Dr Shuker's Casebook

In Pursuit of Marvels and Mysteries

Typeset by Jonathan Downes,
Cover and Layout by ArachnopuSS for CFZ Communications
Using Microsoft Word 2000, Microsoft , Publisher 2000, Adobe Photoshop CS.

Photographs © 2008 CFZ except where noted

First published in Great Britain by CFZ Press

**CFZ Press
Myrtle Cottage
Woolsery
Bideford
North Devon
EX39 5QR**

ISBN: 978-1-905723-33-1

Dedication

To my grandmother, Gertrude Timmins, whose fondly-remembered recollection of how she once witnessed an outbreak of frog rain taught me long ago that some of the world's most extraordinary mysteries can occur right in front of you – all you have to do is keep your eyes, ears, and mind open.

He who can no longer pause to wonder, and stand rapt in awe, is as good as dead; his eyes are closed.

Albert Einstein

Contents

Foreword

by
Nick Redfern

W hether it's due to the fact that we both hail from the mighty and majestic West Midlands (oh yes, it is, I assure you!), or simply that he is a damned good researcher and writer (or both!), I always look forward to a new book from Karl Shuker; and this one most definitely does not disappoint.

And so when Karl asked me if I would write the Foreword for his new book, I quickly said that yes, indeed I would.

Whereas pretty much everything Karl has published so far has been on cryptozoology and weird and wonderful animals, *Dr Shuker's Casebook* is a little different. Yes, it is full to the brim with tales of strange beasts; however, Karl also delves into other areas of Forteana with which he is not generally associated.

This, however, is a very good thing, since we get to learn about a variety of other mysteries that are of interest to Karl. And, indeed, there are enough of them to fill Birmingham's glorious Rotunda and Bull Ring several times over!

So, what do you get for your hard-earned pennies?

Well, I will start with those highly unusual, and almost

legendary, winged cats. Karl provides excellent historical background on these unusual critters, carefully separates fact from fiction, and legend from reality, and gives the reader a fine study of this peculiar phenomenon. Even if you're only vaguely familiar with the winged-cat controversy, this is one you won't want to miss.

And what, I hear you ask, about the mystery animals of Senegambia? Encompassing hairy man-beasts, giant bat-like entities and much more, they collectively make great food for thought as Karl reveals the strange stories of some little-known and elusive beasts of a distinctly exotic and far-off land.

Sky-Beasts is an excellent chapter and one that addresses the theories of researchers such as Trevor James Constable, who suggested the idea that bizarre life-forms might very well soar high in the skies above us, and may perhaps be responsible for at least some UFO reports. Today, Constable's theories are sadly forgotten by many within Ufology, and so it's highly refreshing to see Karl turn his attention to a subject and a theory that perhaps has implications for both the UFO controversy *and* cryptozoology.

Merfolk – strangely humanlike entities that are said to roam the world's oceans and seas - also feature prominently in Karl's book. Of course, any mention of mermaids inevitably provokes furious debate. Fortunately, Karl knows a great deal about this topic and is able to discuss the matter in a fashion that both informs and intrigues. As for how the mysterious Aquatic Ape fits into this story, well...I'll leave that up to you to find out.

The famous 'Green Children' of Suffolk's Woolpit also come under the Shuker microscope, something that leads to such questions as: were they feral children, the denizens of the fairy world, or – if such a thing were possible – something even stranger?

Tulpas – creatures of the mind, thought-forms, or however you want to term them – are a particular obsession of mine. And to see that Karl has devoted a whole chapter of his book to this particularly controversial topic was a fine surprise for me. If you ever mused upon all the fuss that the Tulpa issue provokes, Karl provides the perfect study of the subject, and a look at some of the more significant tales of a distinctly mind-monster kind.

Flying humanoids such as Batsquatch, Mothman, Owlman, Britain's little-known Monkey-Bird, and the Flying Man of Russia also have pride of place. The thorny (or should that be horny?) issue of unicorns is dissected, as are frog-falls, ball-lightning, will o' the wisp, earthquake lights, various other aerial puzzles, and ghosts and specters.

Then there are those weird messages from space: manuscripts, papers, artifacts and even Joe Simonton's famous faerie pancakes. Doppelgangers: they're here too; as are the mysteries of Easter Island, the wonderfully-named Vegetable Man, the Monster of Glamis, and menacing, spectral werewolves.

In other words, *Dr. Shuker's Casebook* is a fantastic and rich collection of oddities that will fascinate everyone with a thing for Forteana. And now, turn the page, turn off the lights and by the flicker of an old candle and a full moon, enjoy the book!

Nick Redfern is the author of many books including *Man-Monkey*, *Memoirs of a Monster Hunter*, and *A Covert Agenda*. He runs the American Office of the Centre for Fortean Zoology, and can be contacted at his website www.nickredfern.com

Introduction

OF MEN IN SCALES, AND UNICORNS,
AND EFFIGIES THAT SING

"The time has come," the Walrus said,
"To talk of many things:
Of shoes - and ships - and sealing-wax -
Of cabbages - and kings -
And why the sea is boiling hot -
And whether pigs have wings."

Lewis Carroll – *'The Walrus and the Carpenter'*,
from *Through the Looking-Glass*

Although I am best known for my cryptozoological writings, I have always been extremely interested in many other anomalous phenomena too, ranging from ancient wonders, religious miracles, and paranormal entities, to geological marvels, extraterrestrial
mysteries, non-cryptozoological animal anomalies, and much more. Indeed, my bestselling book to date, having sold hundreds of thousands of copies worldwide in many different languages, is *The Unexplained: An Illustrated Guide to the World's Natural and Paranormal Mysteries*
(1996) - which, as its title reveals, covers a vast range of unexplained phenomena rather than confin-
ing itself solely to cryptozoological subjects (though it certainly includes a sizeable representation).
Indeed, so outstandingly successful was this book that its publisher, Carlton Books, commissioned a
sequel from me a couple of years later – *Mysteries of Planet Earth: An Encyclopedia of the Inexplicable* (1999), which followed the same format, surveying an equally diverse but this time even more
esoteric spectrum of arcane and uncanny enigmas.

In addition, I have also written numerous magazine and journal articles on non-cryptozoological mysteries. Yet whereas no less than three compilation volumes containing reprinted (and in many cases extensively updated) versions of various of my cryptozoological articles have so far been published – *From Flying Toads To Snakes With Wings* (1997), *The Beasts That Hide From Man* (2003), and *Extraordinary Animals Revisited* (2007) – none of my non-crypto articles have been collected and republished in a compilation volume...until now.

Hence I am delighted to have been given the opportunity by the CFZ Press to assemble in this latest book of mine a selection of these latter articles – which in their original form first appeared in such notable magazines of the present and past as *Alien Encounters, Beyond, Enigmas, Fate, Fortean Times, Goblin Universe, History For All, Prediction, Sightings, Uri Geller's Encounters*, and *The X Factor* - in which I have surveyed or personally investigated a very wide-ranging selection of fascinating non-crypto cases that have long interested me.

As with my cryptozoological researches, I have always been especially intrigued by the less familiar, more obscure cases. So among such major, benchmark subjects as the green children of Woolpit, the Shroud of Turin, and mermaids, and not forgetting my recent close encounters with the stone giants of Easter Island, you will also find plenty of less familiar, more exotic, or relatively unpublicised wonders - such as the singing colossus of Memnon and the living cockerel statue of St Peter, reptilian entities from outer and inner space, a newly-revealed bat-winged monkey-bird from Kent, the Narmer palette, forgotten ancient monarchs of myth and reality, alien artefacts, luminous life-forms, the Porcelain Tower of Nanking, invisible saints, the chirping pyramid of Quetzalcoatl, and the head of Ozymandias, to name but a few.

But what sparked my enthusiasm for such subjects? It is well known that my passion for cryptozoology was ignited by the 1972 Paladin paperback reprint of Dr Bernard Heuvelmans's classic tome *On the Track of Unknown Animals*, bought for me as a birthday present by my mother when I was around 13 years old. However, my interest in mysterious phenomena as a whole stemmed from an even earlier present – a copy of *Stranger Than People*, an enthralling compendium of mysteries from fact and fiction, published in 1968

The author, Dr Karl Shuker, with the two books that sparked his lifelong interest in cryptozoology and other subjects of mystery *(Dr Karl Shuker)*

by YWP, and aimed at older children and teenagers, which I saw one day in the Walsall branch of W.H. Smith when I was 8 or 9 years old, and was duly purchased for me as usual by my mother.

Within its informative, beautifully-illustrated pages I read with fascination – and fear – about Nessie and the kraken, vampires and werewolves, the Colossus of Rhodes and Von Kempelen's mechanical chess player, dinosaurs and the minotaur, witches and zombies, yetis and mermaids, leprechauns and trolls, Herne the Hunter and Moby Dick, giants and the cyclops, feral children, the psychic powers of Edgar Cayce, and lots more. It even included two original – and quite superb - sci-fi short stories: 'Klumpok', about giant ant-like statues found on Mars and what happened when one of them was brought back to Earth; and 'The Yellow Monster of Sundra Strait', in which a giant transparent globe containing an enormous spider-like entity rises up out of the ocean; plus a thrilling (and chilling) fantasy tale, 'Devil Tiger', featuring a royal but malevolent weretiger that could only be killed with a golden bullet.

Needless to say, I re-read the poor book so many times that it quite literally fell apart, and was eventually discarded by my parents. After I discovered its loss, I spent many years scouring every bookshop for another copy, but none could be found. Not even Hay On Wye – world-famous as 'The Town of Books' with over 40 secondhand bookshops – could oblige. A few years ago, however, the Library Angel was clearly at work, because one Tuesday, walking into the bric-a-brac market held on that day each week in my home town of Wednesbury, on the very first stall that I approached I saw a near-pristine copy of *Stranger Than People*! Needless to say, I bought it, and to this day it remains the only copy that I have ever seen since my original one. All of which is a very long way of saying that this is from where my interest in all of the subjects documented in this latest book of mine ultimately derives.

When assembling my cryptozoological compilation volumes, I attempted wherever possible to update all of the articles included in them. Due to the immense range of subjects covered in this non-crypto book, however, and to the fact that some of the articles included in it date back several years, it simply would not have been practical to do this to the same extent here (much as I would have greatly enjoyed doing so!) – otherwise certain of the chapters, such as those dealing with religious marvels, ancient wonders, and balls of fire, for example, would have expanded into book-length treatments in their own right. Obviously, however, if there have been any major developments during the intervening years that clearly necessitated coverage here, these have indeed been incorporated accordingly.

Incidentally, and at the risk of upsetting those readers of a metric persuasion, I have retained here the imperial measurement units used in my original articles. This is not least of all because this book should be of particular interest to the American market, as many of my original articles now comprising it were published in various British magazines not readily available (if, indeed, at all) in the States.

Finally, although the vast majority of subjects documented in the following chapters are non-crypto, it wouldn't really be a true book of mine if it didn't contain at least a few mystery beasts to tantalise and titillate, now would it? So I have also included some fauna from the fringes of cryptozoology and from the eerie realms of zooform phenomena to bemuse and bedazzle, such as a rich assembly of sky beasts, Antipodean mirrii dogs and mini-humanoids, Senegambian terrors, batsquatch and other flying phantasmagoria, a demon hedgehog and a Cambodian stegosaur, necking serpopards, unicorns of fantasy and fact - plus the most comprehensive documentation of winged cats ever published anywhere.

So, with no further ado, welcome to my arcane archive of inexplicabilia, dubitanda, and mirabilia – or, as I prefer to call it, *Dr Shuker's Casebook.*

Imagine yourself an explorer in the jungle. There before you, printed in the soft ground is the footprint of some strange creature you have never even heard of – eagerly you follow, for it is a trail leading to the unknown...
Strange footprints in the jungles of space and time have led explorers and scientists to many strange discoveries. But often the mark of the stranger, be it a footprint or some other clue to the passing presence of something alien, poses a riddle that may never be answered.

'The Mark of the Stranger' – in *Stranger Than People*

Chapter 1

On a Wing and a Purr...

Such Wonderful Things are Cats with Wings

Let not the dream pervade these living hours,
Lest the winged cats of nightmare stalk once more
The haunted spires of distant memory,
Their pinions raised, alert and poised
To deny the conscious dawn.

Patrick Krushenk - *Mirabilis*

lying felids and cats with wings are by no means rare in traditional mythology and folklore. For example, demonic entities are often depicted as cats with bat-wings, just as benevolent, magical beings are sometimes portrayed as angelic cats with white feathery wings. Such images were also used as caricatures by the 19th-Century satirical artist J.J. Grandville. More specifically, the evangelical emblem of St Mark is a winged lion (as extensively featured in Venice and elsewhere in Italy), and the personification of storms and chaos in ancient Sumerian mythology is the imdugud, represented by a lion-headed feathered bird.

Naturally, however, one would not expect to encounter cats with wings in reality - but reality often has a disconcerting habit of turning expectations on their head, and never more so than with winged cats, because, as revealed here, these amazing animals do indeed exist! Indeed, as many aficionados of zoological anomalies will know, over the years I have written several articles dealing with these fascinating felids (which, incidentally, have been shamefully plagiarised since then in uncredited form on various internet websites!). Now, by way of this chapter, which is an updated version of the lead article of mine that originally appeared in the very first issue of the British full-colour magazine *Beyond* (October 2006), I am pleased to present the most comprehensive coverage of winged cats ever published - the culmination of my many years of research into this remarkable subject.

EARLY REPORTS

Kindly sent to me by Paul Sieveking at *Fortean Times*, the earliest record that I have so far encountered of what appears to be a bona fide winged cat exhibits all the characteristics of the more famous examples that would be documented decades later. In 1854, the celebrated American writer Henry David Thoreau published a book entitled *Walden; or Life in the Woods*, which recounted the two years that he had purposefully spent living apart from the rest of the world in a self-built cabin amid woodlands by the shores of Walden Pond in Concord, Massachusetts. In his book, Thoreau recalled that in 1842 a very peculiar cat lived in a Lincoln farmhouse owned by a Gilian Baker close to the pond. The cat's sex was unknown, but was referred to for convenience by Thoreau as 'she', and according to her owner she had first appeared in the neighbourhood during April 1841, before eventually being taken in by the Baker family. She was specifically referred to locally as a 'winged cat' - for good reason:

> ...that she was of a dark brownish-grey colour, with a white spot on her throat, and white feet, and had a large bushy tail like a fox; that in the winter the fur grew thick and flatted out along her sides, forming stripes [often misquoted as strips] ten or twelve inches long by two and a half wide, and under her chin like a muff, the upper side loose, the under matted like felt, and in the spring these appendages dropped off. They gave me a pair of her "wings", which I keep still. There is no appearance of a membrane about them. Some thought it was part flying-squirrel or some other wild animal.

However strange a winged cat might seem, it pales into insignificance beside a crossbreed of cat and flying squirrel, which is truly a zoological impossibility for fundamental taxonomic, genetic, and behavioural reasons. No, the correct explanation for this feline wonder is most likely the same as for all of the other individuals surveyed here, as will be revealed later.

Meanwhile, on 3 August 1894, Cambridgeshire's *Independent Press* newspaper carried the following intriguing report:

> A live cat with wings resembling those of a duckling is now being exhibited in the neighbourhood by Mr David Badcock of the Ship Inn, Reach [near Peterborough]. The cat which is a year old did not until recently expose such a remarkable freak of nature, but being somewhat roughly handled spread out its wings. The owner charges the sum of 2d for callers in the daytime to see such a strange beast and has commented taking it round the neighbouring villages in the evenings to exhibit.

Sadly, however, it seems that Mr Badcock made too much of a show of his marvellous moggie, because a week later the *Independent Press* reported that it had been catnapped!

> The "Remarkable cat" reported in our last issue has been stolen. It is hoped however the thief or thieves will soon be run down, as the animal, our correspondent understands, has been traced to Liverpool.

Nothing more emerged regarding this story, so whether the cat was reclaimed is unknown.

From Cambridgeshire to Derbyshire, and a report from 26 June 1897 in a Matlock newspaper, the *High Peak News*, that described a doubly-strange winged cat. It had been shot by a Mr Roper of Winster, who had seen it on Brown Edge and mistook it for a fox:

> It proved to be an extraordinarily large tomcat, tortoiseshell in colour with fur two and a half inches long, with the remarkable addition of fully-grown pheasant's wings projecting from each side of its fourth rib...

Never has its like been seen before, and eyewitnesses state that, when running, the animal used its wings outstretched, to help it over the surface of the ground, which it covered at a tremendous pace.

Ironically, the most unusual characteristic of this particular cat is not its "pheasant's wings", which is probably no more than a fanciful way of describing long filamentous expanses of furry skin (as opposed to feathers!), but rather its sex. Due to the tortoiseshell condition being a sex-linked genetic mutation, virtually all tortoiseshell cats are female, thus making a male tortoiseshell cat if anything even more extraordinary than a winged cat.

THE WINGED KITTEN OF WIVELISCOMBE

Possibly the most famous of all winged cats, due to the countless times that its picture has since appeared in other publications, was the delightful kitten pictured and described in the November 1899 issue of London's *Strand Magazine*. Under the heading "Can a cat fly?", the kitten in question was reported as follows:

> This sounds very much like a conundrum, and a very absurd one, too. If we look at the picture of pussy which is reproduced here, our question is at least partially answered. This extraordinary cat, it will be perceived, is the proud possessor of a pair of wings. It belongs to a lady of Wiveliscombe, Somerset, to whom it was given when a kitten. There was nothing extraordinary in its appearance at first, but after a time it developed a pair of wings. They are not, of course, covered with feathers, but fur - the same as the rest of the body. Several persons have seen this wonderful cat. The photograph was taken by Mr. G.W. French, of Wiveliscombe, Somerset.

The Wiveliscombe winged cat *(Fortean Picture Library - FPL)*

A later report described this animal's wings as "not flabby, but apparently gristly, about six or eight inches long". Although, as I have already shown in this chapter, the Wiveliscombe individual was not the very first winged cat to have been reported (although it has often, erroneously, been claimed as such by others), it may well have been the first one to have been photographed. Tragically, however, its most famous attributes were also fatal ones, because in a brutal attempt to discover more about them someone subsequently cut its wings off, which caused the kitten's death.

THE SAD SAGA OF THOMAS BESSIE

The year 1900 saw the birth of a kitten whose wings were apparently present from the very beginning. Born at Bramley Workhouse in Leeds, its sex was never determined, and so it became known as Thomas Bessie. It was looked after by the workhouse's relieving officer, William Markham, but while still a youngster Thomas Bessie was stolen. Markham succeeded in tracing the winged cat to a fairground show, and applied for a court order for its return, but before any decision was reached Thomas Bessie was poisoned by person(s) unknown and died. Accordingly, Markham had his deceased pet preserved for posterity by a taxidermist; but after Markham's own death, Thomas Bessie was passed on to his grand-daughter, Mrs Amy Clague, and as a cased specimen began an afterlife of exhibition in various pubs around England.

The most recent news of Thomas Bessie is that in December 1973 it was on display at Mrs Clague's pub, *The Hole in the Wall,* in Scarborough. According to Clague, its wings "seemed to be a kind of rib structure", rather than mere bundles of fur.

This would seem to be the same specimen as the winged cat referred to by S. Peter Dance in his book *Animal Fakes and Frauds* (1976). Dance alluded to a specimen called Thomas-Bessy preserved in a glass-fronted case, which in the early 1960s was being advertised as 'The Famous Winged Cat' by way of an information sheet distributed from an address off New Bond Street, London. Thomas-Bessy's history as described by the sheet closely corresponds with what I have given above for Thomas-Bessie, and also stated that since the death of its owner (not named, but clearly William Markham) it has remained in an attic (presumably Mrs Clague's?) "awaiting an enterprising purchaser". Keen to be that purchaser, Dance wrote an enthusiastic letter to the New Bond Street address, but never received a reply.

BLACK AND WHITE AND WINGED ALL OVER

Like many other mysterious phenomena, whose reports seem to occur in spasmodic 'flaps', i.e. a sudden outbreak of several reports, then nothing at all for quite a time, followed by another 'flap', this pattern emerged with winged cat reports too. Following the Wiveliscombe kitten, no additional cases made headlines until the 1930s, when a new flap of winged cats emerged (incidentally, a flap of winged cats would be a wonderfully apt collective noun for these feline marvels!). One of the most memorable winged cat events took place during the first week of June 1933, and was reported on 10 June by a *Sunday Dispatch* Special Correspondent, who had visited the scene to view this animal for himself:

> I have just seen a cat that has on its back fully-developed fur-covered wings, with which, it is stated, it can fly. It is now housed in the Oxford Zoo, and it is one of the strangest of Nature's freaks - and the most pathetic. All the time it seems to be ashamed of its unusual appearance and tries to be as like a normal domestic cat as possible. When I approached him, the winged cat rolled over on his back and then frisked round an enclosed paddock. Here is the strange history of the animal, which is puzzling eminent zoologists.
>
> A few days ago neighbours of Mrs Hughes Griffiths, of Summerstown, Oxford, saw a strange black and white cat prowling round their gardens. Last evening Mrs Hughes Griffiths saw the animal in a room of her stables.
>
> "I saw it move from the ground to a beam - a considerable distance, which I do not think it could have leaped - using its wings in a manner similar to that of a bird", she said to me.

Mrs Hughes Griffiths at once telephoned to the Oxford Zoo, and Mr Frank Owen, the managing director, and Mr W.E. Sawyer, the curator, went to her house and captured the animal in a net. I carefully examined the cat tonight, and there is no doubt about the wings. They grow just in front of its hindquarters.

The Oxford Zoo no longer exists, but a photo of the cat on a lead, taken during its sojourn at the zoo, still survives and is reproduced here, clearly showing its outstretched black-furred wings, which were 6 inches long. Tragically, the survival of the cat itself was probably much shorter. In a *Daily Mirror* article from the same period, the zoo's curator was quoted as saying: "I have never seen anything like this - it will be kept at the zoo for a short time and then destroyed if it is not claimed - it is not good policy to keep freaks at this zoo, and I do not think our visitors would like to see it". Bearing in mind that the Oxford winged cat is still reported and discussed today, over 70 years later, the general public clearly thinks otherwise.

The Oxford Zoo winged cat *(FPL)*

Three years later, a long-furred female winged cat, white in colour, was discovered on a farm close to Portpatrick in the Scottish county of Wigtownshire. Her wings, which measured 6 inches long and 3 inches wide, were said to fold down into her sides when she was at rest, but rose up when she ran. Adding to her distinctive appearance were her eyes, because one was red and the other was blue.

On 30 July 1939, a photograph of a remarkably similar winged cat to the Oxford Zoo individual, once again handsomely marked in black and white and with long black-furred wings (spanning 2 ft across), was published in London's *News of the World* newspaper.

This particular specimen, a tom, curiously named Sally, belonged to Mrs M. Roebuck of Attercliffe in Sheffield. According to the report, Sally used his wings to assist him in taking long leaps, but couldn't use them to fly. Nevertheless, Sally was striking enough for his owner to sell him to a museum in Blackpool exhibiting freaks.

Sally, the oddly-named winged tomcat originally of Sheffield *(FPL)*

During World War II, yet another black-and-white cat that sprouted wings attracted attention, this time in Ashford, Middlesex. Owned by two pensioners, it was often seen in their garden, and many people came to look at it over their wall, including then-teenager Joyce Harrigan, who spoke to me on the telephone in August 1998, recalling her memories of this interesting animal. A big chubby specimen, its wings sprouted from its shoulders.

In June 1949, a winged cat with a very impressive 23-inch wingspan was shot in northern Sweden after it had supposedly "swooped" down on a child. Weighing 20 lb, its body was later donated to a museum. According to Prof. Rendahl of the State Museum of Natural History who duly examined it, this cat's wings were a deformity of the skin.

A FLAP OF SPANISH WINGED CATS

During May 1950, a grey-furred Angora winged cat called Angolina enchanted the Madrid media, featuring her in several newspaper accounts. Although born in Barcelona, Angolina was now owned by Juan Priego, a porter living close by Spain's houses of parliament in Madrid, and during the course of the year not only attracted great scientific interest but even survived a kidnap attempt. Her eyecatching appearance inspired a wide range of explanations, of which the most charming was that her appearance signalled the return of a race of prehistoric cats originating from before Noah's Great Flood! Her wings had begun as a couple of bulges on her back, but later lengthened into sizeable extensions, the inner surface of each being concave and the outer being convex, which she was reputedly able to stretch and move at will. It was also claimed that she felt pain when the wings were punctured with a needle.

In June 1950, a second Spanish winged cat was unveiled by the media. This one, a thick-furred Angora called Michi, was owned by Rosario Dominguez, a Madrid electrician. A month later, a winged Angora named Lobito was reported from the Spanish province of Granada.

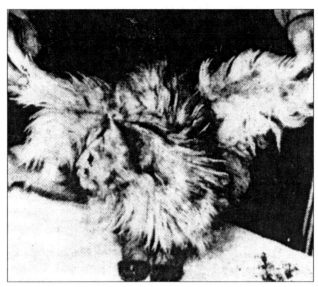

Michi, the winged Angora cat from Madrid *(FPL)*

THE WINGED CAT THAT WENT TO COURT

The most controversial winged cat of the 1950s was unquestionably a female Persian dubbed Thomas-Mitzi. In May 1959, 15-year-old Douglas Shelton caught this cat in a tree near his home in Pineville, West Virginia, and prematurely christened his discovery Thomas, before her true sex was revealed. Her furry wings, which were boneless but seemed to contain gristle and felt gritty towards her body, soon turned Thomas into a celebrity, and she even appeared on television.

As a direct result of her exposure, however, she came to the attention of a local Pineville woman called Mrs Charlie Hicks, who alleged that Thomas was actually her lost cat, Mitzi, and instigated a lawsuit to claim ownership of her. The case was held on 5 October 1959, but when its star exhibit, the winged cat Thomas-Mitzi (as she was now referred to), was brought in, she was soon seen to be lacking the vital prerequisite of any winged cat - wings!

It transpired that she had shed them in July, and the shed wings were duly displayed in the cardboard box inside which they had been placed by Douglas. Following that shock revelation, Mrs Hicks stated that this now-wingless winged cat was not hers after all, and the case was duly dismissed.

Also put on show (but not in a courtroom this time) was the following winged cat. In or around 1950, a fully-grown female tortoiseshell cat called Sandy began to grow a pair of large wing-like flaps of fur on her shoulders, which attracted so much attention that she was eventually exhibited at a carnival at Sutton in Ashfield, Nottinghamshire. Three decades later, according to a *Weekend* magazine report of 12 November 1980, she was still remembered in the area.

WINGED CATS OF THE MODERN AGE

Perhaps the most frequently documented of all modern-day winged cats is an individual that lived for several years during the 1960s in the yard of a building firm called Banister, Walton & Co, based at Trafford Park Industrial Estate, Manchester, and was looked after by the night watchman there. In 1975, the *Manchester Evening News* published a now-famous, much-reproduced photo of this cat (I have seen several other photos of it, but only one of those seems to have been published). Its large fluffy wings sprouted from its shoulders, and began as two furry growths, but matured until each measured 11 inches long, and appeared to be jointed. Moreover, the builders claimed that it could actively raise and lower them. Heightening its unusual appearance was its tail, which was exceptionally broad and flat.

One day during summer 1970, 10-year-old Jay Allen Sanford of Wallingford in Connecticut was passing a neighbour's yard when he spotted a remarkable orange and white long-haired cat that was positively waddling due to a pair of large wing-like growths hanging from its mid-section. Sanford was able to pet this distinctive animal and feel its wings, which he later described as follows:

> The rectangular fur pads were at least five inches long and three or four inches tall. They were attached to its torso nearest to the front two legs and felt like cardboard with fur haphazardly glued to it. The fur was not straight like that on its body - rather, it was matted and a little lighter in color (kind of gray in spots). I thumped the pads and it sounded like thumping light balsa wood though the pads felt sturdy and weighty.

Sanford was informed by the cat's owner that these pads fall off by themselves, and sure enough, when he next saw the cat, over a week later, its wings had gone. What is particularly intriguing about this case is that the wings sprouted from the cat's mid-section rather than the shoulders, haunches, or back as in other winged cats.

During 1975, a verminous, aggressive feral cat, one of a number of supposedly highly-inbred specimens descended from a single original pair, and instantly recognisable by its deformed face and bluish fur, was frequently spied by Russ Williams on his property at Prescott, Arizona. Even more distinctive than its face, however, were its wings - all four of them! One pair hung off its pelvis; each of these two wings was about 6 inches long, 2 inches wide at the base, perhaps an inch wide at the end, and covered in dirty fur. A second, much smaller pair was present on top of its shoulders. Once, when Williams approached too closely, the cat drew back and its pelvic pair of wings began to flail around randomly in all directions, exposing the hitherto-undetected presence of a pair of deformed claws sticking out of the end of each wingtip. After the cat destroyed much of his garden, however, Williams shot it, but due to its parasite-infested state he did not collect or preserve its body.

The Welsh island of Anglesey was the location for a 1980s winged cat report, as this is where a black and white specimen, born in 1980, was encountered by Wyn Williams while visiting a farm there on 22 August 1985. The cat's wings had begun to grow during 1984, sprouting from just in front of its shoulders, but by the time that Williams saw them they were each approximately 7 inches long. Williams took a series of photos of this winged cat, which was fortunate, because by March 1986 it had shed its wings. I have tried on several occasions to contact Mr Williams to request more information and permission to use his photographs (shown to me by the Fortean Picture Library), but have been unable to locate him, so if he is reading this book I would be delighted to hear from him.

In April 1995, Martin Milner spotted a large friendly tabby cat with a very striking pair of fluffy wings

ABOVE: The Manchester winged cat *(FPL)*
BELOW: `Prul`, Martine Smids's winged cat from the Netherlands (p.24)

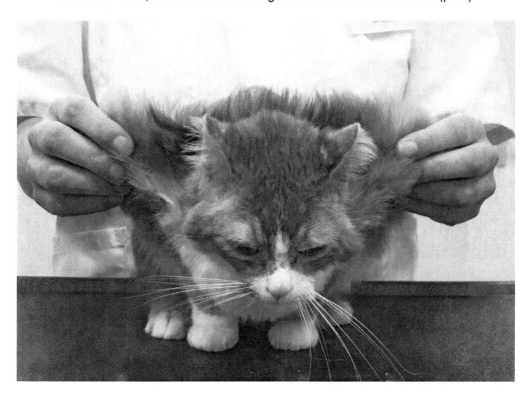

The Anglesey winged cat, photographed by Wyn Williams

while holidaying in Backbarrow, Cumbria, and he learnt that it belonged to the village's retired postman. As the animal was otherwise well-groomed and healthy, Milner doubted that its wings were merely clumps of matted fur. Four months later, Steve Volk of Bradford-on-Avon revealed that some years earlier he had spied a stuffed winged cat displayed alongside a few other freak-type animals in a waxworks exhibition on the Isle of Wight. However, he did not know whether the specimen was genuine or merely a clever-constructed fake.

In April 1997, while visiting Nick Smith of Hersham, Surrey, Jon Downes and Richard Freeman of the CFZ were delighted to find that he owned a winged cat. A mossy-brown Persian cross, its wings were solid but not gristly, and each measured 6-7 inches long.

On 23 May 1998, while taking a cool evening road run in Kumamoto, Kyushu, Japan, Rebecca M.B. Hough encountered two cats sitting on a wall overlooking the road, one of which came over to her when she called. As she stroked its back, Hough was startled to feel and see (by way of the light from an overhead street lamp) a pair of "jutting out, fur-covered wing-like growths...The growths were triangular in shape and covered in soft fluffy fur. They felt like the wings of a chicken, although they were not so long".

Also in 1998, I learnt from postman John Raggatt that earlier that year he had twice encountered a black winged cat while delivering post to Moor Park, a private residential area just outside Northwood, Middlesex. The first occasion, around June, was just a brief sighting of it, walking out of a driveway and into a garden, but during the second sighting, a fortnight or so later, he was only a few feet away, and was able to observe its wings more closely when it paused to groom itself while following precisely the same route as before:

> They appeared to be about 2-3 inches back from the shoulder blades and at about the same height as the shoulder blades. They were about 8-10 inches long, about 4 inches wide and about an inch thick.

The cat ran off when Raggatt attempted to get closer, and as it did so its wings flapped.

Still in 1998 - clearly the Year of the (Winged) Cat - I also received a letter from Lynda Brooks of Burnham-on-Crouch, Essex, in which she recalled seeing a very old stuffed and mounted winged cat while visiting the Niagara Valley and surrounding area on the American-Canadian border back in 1992. Dark reddish-brown with black tabby-like markings and about the size of a one-year-old domestic cat, it had thick coarse fur, a very wild appearance, and was very slim with a thin tail. Its most striking features, however, were its wings:

> ...two bony structures, one on either side of its spine - about shoulder-blade level - and these were covered with flaps of skin which looked much like the ear flaps on a deerstalker hat.

She observed this specimen two or three times during her stay in the area, and inspected it closely for any indications that it had been artificially constructed, but could find no evidence to suggest that it was anything other than genuine. Two years after receiving Lynda's letter, I made enquiries concerning this interesting specimen while visiting Niagara myself, but I was unable to trace it.

21ST-CENTURY WINGED CATS

One of the most recent winged cat reports is also the most tragic, hearkening back to the callous treatment meted out to the Matlock specimen and the Wiveliscombe kitten over a century earlier. In 2004, a very large stray ginger tom with wings that he could stretch out slowly "just like a chicken" entered the yard of Nadezhda Medvedeva, who lived in the village of Bukreyevk, near Kursk in central Russia. She fed him on milk, and her daughter named him Vaska. However, even though Vaska was affectionate and they felt sure that he must be someone's pet, they nevertheless feared that he may be a demon sent by the devil, as did other local people who saw him. News of Vaska soon filtered out as far as Kursk, and a reporter travelled to Bukreyevk to see Vaska for himself, but he was too late. By the time that he had reached the village, Vaska was dead - a local drunk had drowned the poor creature in a sack. The reporter inspected Vaska's remains, which confirmed that he did indeed have wings, but no scientific examination appears to have been conducted.

In May 2007, news emerged of a winged cat in China – the first recorded from that vast country. Owned by Granny Feng of Xianyang city in Shaanxi province, the white four-year-old tom with a handsome black and white face is now the proud possessor of a pair of hairy 8-inch-long wings, and has been pictured in media accounts worldwide. His wings began as a small pair of bumps in April 2007, but within a month had quickly grown into their present form. According to Feng, they contain bone, but this is more likely to be gristle, or even hard pads of matted fur. Intriguingly, Feng also claims that her tom grew his wings after being harassed by many female cats in heat.

The most recent winged cat case known to me at the time of writing in the spring of 2008, was brought to my attention by Martine Smids, a small animals veterinarian from the Netherlands, who happens to own one, a male called `Prul`. In an email to me of 24 January 2008 (the first of several that we subsequently exchanged, discussing Prul), Martine provided the following information:

> It happens to be, that I own a so called 'winged cat'. It was brought to our practice in November 2005 at the age of 6 months. He was brought in because he had lost the complete skin of his tail, in a fight with a dog. The tail needed to be amputated. The owner didn't want to make a lot of costs and decided to leave the cat with us (this instead of euthanasing the cat immediately). The cat didn't seem to be sick or unhappy, so I decided to keep the cat, although I didn't know what was wrong with him at that point. After weeks of diagnostic investigations and talking to veterinarian dermatologists, the diagnosis of Cutis [sic] Asthenia was made.
>
> Now, 2 years later, the cat is doing fine. Of course he is an indoor cat and he wears most of the time a little baby-sweater to protect his skin. I have 3 more cats and a dog, this doesn't seem to be a problem. But this is probably due to the good character of the cat. Whenever he has skin lesions, I treat them very easy with agraffes (staples), he doesn't need sedation for this and even the largest wounds heal within a week. His wounds don't bleed and don't seem to hurt really. I've stopped giving him antibiotics for every wound that he has, because it didn't seem to make a lot of a difference. In the last 2 years, he only had one infected wound, that formed a small abscess. He also has hip dysplasia. The first months that I had him, he seemed to have problems with this. His hips subluxated [partially dislocated] spontaneously when he walked. This causes him to walk lame, non responsive to NSAIDs. After a few months, he started to walk normally again and he doesn't really have problems with this ever since. The last year he even seems to be getting stronger, he can jump up much better, although not as good as a normal cat.
>
> Now I am writing a case-report about him for the Dutch Veterinarian Magazine (*Tijdschrift voor Diergeneeskunde*).

In her next email to me, of 6 February, Martine included some additional details:

> Prul hasn't always 'wings', he only really has them, when he has been licking on a certain spot for a long time, his skin stretches then into folds, sometimes his skin tears. These folds usually disappear after a while, when he stops licking. I've made some pictures with his skin stretched out.

In later emails, Martine did indeed enclose a number of photos, including the one reproduced here with Martine's kind permission.

To end this chronology of winged cat reports are a few cases that seem too bemusing, and bizarre, to qualify for inclusion within the main body of data, warranting separate categorisation as follows:

WINGED CAT, OR FLYING LEMUR?

Several years ago, I uncovered the following intriguing but previously-unpublicised report, entitled 'Flying Cat', published in the volume for 1868 of a long-forgotten British journal entitled *The Naturalist's Note Book*:

> A nondescript animal, said to be a flying cat, and called by the Bhells *pauca billee*, has just been shot by Mr. Alexander Gibson, in the Punch Mehali [India]. The dried skin was exhibited at the last meeting of the Bombay Asiatic Society. It measured 18 inches in length, and was quite as broad when extended in the air. Mr. Gibson, who is well known as a member of the Asiatic Society and a contributor to its journal, believes the animal to be really a cat, and not a bat or a flying-fox [fruit bat], as some contend.

Today, 140 years later, the identity of this animal is still undetermined. Could it be an early example of a winged cat, or was it truly nothing more than a species of bat? There is even the interesting possibility that it was an extraordinary creature known as a colugo or flying lemur (despite neither flying nor being a lemur, though it is a distant relative). Possessing a large web-like membrane that connects their limbs and enables them when outstretched to glide passively through the air, colugos are southeast Asian species not native to India. However, perhaps the Gibson specimen was not actually shot in India after all, but merely preserved or exhibited there.

Athanasius Kircher's bat-cat engraving

Moreover, the grotesque bat-cat depicted by the eminent Jesuit scholar Athanasius Kircher in 1667 may conceivably have been an early attempt to portray a preserved colugo, because the bat-cat's wings are actually pictured as a membrane extending from the forelegs to the hind legs and onto the tail, exactly mirroring the gliding membrane of colugos.

A distorted, secondhand (or more) account of a colugo may even explain traveller Marco Polo's curious mention of a still-unidentified beast from the Far East known as a cat-a-mountain, which was said to be a predatory cat with the body of a leopard but also with a strange skin that stretched out when it hunted, enabling it to fly in pursuit of its prey.

COLUGOS OR FLYING LEMURS

Native to the tropical forests of southeast Asia, most famous for the extensive gliding membrane (patagium) connecting their limbs, tail, and even the digits of their paws, and as big as a medium-sized possum or very large squirrel, the two modern-day species of colugo are the only surviving members of the mammalian order Dermoptera. Widely believed to be the primates' closest living relatives, these extraordinary yet surprisingly little-known gliding mammals are also called flying lemurs - even though they glide rather than fly and are not lemurs! Perhaps the most memorable description of a colugo that I've ever read appears in Bill Garnett's book *Oddbods!* (1984): "*Imagine a floppy shopping bag with an avocado sticking out sideways at the top; hang it by claws beneath a branch; put a huge round eye on the avocado - and cover the lot in a soft furry pelt, mottled fawn and grey. You've now got yourself a colugo*".

VAMPIRE CATS

According to traditional Eastern European legends and folklore, one of the guises that a vampire can adopt is that of a cat. If not a colugo, then perhaps this belief was the inspiration for Kircher's above-mentioned bat-cat engraving - which may well be the earliest-ever depiction of a winged cat.

However, such legends certainly cannot explain a supposedly airborne vampire cat that terrorised the citizens of Alfred, a village in Ontario, Canada, during 1966. According to media reports, this bizarre black-furred creature possessed a pair of long needle-like fangs as well as a pair of large furry wings, which enabled it not only to become airborne but also to swoop down savagely upon the village's more mundane, terrestrial moggies. Its reign of terror ended on 24 June when confectioner Jean J. Revers shot it, and its carcase was buried.

Due to the media attention that its alleged exploits had attracted, however, its remains were soon unearthed again and were examined at Kemptville Agricultural School's veterinary laboratory, but its famed wings were found to be nothing more than matted fur, which clearly could not have enabled it to fly, though the cat itself was rabid, explaining its vicious behaviour. What remains unexplained, conversely, is how the extraordinary claim that it could fly had originated.

Equally strange is a brief UPI report from the same period that a flying cat had also been shot near the village of Lachute, roughly 24 miles north of Montreal, and that yet another one had been shot near Ottawa.

WINGED CATS – IN SEARCH OF AN EXPLANATION

Over the decades since winged cats first began to attract widespread public attention and interest, a number of explanations for their unexpected appendages' development have been offered. The most basic, conservative suggestion is that their wings are nothing more than clumps of matted fur, which can develop if a long-haired cat is not satisfactorily groomed.

There is little doubt that this may indeed explain a few cases, namely those in which the wings are clearly composed entirely of fur with no internal tissue or flesh present. Moreover, as recently mentioned to me by cat worker Sarah Hartwell, cats suffering from an increased metabolic rate due to feline hyperthyroid condition grow fur at an accelerated rate but can become listless and unwilling to groom themselves, again resulting in matted fur. Sometimes these furry mats can become so contaminated with dirt and hardened saliva that they acquire the form of hard wing-like appendages, which even flap passively as the cat runs.

However, these conditions cannot explain the majority of winged cat cases for a number of reasons. Firstly, if the wings of all winged cats were indeed nothing more than clumps of matted fur, as this is by no means an uncommon occurrence and is readily recognisable as such (especially by experienced cat owners, breeders, and groomers) there would never have been a winged cat mystery in the first place. Yet despite various winged cat cases having featured in newspapers and magazines, I have never seen follow-up letters from experienced cat people confidently explaining these animals as merely ungroomed, unkempt cats.

Moreover, whereas clumps of matted fur can occur anywhere on the body, the wings of winged cats are consistently reported only on the shoulders, back, and haunches. Also of note is that some winged cats have been short-haired, not long-haired, and therefore much less likely to suffer from matted fur. Furthermore, whereas clumps of matted fur may indeed flap passively as the affected animal moves, in bona fide winged cat cases the cat can raise and lower its wings actively, clearly demonstrating that they contain neuromuscular tissue and are not just furry mats.

A very different but pertinent explanation is that some wings may actually be supernumerary limbs, i.e. freak extra legs - the result of a pair of twins incompletely separating or rejoining during embryonic development, or, less dramatically, a freak duplication of limbs occurring during development. The most obvious example of a winged cat exhibiting such a condition is the deformed specimen shot in 1975 by Russ Williams in Arizona, which possessed four wings, two of which had claws at their tips. These were clearly abnormal, partially-formed supernumerary limbs. However, in the vast majority of winged cats, there is no suggestion that the wings are bona fide limbs.

As someone who has always been passionately interested in - and also, for a long time, perplexed by - winged cats, during the early 1990s I made a concerted attempt to seek out the true, hitherto-undisclosed answer to the mystery of these enigmatic creatures. Prompted and encouraged by renowned feline geneticist Roy Robinson, I decided to comb the veterinary literature in search of inspiration and possible clues, and eventually, after scanning countless specialised and often obscure papers, I came upon a remarkable case history.

In 1977, the International Academy of Pathology's journal, *Laboratory Investigation*, published a paper by two American veterinary medical scientists, Drs Donald F. Patterson and Ronald R. Minor, which documented a domestic short-haired tom cat exhibiting a little-known but quite extraordinary genetically-inherited skin condition called feline cutaneous asthenia (FCA). In FCA, collagen, the protein functioning as binding and packing tissue in the dermis of the skin, is defective. As a result, the cat's skin was exceptionally elastic (hyperextensible), so much so that it could be readily stretched to yield furry wing-like extensions. Indeed, when, during an examination, the fur on the upper portion of the cat's lower back was gently lifted, it extended to a distance above the spine equal to about 22 per cent of the body's entire length. Moreover, their paper included a photo taken of the cat while this action was being demonstrated - revealing it as a classic winged cat, identical in appearance to those previously depicted in newspapers and magazines.

Further detective work by me in the veterinary literature unearthed several additional examples of cats exhibiting FCA, and it was evident that this was unquestionably the long-awaited explanation for the winged cat mystery, which I formally published in 1994. Just as described for winged cats, the skin of FCA cats is most elastic on the shoulders, back, and haunches, and the mere act by an FCA cat of brushing or rubbing these body regions against something hard or rough is enough to stretch the skin out into wings. Moreover, if these wings of elasticated skin contain sufficient musculature, the cat can raise and lower them actively, exactly as described for various winged cats.

In addition, the skin of FCA cats is not only abnormally stretchable but also aberrantly fragile, so that elasticated skin extensions will readily break off, yet, remarkably, without causing bleeding. Instead, the extensions simply peel away and drop off - thus giving the impression that the cat has moulted them, once again matching the descriptions of winged cats that have moulted their wings. (Martine Smids's emails to me in early 2008 concerning her winged cat Prul describe this very same phenomenon.)

There could no longer be any doubt. FCA was the primary answer to the longstanding mystery of winged cats - but why had it not been revealed earlier? The simple answer was that although veterinary scientists were aware of FCA, they had not encountered reports of winged cats in popular-format publications, whereas investigators of mysteries who were familiar with winged cat reports had not encountered FCA cases in the veterinary literature. However, because I was not only a zoologist but also an investigator of mysteries, my interests overlapped both fields, so I had been uniquely placed to make the crucial connection - and thereby bring down to earth in every sense of the expression an amazing phenomenon that had until then been dismissed by many as a veritable flight of feline fancy.

And finally: although its subject is not a cat, the following, seemingly unique report is sufficiently allied to the main theme of this chapter to warrant inclusion here.

HARE TODAY...THE WINGED RABBIT OF COLCHESTER

The only known winged rabbit was featured in an article by the *Essex County Standard, West Suffolk Gazette and Eastern Counties Advertiser* newspaper on 16 December 1911:

> Mr William Walton, of Morant Road, Colchester, is the owner of an extraordinary freak of nature - a winged rabbit. This animal, which is about six months old, is a great pet in the home. Its wings, which seem to take no regular form, proceed from near the two fore feet, and are about seven inches in length and in width about two inches at their widest part. They are fed with blood and resemble pieces of flesh. These wings are covered with white fur similar to the fur of the animal. At times this wonderful rabbit will sit in front of the fire and flap its wings, and it has been known to ascend on to a chair. When the fur which covers the head is ruffled, it somewhat resembles an owl.

Clearly, its wings were more than mere clumps of matted fur, and suggest a condition comparable to FCA. Indeed, conditions featuring hyperextensible skin have been recorded from several other types of animal, including cutaneous asthenia in dogs and mink, and dermatosparaxis in cattle and sheep. There is even a human equivalent, known as Ehlers-Danlos Syndrome, which occurs in at least seven different forms.

In any case, one thing is for sure. However powerful its wings may have been, the Colchester rabbit's "ascent" onto a chair owed more to its species' natural jumping ability than to any unnatural flying capability!

J.J. Grandeville's bat-winged and angel-winged cat caricatures

Chapter 2

TALES OF THE UNINVITED

I Thought I Saw the Strangest Ghost...

Crispin felt very lonely there in the dim dusk of the old trees, with the damp leaves underfoot and the heavy, brooding stillness all around. He moved noiselessly and aimlessly, for the spirit of this pathless glen was upon him. And as he thus walked, a breath of wind from above the dripping pine plumes parted the ghostly mist; and he saw a still, dark pool surrounded by sombre yews and ivy-girt trunks; and a woman crouched by the pool and peered into its motionless depths. He could see nothing of her face, save a pallid cheek which stood out with startling clearness against the black, hooded cloak that enveloped her.

While he stood and gazed at the strange sight, the fog closed down once again. The boy went forward, silent and wondering. When he came to the place, no one was there.

W.R. Calvert – *The Secret of the Wild*

Ghosts come in all shapes and sizes - but few are any stranger than the veritable phantasmagoria of weird apparitions and bizarre wraiths conjured forth herewith from every corner of the British Isles (allegedly the most haunted locality in the world) and beyond, for your appreciation...and apprehension. From radiant boys and spectres of smoke to phantom frozen chickens and fountains of blood - all of these and many more await your command to tantalise and terrify. So draw a little nearer, and you will see what you will see...maybe.

A PHANTOM LAKE AND A FOUNTAIN OF BLOOD

On 10 November 1969, the *Daily Mirror* reported the curious case of Beaulieu Abbey's phantom lake, and the tenacious if unsuccessful search for this watery wraith by John and Christine Swain from Ilminster, Somerset. At the time of the newspaper report, they had already made about 250 100-mile visits to the location near Beaulieu Abbey in Hampshire's New Forest, where they had chanced upon the mist-enshrouded lake 17 years earlier, but had never been able to relocate it.

The reason why this body of water had made such an impression upon them was that at its centre stood a large boulder in which was embedded a sword. Not surprisingly, they assumed that this was some form of Arthurian memorial, but all their attempts not only to find it again, but even to find out anything about it, have failed. Undaunted, however, their search has continued, every three weeks, ever since.

William the Conqueror marked the site in East Sussex where he defeated Harold II at the Battle of Hastings in 1066 by the erection of Battle Abbey. The church is long gone, but many sightings have been reported here of a terrifying fountain that gushes forth scarlet cascades of blood. Other spectral sights witnessed in this vicinity include a phantom Red Lady, Grey Lady, monk, Norman knight, and even the ghost of King Harold himself, with an arrow protruding from his eye.

ANTIPODEAN APPARITIONS

Some of the oddest phantoms have been reported from Australia. In 1905, for instance, two holiday-makers trekking through some mountain countryside in Queensland came upon a large waterfall. As they stood close by to admire its beauty, however, they were greatly alarmed to see two translucent, ghostly hands emerge from the waterfall, and beckon towards them! Despite their alarm, the men were intrigued enough to peer behind the waterfall, to see if anyone - or anything - was there, and spied a cave, which was found to contain three human skeletons. The disembodied hands were not seen again; evidently, their task was complete - the skeletons had been brought to public attention.

Australian ghosts are not exclusive to this continent's European settlers either. Long before they first arrived here, its aboriginal peoples were seeing phantoms of their own. In New South Wales, the Wiradjuri tribe refer to ghosts as the gunj, but few are feared, as they are looked upon as part of the natural course of existence.

A notable exception, however, which the Wiradjuri do fear very considerably, is a terrifying spectral entity nicknamed Old Red-Eye, documented in Frank Povah's fascinating book *You Kids Count Your Shadows* (1990), a unique account of traditional Wiradjuri lore. Sometimes vaguely human in form, but often virtually amorphous, this apparition can materialise anywhere, at any time, and is named after its bright red eyes. Anyone whose own eyes are caught in their fixed gaze is mesmerised for however long Old Red-Eye chooses - and as time stands still for his victim, there is no hope of rescue by crying out, because no-one will be able to hear the cries.

Perhaps the most famous of Australia's photographed phantoms is the Ghost of the Outback. Featuring what appears to be a short, dark-skinned spectre, the photo was snapped one day in May 1959 by Rev. R. S. Blance from Adelaide's Tusmore Presbyterian Church, while walking through a clearing at Corroboree Springs in the Northern Territory. Intriguingly, the Reverend did not see the figure when taking the photo - but it is interesting that this area was formerly used by the Arunta aboriginals as a site for sometimes fatal initiation rites. Could the Ghost of the Outback be the spirit of one of these rites' victims?

OUT OF THE MOUTHS OF BABES?

Children have long been said to be more sensitive to supernatural phenomena than adults, amid suggestions that reports by infants of invisible entities popularly dubbed 'imaginary friends' might sometimes actually refer to spirits unseen by adults but which are visible to the young. Occasionally, however, these 'imaginary friends' may be discerned by a wider age-group - as seems to be true in the case of Old Nanna.

In 1991, two-year-old Greg Sheldon Maxwell from Ruislip, Middlesex, began saying, for no apparent reason, "Old Nanna's here". His family would ask him to point to 'Old Nanna', and he would point to the lounge, or the kitchen, or up into the air, but no-one could ever see anything. Once, however, a photograph was taken, and when developed it depicted Greg sitting on a cushion on the kitchen floor, looking up at a large white misty object seemingly materialising just in front of him. The object is undefined, but Greg's description of it is notable - 'Old Nanna' is the name that he called his great-grandmother before she died.

MALIGNANT MISTS AND SMOKE SPECTRES

One of the most feared types of British apparition is variously referred to as 'Boneless' or simply 'It', because it has no constrained shape, merely assuming the form of a ghostly mist. Nevertheless, it is usually tangible and often exudes a palpable sensation of evil.

Folklorist Ruth Tongue was given details of a Somerset policeman's allegedly true-life encounter with one of these 'boneless' entities by the terrified man's sister-in-law, which was also substantiated by Colonel Luttrell and a Mr H. Kille. The policeman had been cycling on his beat along the Minehead-Bridgwater road one night sometime after World War I when his bicycle's lamp lit up something white lying across the road ahead. He could see that it was neither a patch of low-lying fog nor a sheep, but it did seem to be woolly - and alive.

Before he could ride away from this uncanny entity, however: "...it slid up and all over him on his bike, and was gone rolling and bowling and stretching out and in up the Perry Farm Road. It was so sudden he didn't fall off - but he says it was like a wet heavy blanket and so terrible cold and smelled stale". So frightened was the policeman that he insisted upon being transferred to another district, far away from the scene of his eerie encounter.

Back in 1954, while on guard in the much-haunted Tower of London, a sentry spotted a strange cloud of smoke that appeared to be moving quite independently of any prevailing breezes or other propulsive agents. After watching it change shape and drift in what looked very like a deliberate, purposeful manner for several minutes, the sentry decided to inspect this anomaly directly, but when he approached it, the smoky spectre vanished.

Further examples of amorphous apparitions and foggy phantoms can be found in my book *Mysteries of Planet Earth* (1999).

HERNE THE HUNTER

One evening in September 1976, while on duty at the East Terrace of Windsor Castle, a young Coldstream guard happened to gaze at a statue in the Italian Garden - and, to his horror, the statue began to move, and sprouted a pair of horns! This, at least, is the story that he told his superior officer, after being discovered unconscious by the relief guard. When he was finally roused at the garrison's medical centre, he was found to be stone cold sober, but in a state of acute shock, and vehemently swore that his bizarre claim was true.

Moreover, in 1962, some youths found a strange hunting horn in Windsor Park, and when one of them picked it up and blew it, a terrifying figure with deer's antlers growing out of his head appeared, riding a huge black horse. Not surprisingly, the youths decided to leave the horn where they found it and fled with all speed!

These are just two of many reports attesting to the modern-day persistence of an ostensibly mythical phantom figure known as Herne the Hunter (comically impersonated, incidentally, by Falstaff in Shakespeare's *The Merry Wives of Windsor*). According to one legend, Herne was a deer keeper in Windsor Great Park during the reign of Richard II (1377-1399), who committed some heinous crime and then, in

shame, hung himself from a huge oak tree in the park. According to another legend, Herne hung himself from the tree when, after valiantly saving the king from being gored by a stag, he was dismissed by him from his service not long afterwards. In any event, following Herne's suicide the tree became known as Herne's Oak, and was allegedly haunted by his ghost - taking the form of a wild, rampaging man with a pair of stag's antlers sprouting from his head, riding through Windsor Park at the head of a pack of phantom baying hounds, and blowing upon a great horn.

In 1863, it seemed as if the Herne tradition had met its end, when, during a violent storm, a bolt of lightning struck a Windsor Park oak tree claimed to be Herne's and blasted it apart. However, Queen Victoria duly planted a new young oak in its place, and the legend continued.

An engraving of Herne the Hunter

Having said that, some historians claim that the real Herne's Oak was actually an aged specimen, sited elsewhere in the park, that was felled in spring 1796; this one was replaced on 29 January 1906 by Edward VII. In 1915, the eminent British folklorist Dr Katharine Briggs learnt that the father of one of her teachers, a retired colonel with apartments in Windsor Castle, used to espy Herne on moonlight nights, standing underneath his oak tree, but the precise locality of the tree in question was not specified.

During the 20th Century, Herne reputedly appeared at times of great national distress. For instance, sightings of him in the Park were reported just before the abdication of Edward VIII, the onset of World War II, and the death of George VI. He was also seen by a team of workmen renovating Windsor Castle during the mid-1930s, but as far as I am aware he did not appear before the more recent fire here in 1992. Nevertheless, if his restless spirit can linger for 600 years, who can say with certainty that the horned figure of Herne will never be seen or heard again within the Great Park at Windsor?

RADIANT BOYS

Also termed 'enfants brillant' (even though few, if any, are girls), phantoms known as radiant boys appear surrounded by intensely bright, glowing light, and are generally believed to be the ghosts of young boys that have been murdered. One of the most spectacular and oft-spied examples was witnessed by the rector of Greystoke on 8 September 1803 while visiting Corby Castle in Cumbria, home of the Howard family. Waking up at between one and two o'clock in the morning during his first (and only!) night at the castle, the rector saw:

> ...a glimmer in the middle of the room, which suddenly increased to a bright flame. I looked out, apprehending that something had caught fire; when, to my amazement, I beheld a beautiful boy clothed in white, with bright locks resembling gold, standing by my bedside, in which position he remained some minutes, fixing his eyes upon me with a mild and benevolent expression. He then glided gently towards the side of the chimney; where it is obvious there is no possible egress, and entirely disappeared. I found myself again in total darkness, and all remained quiet until the usual hour of rising. I declare this to be a true account of what I saw at Corby Castle, upon my word as a clergyman.

As a visit from a radiant boy usually portends bad luck and violent death, it is not surprising that the rector curtailed his visit to Corby, leaving the castle before needing to spend a second night there. Nevertheless, he was still alive and well 20 years later - and still telling people of his amazing encounter with the glowing ghost.

Someone with whom the appearance of a radiant boy is indeed associated with violent death, however, was a renowned English statesman - Lord Castlereagh, second Marquis of Londonderry (1769-1822). Although details of his encounter vary from one source

Engraving of Lord Castlereagh and the radiant boy

to another, it seems that when still a young man, and known then as Captain Robert Stewart, he had been visiting an acquaintance in Northern Ireland when, upon waking in the bedroom allocated to him by his acquaintance's butler, he saw a naked boy surrounded by blazing light. According to local tradition, anyone who saw this apparition was destined to achieve greatness but would die violently and unexpectedly.

Sure enough, Stewart's career became little short of meteoric. In 1796, he was created Viscount Castlereagh, and became Irish Secretary in 1797, followed by President of the Board of Control (1802), Minister of War (1805-6, 1807-9), and, from 1812, Foreign Secretary. In 1821, he succeeded his father as Marquis of Londonderry, but during the following year he began to suffer from ill-health and increasing mental turmoil, culminating in an obsessive fear of being blackmailed. On 12 August 1822, Lord Castlereagh unexpectedly committed suicide by cutting his throat with a razor.

THE GROWLING GHOST OF A WELSH WEREWOLF?

Encountering a mortal werewolf would be terrifying enough - so imagine the horror of confronting the snarling visage of a werewolf's ghost! Yet one such macabre event may have occurred in the unremarkable backdrop of the Welsh countryside little more than a century ago. According to Rev. Montague

Summers, an authority on werewolves and vampires, the following true-life vignette took place in summer 1888, and featured a professor from Oxford who had rented a holiday cottage in a mountainous area of Merionethshire.

During his stay there, the professor decided one day to go fishing in a local lake, and while doing so he hooked an unusual object that proved to be the skull of an extremely large dog-like beast. Curious to discover more about his unexpected catch, he took the skull back to the cottage, but left it in the kitchen when he went out for the evening with a friend, leaving his wife alone in the cottage. During their absence, his wife suddenly heard a strange snuffling sound outside the kitchen door, but when she went into the kitchen she saw to her horror the face of a terrifying red-eyed beast, seemingly part-human and part-wolf, outside the window, grasping the window ledge with paws that resembled hands.

Greatly frightened, she ran back to the front door of the cottage and bolted it...just in time. Moments later, the monster was panting outside, and rattling the door's latch. Unable to open the door, it paced round and round the cottage, snarling and growling with rage as it vainly sought some way to enter, until eventually it departed, leaving the petrified woman shaking with fear inside.

When her husband returned with his friend, they made absolutely sure that the house was totally secure, and then sat waiting with stout sticks and a gun at the ready, in case the wolf-man, werewolf, or whatever it was, returned - which it did.

The Welsh werewolf...or wulver...or wolfen?
(Richard Svenssen)

Later that same night, as the cottage lay enshrouded in a still darkness, its three alert occupants heard the soft crunching of paws upon the gravel outside, followed by the scratching of nails or claws against the kitchen window. And as they peered towards the sound: "...in a stale phosphorescent light they saw the hideous mask of a wolf with the eyes of a man glaring through the glass, eyes that were red with hellish rage". They raced to the door, but their quarry had heard them, and as they opened it they could just discern a huge form racing into the lake, and disappearing from view beneath the surface

There seemed only one way of bringing this living nightmare to a close, so as soon as it was light the professor left the cottage, rowed out into the lake, and hurled the mysterious skull as far as possible into the lake's depths.

Returning the skull from whence it had come evidently restored an equilibrium of sorts, because the werewolf's ghost never returned to the cottage.

Intriguingly, when discussing the above case (together with a remarkably similar, Hebridean incident) in a detailed *Fate* article (November 2000), I pointed out that a hairy humanoid with the head of a wolf occurs in Shetland mythology, and is known as a wulver.

However, in stark contrast to the terrifying lupine entity from Wales, the wulver is said to be totally harmless and shy. In fact, a much closer counterpart to the Welsh werewolf would be the wolfen, a highly intelligent species of humanoid wolf that has eluded scientific discovery but is responsible for werewolf reports down through the ages. Of course, the wolfen exists only in a 1978 novel of that name by Whitley Strieber and the subsequent 1981 film...doesn't it?

WHAT'S IN A NAME?

Today, chagrin, nightmares, and larvae are terms used, respectively, for disappointment, bad dreams, and juvenile insects - but long ago, all three had distinctly spectral associations.

Perhaps the most bizarre member of this etymologically-transformed trio is the chagrin. This name originally referred to an evil ghost dreaded by gypsies, in whose presence it assumes the decidedly unusual guise of a giant yellow hedgehog! Like many other animal ghosts, its appearance is believed to be an omen of impending doom for those who espy it.Equally malign is the mara - a dreadful hag or incubus-like entity of misty, insubstantial form occurring in traditional French and even modern-day Scandinavian lore (and not to be confused with a harmless Patagonian guinea-pig relative of the same name!). A mara torments people at night by sitting upon them while they are asleep and infesting their minds with horrific dreams. It is from this nocturnal wraith that the term 'nightmare' is derived.

Showing that ghosts are not a modern-day phenomenon, larvae were malevolent spirits greatly feared by the ancient Romans. Also called lemures (after which Madagascar's nocturnal primates, the lemurs, are named) if spied at night, some were unidentifiable wraiths, but others were the recognisable ghosts of evil persons. One of the most (in)famous larvae was that of the insane emperor Caligula, which was often seen lurking in his palace after his death. Larvae were notably abundant during the month of May, but could be dispelled by loud noises, and also by the smell of black beans burning on a fire!

THE MONSTER OF GLAMIS

One of the most macabre ghostly tales of the very unexpected derives from Glamis Castle, Scotland's oldest inhabited castle, which is renowned for its varied assortment of spectres and other paranormal entities reported down through the centuries.

According to local myth, during the early 1800s the 11th Earl of Strathmore, owner of Glamis Castle, was presented with his first son, who was therefore heir to the estate. Tragically, however, the infant was grossly deformed, with a hideous egg-shaped body and hairy torso, no neck, and disproportionately small limbs. Due to his monstrous form, the child was not expected to live, and was confined in a secret room, but instead of dying he thrived, and became immensely strong. In time, the Earl was presented with a second son, who was duly designated his heir, while the 'monster' remained concealed in his room.

When he reached the age of 21, however, the second son was informed of his older brother's existence by his father and sworn to absolute secrecy - a tradition persisting from generation to generation thereafter, for the Monster of Glamis is said to have been incredibly long-lived. Indeed, some reports claim that he did not die until as recently as 1921, though, as this would mean that he was over 200 years old, such reports cannot be taken seriously - but what about the principal component of the story? Had there really once been a hideously misshapen, disinherited heir to Glamis Castle, locked away from the world's prying eyes?

Unconfirmed rumours abound concerning a hidden, sealed-off room, as do tales of how a grotesque figure had occasionally been seen in the grounds at night. Moreover, in 1880 one Scottish newspaper even carried an account claiming that a workman in the castle had accidentally knocked through a wall and discovered a secret passage leading to a locked room, but that after reporting this to the castle's steward the workman had disappeared, and was believed to have been paid a sizeable sum of money and dispatched to Australia!

The headless phantom dog of Tingewick *(FPL)*

A HEADLESS DOG, A DOGLESS HEAD,
AND THE FRIGID PHANTOM OF A FROZEN CHICKEN!

Countless animal ghosts have been reported, but few are stranger than the following ones. Take, for example, the canine phantom inadvertently photographed at a garden party in Tingewick, Buckinghamshire, in c.1916 by Arthur Springer, a retired CID inspector from Scotland Yard. The image in his viewfinder as he clicked the camera's button was that of three ladies sitting around a small table, taking tea.
When the photo was developed, however, a slightly hazy, ghost-like form of a medium-sized terrier-type dog with medium-dark fur, broad collar, thin tail, and white front paws could clearly be seen standing next to one of the women. Yet no dog fitting its description, living or dead, was known from that area. The most remarkable aspect of this case, however, was that this animal spectre was not complete - above the region of its neck encircled by the collar, much of its head seemed to have faded away, leaving behind no clear outline.

The converse situation arose in September 1926, when Lady Hehir's Irish wolfhound, Tara, was photographed by a Mrs Filson. Six weeks before, Tara's devoted friend, a cairn terrier puppy called Kathal, also owned by Lady Hehir, had died, and the spot where the photo was taken had been a favourite haunt of the two dogs. Judging from the picture, the spot had not lost its canine appeal - for when Filson's photo was developed, the readily-recognisable head of Kathal could be clearly perceived peeping above the back end of Tara.

Perhaps the most bizarre animal ghost of all, however, allegedly haunts Pond Square in Highgate, London, for it is said to be the spectre of a half-naked, half-frozen chicken! Its origin lies in a pioneering if ill-fated experiment in refrigeration made one cold snowy day in March 1626 by the renowned English philosopher-scientist Francis Bacon, seeking to preserve fresh meat using snow. He purchased a chicken from a nearby farm, killed it, plucked off most of its feathers, cleaned out its abdominal cavity, and then proceeded to fill it with snow.

After completing that task, he placed the chicken's snow-filled corpse in a sack, which he then filled with more snow. While doing all of this, however, Bacon had become extremely chilled - so much so that he collapsed, severely ill, and he died a few days later.

Remarkably, in the centuries that have followed, many people claim to have witnessed the grotesque spectacle at Pond Square of a phantom, semi-plucked, shivering chicken, half-running and half-flapping its wings. One eyewitness who claimed to have seen it several times here on moonlit nights during World War II was Mrs John Greenhill, who described it as a "big, whitish bird". Aircraftman Terence Long also spied it once during the war, and was informed by an Air Raid Precautions fire-watcher that it was often seen here.

As recently as January 1969, a passing motorist stopped and actually attempted to catch this unique example of phantom poultry, mistakenly assuming that it was a living chicken whose feathers had been plucked by some mindless hoodlums, but the bird eluded him by vanishing into thin air.

THE MONSTROUS SPECTRES OF BEN MACDHUI AND BRAERAICH

Not for nothing is Ben MacDhui, Scotland's second highest mountain, known as the haunted mountain. This gloomy Cairngorms peak is said to conceal an evil, panic-inducing, phantasmal entity dubbed 'the big grey man', which has long been blamed for inducing many experienced, level-headed climbers to flee down its rugged slope in fits of blind terror, racing away from some unseen but tangibly foreboding presence. Sometimes its heavy crunching footsteps have been heard pursuing its hapless victim, and strange laughter or ghostly music has also been reported from Ben MacDhui.

Occasionally, a giant hairy yeti-like form has been spied here too, even though it is surely impossible for a corporeal entity of this nature to exist in such a well-explored region yet remain unknown to modern science. A more detailed examination of the Big Grey Man will appear in my second casebook, currently in preparation.

The same is true for nearby Braeraich, yet this does not prevent sightings of bizarre beings occurring spasmodically here. Notable among these was an undeniably weird encounter experienced by mountaineer Tom Crowley, while coming down from Braeraich to Glen Eanaich one day in the early 1920s. As in a number of cases reported from Ben MacDhui, Crowley suddenly heard the sound of footsteps behind him - and when he looked around, his eyes met with a spine-chilling sight. Standing before him was an

Is the Big Grey Man of Ben MacDhui a corporeal or a supernatural entity?

enormous figure partly shrouded by mist, but sufficiently visible to reveal a pair of pointed ears, long legs, and bizarre finger-like talons protruding from its feet!

Moreover, unaccountable man-beast entities have been reported from many other parts of the U.K. too, particularly in recent years, as investigated and documented by Nick Redfern in his fascinating book *Man-Monkey: In Search of the British Bigfoot* (2007).

PHANTOMS OF THE PAVILION, AND OTHER STRANGE GHOSTS OF BRIGHTON

It is said that people who have been closely associated with a certain building or locality in life sometimes continue this association even beyond death, as ghosts. The Royal Pavilion in Brighton seems to be a case in point. This spectacular edifice was the brainchild of the then Prince of Wales ('Prinny'), later to become Prince Regent and, ultimately, King George IV (1820-1830). Responsible for its initial creation in 1784, he engaged architect John Nash in 1811 to redesign many architectural masterpieces, a task that included rebuilding the Royal Pavilion in a sumptuously-elaborate Oriental style, mirroring a Moghul emperor's palace.

When complete, the Pavilion included a subterranean passage, and it is here that during the years following his death, the familiarly corpulent, ornately-attired spectre of Prinny was allegedly encountered on numerous occasions by visitors. Intriguingly, however, he does not seem to have been seen in any of the Pavilion's state-rooms - even though these, surely, are more appropriate localities to be haunted by the ghost of this building's royal creator.

During the 20th Century's inter-World-War period, conversely, a phantom was indeed spied in a Pavilion state-room - but the figure was definitely not Prinny. A caterer visiting this room one night to check that everything was laid out for a banquet to be held there the following day was startled to see the figure of a dumpy, elderly lady come from the direction of the kitchen, move up and down the aisles between the tables, and then disappear through a door. She was wearing what the caterer later described as a long "bunchy" skirt, a large bonnet, and a triangular shawl - all typical of the Regency period.

The caterer pursued this unexpected visitor, but when he opened the door through which she had passed, he found the corridor behind it empty - yet no-one reported seeing anyone matching her description leaving the building. Afterwards, while flicking through a collection of old Brighton pictures, the caterer spotted one that precisely matched the figure that he had seen. It was a print of Martha Gunn, better known as the Brighton Bather due to her pioneering popularising of sea-bathing. She was even known to have bathed Prinny once when he was a child, but she had been dead for many years. Yet even if it was her ghost that the caterer had seen, why was it interested in the state-room banquet? No-one knows.

The Royal Pavilion is not the only locality in Brighton linked with encounters of the spectral kind. Meeting House Lane is said to be frequented by the ghost of a pre-Reformation nun in a grey habit, often sighted disappearing through a bricked-up doorway in a wall next to the Friends' Meeting House. In the 12th Century, the priory of St Bartholomew stood behind this wall, and was guarded by soldiers - one of whom fell in love with a nun. They eloped, but were soon caught, and the nun was walled up alive. Moreover, while walking in the churchyard at St Peter's Church in Preston Park, North Brighton, in the 1970s, a couple saw a lady in medieval costume; they assumed that she was wearing fancy dress for a pageant - until, after ignoring them, the lady simply faded away while next to a large tomb.

During the summer of 1922, Albert D. Glovert from Kensington stayed at a boarding-house in a street near to the Royal Pavilion. During the third day of his stay, he was a little surprised to see a tall slender girl wearing a very old-fashioned green dress walk down the stairs while holding a handkerchief in front of her face, so that her face was completely obscured. He was even more surprised when he saw her again the next day, coming out of one of the bedrooms and entering another one, but wearing the same dress and still holding a handkerchief in front of her face. Yet when he asked one of the maids about her, the maid replied that she had never seen anyone fitting that description in the house. Nevertheless, two days later, Glovert saw her for the third time - and on this occasion, she was walking ahead of him, down a corridor leading directly towards his own bedroom!

When she reached his door, however, she stopped, turned around to face him, and slowly removed her handkerchief. To Glovert's horror, the sunlight streaming into her face revealed a nightmarish mouth of grotesquely large size, filled with long yellow teeth that projected like tusks - but that was not all! As he gazed in macabre fascination at her repellent visage, the girl in green leered lasciviously at him, then entered his bedroom, leaving the door half-open in evident expectation that he would follow her! Shocked out of his stupor, Glovert threw the door open wide in rage, ready to tell her in no uncertain terms to leave his room at once - but when he looked inside, his self-invited 'guest' had utterly vanished!

When he challenged the boarding-house's owner, a Mrs Dealman, she vehemently denied the existence of such a person, so Glovert left at once. Two years later, however, he met Mrs Dealman again, and while chatting she confessed to him that her boarding-house had indeed been haunted by an unidentified ghost fitting the green girl's description. At first, she had tried to silence the rumours by denying them, but so many visitors and staff reported seeing the apparition that she had finally closed the boarding-house down and moved away.

Two very different but equally odd spectres are linked to an ancient stone plinth within St Nicholas's churchyard in central Brighton. According to one legend, a knight and his steed, both attired in full armour, are buried beneath it, and on moonlit nights the steed can sometimes be spied galloping alone around the churchyard. Alternatively, it is said that the plinth marks the grave of a beautiful but tragic medieval maiden known as Lady Edona. Local lore claims that Edona died here from shock and grief after seeing the galleon bearing her betrothed lover, Lord Manfred (eldest son of the 4th Earl de Warrenne), run aground upon a concealed rock off Worthing Point on 17 May. Lord Manfred was returning home from a Byzantium pilgrimage to marry her; but instead, his vessel swiftly overturned, drowning him and almost everyone else aboard. Since then, however, an eerie glowing phantom facsimile of this ill-fated ship can allegedly be seen here every 17 May at midnight, sailing towards the harbour, only to keel over and sink when it collides with the fatal rock.

In 1950, a long-established Brighton stationers shop in East Street, called John Beal's, was plagued by a series of unheralded (and unwelcome) appearances by a mysterious dark-robed entity wearing a cowl. Fortunately, these visitations ceased, and the figure is no longer seen here - but perhaps it simply moved elsewhere.

So keep your eyes open the next time you visit Brighton, and if someone there says that you look like you've just seen a ghost - who knows, they could be right!

HAIRY HANDS AND DISEMBODIED LEGS

One of the most malevolent apparitions on file must surely be the terrifying 'hairy hands' of Dartmoor. As far back as the days of horsedrawn carriages, these spectral hands were a source of terror to travellers journeying along the lonely road crossing the moorland between Two Bridges and Postbridge, but it was not until the 1920s that news of their deadly deeds attracted widespread attention beyond Dartmoor. Perhaps the most notable event took place on 26 August 1921. This was when a young army officer riding a motorbike along the hands-haunted road suffered, but survived, a very unpleasant - and uncanny - motoring accident. When later questioned by friends, he made the following, extraordinary statement:

> You will find it difficult to believe, but something drove me off the road. A pair of hairy hands closed over mine - I felt them as plainly as ever I felt anything in my life - large, muscular, hairy hands. I fought against them as hard as I could, but it was no use, they were too strong for me. They forced the machine into the side of the road, and I knew no more until I came to my senses lying on my face on the turf a few feet away from the bicycle.

What makes this incident even more chilling than it already seems is that only two months earlier, at a spot very close to where the officer's inexplicable 'accident' had taken place, another motorcyclist, medical officer Dr E.M. Helby, had also suffered a mysterious accident - but he was less fortunate, and was killed.

Moreover, years later, 28-year-old Florence Warwick, who was only visiting Devon on holiday and knew nothing about the hairy hands stories, was driving along that fateful road between Postbridge and Two Bridges one evening when her car unexpectedly broke down. As she sat inside, attempting to read its manual, she was horrified to see a pair of huge, disembodied, hairy hands press themselves against the outside of the windscreen, and crawl along it, as if seeking a means of entry!

Too frightened even to scream, Florence sat in fear-frozen silence, staring up at these twin terrors for what seemed like a lifetime, until she finally summoned up enough courage to cry out - and as she did so, the hands disappeared. Remarkably, the car promptly started again too, enabling her to drive away from this accursed spot.

Far less threatening, if no less bizarre, surely, is the sight of a pair of ghostly disembodied male legs racing downstairs! Yet this unlikely-sounding, comical spectacle has indeed been witnessed, and on more than one occasion, in the landlord's private quarters of the Black Swan, the oldest public house in York. Who knows? Perhaps the explanation for this amazing apparition is that some long-deceased frequenter of the Black Swan took the alcohol-related expression 'becoming legless' a little too literally for his own good!

BIRDS OF ILL OMEN

Although not strictly spectral, the following subject is certainly uncanny enough to warrant inclusion in this chapter. In their book *Phenomena: A Book of Wonders* (1977), John Michell and Robert J.M. Rickard devote a chapter to what they refer to as "Family retainers, weird and ominous". These are animals whose appearance presages a death, particularly within a specific family. Michell and Rickard provide details of several examples, the most famous being the large white bird of unidentified species that allegedly has mysteriously appeared inside the room of several different members of the Oxenham family of Devonshire origin just before their respective deaths.

The first recorded instance featured the death of Grace Oxenham in 1618, and Michell and Rickard mention a number of other cases down through the generations until as recently as 1873, with the death of G. N. Oxenham.

In another remarkable example, on the day of amateur ornithological enthusiast Reverend W.W. Fowler's death in Lincoln a surprising number of owls were seen to gather around his house, disappearing when

Miss Edith Griffin with her great-nephew,
the author Dr Karl Shuker (aged 7 months)

he died. On the day of his funeral, however, a single barn owl abruptly appeared, swooping down low in broad daylight over his coffin as it was being carried under the lychgate to the church, as if in deference, before flying away again.

These and other similar cases echo the traditional English folk-belief that birds will come to tell a person when a relative is about to die. Moreover, not everyone is convinced that it is just folklore. The following incident involved members of my own family.

During the late 1960s, Miss Edith Griffin, my very elderly great-aunt, suffered a severe stroke from which there was no hope of recovery. Each day, her niece, my mother Mary Shuker, visited her at her home, which she shared with my grandparents (one of whom was her younger sister Mrs Gertrude Timmins, whose recollection of a shower of frog rain is included in Chapter 18). Back at our own house one night, after spending the day with my great-aunt, my mother was unable to sleep, and lay awake in bed for some time. Suddenly, she heard a bird land on her windowsill and begin chirping loudly, even though it was now approximately 2 am the next morning and totally dark outside. This continued for quite a time before the bird eventually flew away, leaving my mother feeling very ill at ease by this strange, unexpected visitation. So much so, in fact, that she called in at my grandmother's house just a few hours later, earlier than normal - where she learnt that my great-aunt had passed away, at around 2 am that morning. Just a coincidence?

There are countless phantoms, spectres, and suchlike whose veracity has never been disproved, but that does not prevent sceptics from denying the existence of *every* type of ghost. Yet if ghosts are entirely fictitious, without any basis whatsoever in reality, why are there so many reports, from every corner of the globe, dating back into the far-distant mists of antiquity? This paradox is, perhaps, the greatest ghost-related mystery of all.

Chapter 3

VISITING THE LAND OF THE STONE GIANTS

Moai, Manbirds, and Other Mysteries of Easter Island

Wherever we climbed and wherever we halted, we were surrounded, as in a hall of mirrors, by enormous faces circling about us, seen from in front, in profile, and at every angle...We had them above us, beneath us, and on both sides. We clambered over noses and chins and trod on mouths and gigantic fists, while huge bodies lay leaning over us on the ledges higher up. As our eyes gradually became trained to distinguish art from nature, we perceived that the whole mountain was one single swarm of bodies and heads, right from the foot up to the very top of the precipice on the uppermost edge of the volcano. Even up here, five hundred feet above the plain, half-finished giants lay side by side, staring up into the firmament, in which only the hawks were sailing. But the swarm of stone men did not stop even up here on the topmost ridge, they went on side by side and above one another in unbroken procession down the inside of the crater. The cavalcades of stiff hard-bitten stone men, standing and lying, finished and unfinished, went right down to the lush green reed-bed on the margin of the crater lake, like a people of robots petrified by thirst in a blind search for the water of life.

Thor Heyerdahl – describing the quarry at Rano Raraku, in *Aku-Aku*

The travel guides had not been joking when they described it as the world's most remote inhabited locality - a minute triangular speck in the middle of the South Pacific, almost equidistant between Tahiti to the west and Chile to the east, and more than 2,200 miles away from both of them. It had already taken me well over a day and who knows how many thousands of miles of air travel just to reach Chile from my home in England, on the other side of the world. And then I had yet another 5 hours of flying time across a blue expanse of ocean awaiting me before I would reach my intended destination. But at last, just before mid-day local time on 5 April 2008, after poring over countless books and perusing untold television documentaries detailing its unique history and mysteries for as long as I could remember, I finally arrived, stepping down from the aeroplane into the extraordinary land of the stone giants – or, as it is better known, Easter Island.

FIRST IMPRESSIONS

Also called Rapa Nui, and with a population of roughly 3500, Easter Island has been an overseas territory of Chile since 1888, and was first brought to the attention of the Western world on Easter Sunday (hence its name) in 1722, when it was formally discovered by Dutch explorer Admiral Jacob Roggeveen. Today, it has a single town, Hanga Roa, which, with dirt-track roads and dusty stores juxtapositioned amidst leafy groves and luxuriant blooming flora of every conceivable kind, recalled to mind a 19th Century American frontier town that had somehow been dropped headlong onto a tropical island. However, it is the pre-European contact history of Easter Island that makes it so fascinating and mystifying.

Where else on earth could you find a tiny island (less than 20 miles across at its widest) that was once ringed by hundreds of enormous inward-facing monolithic statues and home to a thriving birdman cult, with archaic stone houses representing the human womb, a bleak and almost treeless vista that had formerly been swathed in dense forests, and a cryptic hieroglyphic script that continues to defy all attempts at deciphering it?

My initial assumption upon arriving on Easter Island – whereupon I was promptly garlanded with a lei of exotic flowers by the welcoming locals meeting the latest influx of tourists at the tiny airport – was that it was a typical Polynesian island, brightly carpeted by multicoloured blossoms and lush vegetation...until I learnt that almost every flower, tree, and shrub that I could see, and even the butterflies and dragonflies flitting close by, had been imported here from elsewhere. Hardly anything was native to Easter Island – it was as if it had been created as a massive film set, with each feature carefully planned and meticulously planted to produce a completely new world.

To understand why this was so, we need to understand how Easter Island's civilisation came about, evolved, and devolved – and, during this complex process, engendered the multitude of mysteries and controversies for which it has become so rightly famous in modern times.

HOW IT ALL BEGAN...AND ENDED

Thanks to modern archaeological research, we now have at least a basic idea of Easter Island's history prior to its discovery by Europeans, although there is still much dispute concerning the finer points. It seems to have been colonised during the 4th Century AD (though some claim as late as the 7th or even the 8th Century), by Polynesian seafarers, probably from the Marquesas Islands, who would have been confronted by an Easter Island very different indeed from the one that I encountered a few months ago. Back in those long-gone colonisation days, the entire island was clothed in subtropical moist broadleaved forest (confirmed by pollen studies and archaeological work), which most sensationally included among its many botanical endemics the world's largest species of palm tree, *Paschalococos disperta*.

The elders and leaders of the independent clans (kin-groups) that gradually emerged following colonisation were greatly respected and admired by their clans' members. So much so that a tradition began whereby each clan would carve a stone statue of its leader, and erect it upon a platform near the coast, but facing inwards, in order to look over the clan as a symbol of protection. These stone statues became known as moai, and the platforms were called ahus, the earliest known examples of which date back to 690 AD. Originally, the statues were little larger than their human models, but over the centuries, as more and more were created, they became ever bigger. By the 15th Century, at which time moai production had reached its peak, with hundreds in existence all over the island, the moai had become colossal, over 20-30 ft tall in some cases, and were unlike anything to be seen anywhere else in the world.

The author Dr Karl Shuker (5'10" tall) alongside a half-buried moai on the slopes of Rano Raraku (*Dr Karl Shuker*)

A moai with restored eyes at Ahu Kote Riku, Tahai *(Dr Karl Shuker)*

They were hewn from tuff, an igneous rock ash present within a huge volcanic crater called Rano Raraku in the island's eastern half, which became a moai quarry where the statues were fully carved before being moved to their required locations elsewhere on the island. Some even bore red topknots (pukao) on their heads (carved from scoria rock hewn from a quarry called Puna Pau), and their carved eyesockets originally contained eyes, made from white coral and black obsidian.

There has been much dispute as to how these stupendous statues were moved, bearing in mind their size and immense weight (averaging 14 tons, but sometimes very considerably more), with some of the more intriguing suggestions including air-lifting by aliens, levitation by the harnessing of electromagnetic forces, and the statues walking by themselves using a special life force called mana. It is now widely believed that they were transported on rollers made from the trunks of the palm trees. Needless to say, however, on such a small island there was a very finite number of trees, and with moai production reaching almost frenzied proportions as clans actively competed with one another to see who could produce the largest and most spectacular examples, Easter Island ultimately became entirely deforested, its giant palm tree now extinct.

This in turn brought calamity upon the entire civilisation, because once the trees had gone, together with shrubs and other woody plants, substantial soil erosion occurred that in turn limited attempts at crop cultivation; spears used for hunting sea fishes could no longer be made; and even timbered fishing boats became a thing of the past with only reeds available for binding together to create canoes. To sustain themselves, the people killed the seabirds that once nested in great quantities around the coasts, as well as the various native land birds that existed here up until then, until they were all wiped out except for a few that escaped to some minuscule offshore islets.

Now in rapid decline, and without the trees needed for manufacturing rollers, the clans quite suddenly stopped creating moai, and hundreds can be found today in varying states of completion still attached to the inner walls of Rano Raraku's crater, which, as so evocatively described by Norwegian explorer Thor

The author alongside a toppled moai with Rano Raraku in the background *(Dr Karl Shuker)*

The 15 upraised moai at Ahu Tongariki
(Dr Karl Shuker)

Heyerdahl (see this chapter's opening quote) in his book *Aku-Aku: The Secret of Easter Island* (1958), is a truly breathtaking, surrealistic sight to behold. They include the biggest moai ever produced, the unfinished but aptly-named El Gigante, measuring a colossal 71.93 ft, and of such stupendous weight (estimated at 145-165 tons) that even if it had been completed, it seems very unlikely that it could ever have been transported out of the quarry. Now, instead of friendly moai-creation competition, rivalry between the clans became decidedly unfriendly as they viciously turned against one another, producing obsidian daggers and other weapons with which to wage savage ongoing bouts of civil warfare, and even incidents of cannibalism.

Finally, not long after the first Europeans had reached the island, the native clans defied what until then had been their ultimate taboo. Losing faith in the protective powers of their moai, as well as seeking revenge on rival clans, they began overturning the giant statues that they had once carved and erected with such zeal and pride. In 1722, Roggeveen had declared that all of the moai observed by him and his crew were standing, but when famed British explorer Captain James Cook arrived in 1774 he reported seeing many overturned statues lying beside their ahus, and in 1868 Linton Palmer, a visiting English doctor, recorded that not a single moai on the island remained standing, with a great number of them lying face down and broken.

From veneration to vilification, the rise and fall (in every sense) of the giant stone statues and the entire Easter Island civilisation, as well as the wholesale destruction of the latter's ecosystem, had occurred within the space of a single millennium.

Happily, during the last half-century, several funded projects have helped to raise up quite a number of unbroken moai onto remaining ahus, and today there are several sites featuring spectacular series of upright moai once again dominating the landscape. These include Ahu Tongariki, comprising no less than 15 moai on a huge ahu, restored to former glory after having been flattened in modern times, though not by human agency on this occasion but by the savagery of nature, when a huge tsunami hit the island's southeastern coast in 1960; Ahu Nau Nau at Anakena on the east coast, bearing seven upright moai on an ahu; Ahu Akivi, the only inland moai-bearing ahu, bearing seven moai that uniquely face out towards the coast instead of inward; and Ahu Kote Riku at Tahai, which has a moai with restored inset eyes.

Currently, a total of 887 moai have been recorded on Easter Island (of which 397 remain in the Rano Raraku quarry), but it is suspected that many more remain buried, especially on the slopes of Rano Raraku, covered over by encroaching soil and vegetation in the centuries since they were toppled during the outbreaks of inter-clan war.

In addition, the island's devastated landscape has been repopulated via a huge programme of biological introductions, including many different species of plant, insect, and even a few species of land bird (most notably the very impressive Chilean caracara hawk).

A fully-revealed moai at Ahu Tongariki *(Dr Karl Shuker)*

This, then, is the official history of Easter Island – but as I was to discover during my visit here, there is still much that is shrouded in controversy, confusion, and mystery.

MOAI MYSTERIES

Take, for example, Easter Island's iconic moai. They are commonly referred to in many publications as stone heads, but this is wholly incorrect. True, when walking upon the grassy slopes of Rano Raraku, one would be forgiven for assuming this description to be accurate, because there are indeed monstrous heads standing upright everywhere - staring imperiously but sightlessly ahead through empty eyesockets, with thin lips and haughty visage radiating Ozymandian disdain.

However, just like icebergs, much of their total form is hidden from sight. As Thor Heyerdahl revealed back in 1955 when he and a team of archaeologists spent several months here conducting the first, pioneering studies of Easter Island's enigmatic archaeology, hidden beneath the layers of shifting soil that have gradually accrued around them down through the centuries since they were originally erected are these statues' torsos. Each comprises a long body down to the hips, a pair of spindly arms pressed closely to the sides of the body (which has nipples), and a pair of hands with very elongate fingers splayed across the bulbous stomach.

These features can be readily perceived in specimens that have been re-erected on ahus elsewhere on the island. Furthermore, one torso-exposed Rano Raraku moai has an early three-masted European sailing ship skilfully carved upon its stomach, suggesting that although by the time of Western arrival here the age of moai manufacture had passed, there were still some talented local craftsmen in existence.

A particular revelation for me, after reading in several publications that the moai all shared the same face, was that they were actually all recognisably different. This supports the belief that they represent the elders of different clans. Moreover, whereas it is widely claimed that they are all male, my guide pointed out one moai on the slopes of Rano Raraku that is nowadays believed to be female. Even more intriguing is Tukuturi - the enigmatic round-faced, bearded, kneeling moai with squat body and well-formed arms, discovered in 1955, and which can be found on the outer flank of this volcano's quarry. The most popular view is that it is an extremely primitive, prototype moai, the earliest still in existence, in fact, from which the more familiar gaunt, lantern-jawed, taller moai style of carving subsequently evolved. In contrast, a differing school of thought has postulated that it may be the most recent moai to have been carved, some time after all of the others.

Equally mystifying are the topknots or pukao originally present on the heads of the moai – for the simple reason that no-one has any idea how they were placed there. As many of the larger moai are over 20 ft tall, and their correspondingly-sizeable topknots are exceptionally heavy, it required a crane to lift them up when replacing them on the heads of moai re-erected by contemporary researchers. How, therefore, lacking the convenience of such modern-day machines, did the early Easter Island clans accomplish this formidable task?

The leaning moai of Rano Raraku *(Dr Karl Shuker)*

Tukuturi, the unique round-faced, kneeling moai *(Dr Karl Shuker)*

One might also ask, when first observing it, how the famous leaning moai of Rano Raraku, tilting forward at an extremely precarious angle, continues to avoid toppling over onto its face. The answer of course is that the rest of its form, buried beneath centuries of soil and vegetation, anchors it firmly in place. Even so, I resisted the temptation to walk underneath it - just in case!

LONG EARS...OR TALL TALE?

Another oft-quoted Easter Island claim is that there were once two totally different tribes here. One was the aristocratic Long Ears, named after their purposefully elongated earlobes, a very distinctive characteristic faithfully reproduced on the moai. The other was the menial Short Ears. According to lore reported by locals to Thor Heyerdahl, and since reproduced in countless books, magazines, and other publications, during the late 17th Century the Short Ears rose up and rebelled against their long-eared overlords, massacring all but a single Long Ear, thereby marking the end of the latter as a separate tribe.

When I mentioned this to Karen, my guide on Easter Island, however, she was very amused, informing me that this was totally fictitious. In reality, she stated, every clan had a Long Ear as its leader (thus explaining why the clan leaders' stone effigies, the moai, were long-eared) rather than there being a discrete Long Ear tribe, and that the story of the Long Ear versus Short Ear battle had been nothing more than a tall tale spun to a gullible Heyerdahl by locals as a joke, but which had since taken on a life of its own until it was now widely, albeit wrongly, accepted as fact.

Speaking of earlobes – just a week before I arrived on Easter Island, I was horrified to read in the media that a Finnish tourist had clambered onto one of the magnificent moai at Anakena and had broken off one of its earlobes to keep as a souvenir! Not surprisingly, he was swiftly arrested and jailed on the island, but it is unclear whether the earlobe can be reattached to the damaged moai. Bearing in mind that it is a strictly-observed taboo on the island even to touch a moai, let alone climb onto it and vandalise it, the shock waves generated by this incident were still reverberating at the time of my visit there, and it beggars belief how anyone could even think of committing such desecration.

Incidentally, a number of travellers to Easter Island have reported a dark feeling of oppression and apprehension when in the presence of the moai, as if the statues considered them interlopers or threats to their island sanctuary. Yet I experienced no such sensation whatsoever, and was in close proximity to moai on many occasions during my stay here. Who knows, perhaps they could somehow sense my respect for their status and antiquity, and were mindful that I was here to marvel at but never to malign their unique, lonely realm, whose silent guardians and sentinels they have been for so very long.

EAST OR WEST - WHICH IS BEST?

Perhaps the thorniest subject of contention relating to Easter Island is the origin of its initial, pre-European colonists. It was Thor Heyerdahl who put forward the dramatic hypothesis that instead of reaching the island from Polynesia to the west (as had traditionally been assumed), perhaps its inhabitants had come from the east, making their way from South America across the Pacific on huge balsa wood rafts, as found in Inca culture. To substantiate this very radical claim, he built such a raft, dubbed it the Kon-Tiki after an Incan deity called Kon-Tiki Viracocha, and in 1947 he and five companions successfully sailed 4300 miles on it right across the Pacific, from Peru to an island east of Tahiti, taking 101 days. In 1979, almost three decades before finally reaching Easter Island, I had visited the Kon-Tiki Museum in Norway, housing a reconstruction of this historic craft – alongside the world's only life-sized giant moai replica outside Easter Island itself and Chilean museums – and at that time I was very impressed by the boldness of Heyerdahl's notion.

A Rano Raraku moai clearly showing an elongated earlobe *(Dr Karl Shuker)*

Nevertheless, as more recent research has intensified into Easter Island's history, coupled with very extensive technological advances, it has become increasingly clear that although Heyerdahl's own voyage confirmed that such a journey was indeed possible, it is not supported by other, independent evidence. Studies in disciplines as diverse as linguistics, palaeobotany, genetics, and archaeology have all revealed and emphasised links between Easter Island and Polynesia (even the DNA obtained from a series of ancient human skeletons found on Easter Island in 1994 was shown, when analysed, to be of Polynesian affinity), whereas comparable links between this island and South America are conspicuous only by their absence. Having said that, two issues that had been offered in the past by Heyerdahl as particularly convincing evidence of a South American origin for the Easter Island colonists were the Vinapu ahus and the presence on the island of the sweet potato *Ipomoea batatas*.

On Easter Island's southwestern coast, Vinapu is the site of two large ahus. The bigger of the two is known as Vinapu I, and it is constructed from huge blocks of skilfully-shaped, precisely-carved stones fitting very tightly together yet without the presence of mortar, and thus recalling the masonry of various Inca ruins in Peru. The smaller ahu, Vinapu II, is of very inferior construction, and modern radiocarbon testing has shown that it was built in 857 AD, whereas Vinapu I dates from 1516 AD. However, when Heyerdahl encountered them back in the 1950s, he believed (wrongly) that Vinapu I was the earlier of the two, and he promoted its apparent similarities to Inca relics as proof for his theory that Easter Island had been originally colonised not by Polynesians but by advanced South American voyagers from the Tiahuanaco culture.

Moreover, Heyerdahl believed that the leader of the South American colonists had been none other than Kon-Tiki Viracocha, venerated as a deity by the Incas but claimed by Heyerdahl to have been a real man. Unfortunately, although undeniably fascinating, this entire scenario was riddled with anachronisms, which were later exposed in succinct style by Polynesian anthropology expert Dr Robert Suggs:

> Heyerdahl's Peruvians must have availed themselves of that classical device of science fiction, the time machine, for they showed up off Easter Island in A.D. 380, led by a post-A.D. 750 Incan god-hero, with an A.D. 750 Tiahuanco material culture featuring A.D. 1500 Incan walls, and not one thing characteristic of the Tiahuanaco period in Peru and Bolivia.

Also, Heyerdahl was apparently unaware that deceptively Inca-like Polynesian stone-working traditions such as the Marae exist. And although it might appear superficially similar, the design of the Vinapu I ahu's retaining wall is fundamentally different from those of Inca relics.

As for the sweet potato: Easter Island grows a number of important food plants, including the banana plant, taro root, sugar cane, and sweet potato, which sustained the first colonists here. However, none of these plants is native to the island, so they were all evidently brought here by the colonists. The only problem with this is that whereas the banana, taro, etc, are known to be of Polynesian origin, the sweet potato is South American. Once again, this geographical anomaly was not lost upon Heyerdahl, who believed it provided strong support for his theory of a South American origin for Easter Island's original colonists. Even today, moreover, it is widely believed that this riddle has not been solved.

However, my Easter Island guide informed me that according to Pacific botanist E.D. Merrill, writing in his book *Plant Life in the Pacific* (1945), there was evidence for believing that the sweet potato had actually originated not in South America but rather in Africa or South Asia, and had only later been transported across the Pacific, ultimately reaching the New World. Other researchers have subsequently suggested that this plant may indeed have been brought to Easter Island from South America, but not by its original colonists, which were Polynesian, but instead during occasional later visits by South American seafarers.

CULT OF THE BIRDMAN

Second only in fame to its gargantuan monolithic moai is Easter Island's extraordinary birdman cult. We still do not know exactly when, how, or why it arose, and there are many conflicting opinions. Some believe, for instance, that the cult developed in tandem with moai production; others that it began as late as the 18th Century. Nevertheless, this remarkable annual ceremony, occurring each September, was still being practised as recently as 1878, and would begin at the village of Orongo, sited on the slopes of Rano Kau - the island's spectacular westernmost volcano. Although long-abandoned, this village can still be explored today, where, as I was to discover during my own visit, many of its very strange, low, stone-slab houses have been partially restored. Containing only a tiny door through which the occupant had to crawl in order to enter, an Orongo house could only be slept in; all other activities, including eating, had to be done outside, because in Orongo tradition the house symbolised the human womb, and its internal space was therefore sacred.

Traditional stone-slab house restored at Orongo *(Dr Karl Shuker)*

The birdman ceremony was basically an inter-clan competition. Each clan chief would have a representative whose goal was to obtain the first egg laid that year by a small migratory species of seabird called the sooty tern, which nested on Moto Nui, the largest of three small offshore islets. The egg was deemed to be an incarnation of the creator god, Make-make, so it was very precious. The clan chief whose representative succeeded would then become the Birdman or Tangata Manu for the next year, bringing great glory and status to his clan, but he would have to live apart from them, in a special cave at the foot of

Moto Nui, with two smaller islets in front *(Dr Karl Shuker)*

Petroglyph of the Birdman, lying on its back *(Dr Karl Shuker)*

Rano Raraku, at the other end of the island.

To reach Moto Nui, the competitors would have to scale down the treacherous, near-vertical, ocean-overlooking cliff of Rano Kau, risking death if they fell and hit the jagged rocks below, then swim through shark-infested waters until they reached the islet. Once there, they would seek a newly-laid tern egg (though this could require several weeks of waiting if none had already been laid), then swim back with it, ensuring that the sacred egg remained unbroken, and present it in triumph upon reaching Orongo.

Commemorating this ceremony are hundreds of Birdman petroglyphs on rocks around Orongo and on the cliff, always depicting a bird-headed human or manbird, lying curled up on its back, sporting a long curved beak and a crest. Even today, the Birdman has not entirely vanished, as the statues of Christian saints in the island's church have the heads of birds.

THE RIDDLE OF RONGORONGO

Perhaps the most perplexing mystery of all associated with Easter Island is its unique and seemingly indecipherable hieroglyphics. It was French missionary Father Eugène Eyraud, writing in his journal of 1864, who first reported that most Easter Island villagers' homes contained slabs or tablets (occasionally staffs) of wood (never stone) bearing an extraordinary form of picture writing that they greatly revered but which none of them could understand. Later investigations revealed that only the clan elders and priests had been able to read this script, known as rongorongo. Tragically, however, all who had possessed such knowledge had been transported off the island during the 1850s-60s, taken as slaves to work – and die - in Peruvian guano mines after Easter Island was seized by Peru (prior to Chile assuming control in 1888), or had died in a later smallpox epidemic.

In short, the ability to read rongorongo had been lost. Moreover, many of the slabs and staffs were soon to be lost too, when villagers, fearing that they would fall into the hands of foreigners and thereby invoke their ancestors' displeasure, destroyed them or hid them away in caves (where they were eventually forgotten about and/or ruined by the damp conditions). Today, less than 30 examples of rongorongo exist, and, ironically, not a single one of them is on Easter Island itself, as they are now all housed in various Chilean and other overseas museums.

A rongorongo replica tablet *(Dr Karl Shuker)*

The picture writing, containing images of birds, plants, humans, and fishes alongside strange symbols, has defied every attempt to decode it fully, although linguistics expert Steven Fischer, assisted by a wooden rod known as the Santiago Staff - once owned by an Easter Island clan chief and bearing rongorongo script uniquely divided up into sections - has recently achieved a limited degree of success. He now believes that at least some of the rongorongo inscriptions are creation chants, and that another set comprises a calendar, but there are still many others of wholly unknown meaning.

SEEING A COELACANTH?

A final, much less famous, but no less interesting, Easter Island enigma can be seen in the form of local wood carvings – a strange fish with leg-like fins, known as the patuki. In his book, *Mysteries of Easter Island* (1969), Francis Mazière stated that according to island tradition, man descended from this somewhat frog-like entity after ten transformations, brought about by alterations in the climate and therefore in his food, and by various "man-like reactions". This is certainly intriguing, but what is even more so, especially for those of a cryptozoological persuasion, is Mazière's claim that this metamorphic legged fish has marked analogies with the coelacanth.

What needs to be explained at this point, however, is that the patuki is actually a small species of blenny, *Cirripectes alboapicalis* (=*C. variolosus patuki*), whose fins are not leg-like, and which bears no resemblance whatsoever to a coelacanth.

The entity that Mazière was describing, and which was depicted in the carvings that I saw while visiting Easter Island, is actually the supposed appearance of this blenny once it has begun (according to traditional legend) its series of transformations that ultimately convert it into a man. So it is during this mythical metamorphosis that the blenny resembles a coelacanth – but from where does such an unexpected

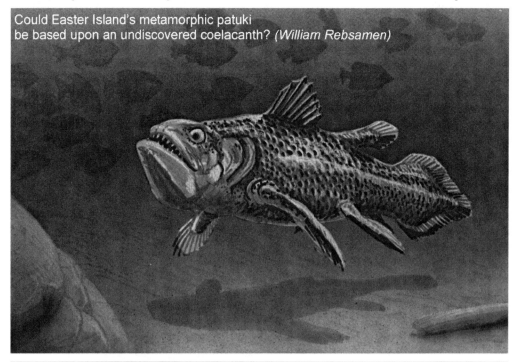
Could Easter Island's metamorphic patuki be based upon an undiscovered coelacanth? *(William Rebsamen)*

similarity derive? Could it be more than just coincidence? Believed to have died out many millions of years ago, the coelacanth line was sensationally resurrected in 1938 when a living species, dubbed *Latimeria chalumnae*, was discovered off South Africa, with specimens also later being obtained off the Comoro Islands, north of Madagascar.

During the late 1990s, a second, closely-related living species, *L. menadoensis*, was discovered many thousands of miles to the east, in waters off the Indonesian island of Sulawesi. Moreover, there have been unconfirmed reports of coelacanth scales having been found in other far-flung localities, such as the Mexican Gulf, and Australia. Consequently, some cryptozoologists have speculated that living coelacanths may be much more widespread than the two currently-known, geographically-disparate species would suggest.

Could it be that the mythical patuki is based upon sightings or even captures of coelacanths off Easter Island by its early colonists? Separated from all other inhabited islands by thousands of miles, the waters around Easter Island would have been a longstanding sanctuary for coelacanths, at least until the arrival 1600 years ago of the first human colonists. Yet who can say? Like so much about this extraordinary mid-oceanic microcosm, the answer lies shrouded in the mists of Easter Island's distant past.

THE MOTHER OF ALL JOURNEYS

My visit to Easter Island will undoubtedly remain one of the most memorable that I am ever likely to experience.

As for its mysteries and controversies, I returned home to England with many more questions than answers – not least of which is: having now visited the world's most remote inhabited locality, where on earth can I journey to the next time I decide that I need to get away from it all?!!

And finally: just in case anyone reading this chapter imagines that journeying from England to Easter Island would surely be much too tortuous or traumatic for them even to consider attempting it, let me point out that throughout my journey to this remote island and to all of its major sites once I arrived there, I was accompanied by my intrepid, indefatigable mother, Mrs Mary Shuker, aged 87.

Who said that Britons no longer have any spirit of adventure?!

The author's mother, Mrs Mary Shuker, alongside a mini-moai in the lush gardens of Hotel Otai on Easter Island (*Dr Karl Shuker*)

Chapter 4
MUMMIES FROM SPACE
AND VEGETABLE MAN
Close Encounters of the Bizarre Kind

Either the eyewitness reports can be trusted, or they can't. If they are reliable, then the aliens certainly display a bewildering array of anatomical forms and show a great diversity of shape, size, skin color, and other features. Through the years there have been aliens of all colors: black, white, red, orange, yellow, blue, violet, and of course, gray and green. They can be minuscule, just a few inches tall, or tower above the witnesses, standing 10 feet tall or more. They range from small hairy dwarfs to bald giants. Some look nearly human, others comically alien. A few are living manifestations of a nightmare. While they can often look like flesh-and-blood or metallic beings, many can perform ghostlike feats such as walking through walls. They display various eccentricities in their dress, behavior, and speech content. Some act like saints, others like demons. And when it comes to telling fibs, it has been noted, no politician on Earth could do better.

Patrick Huyghe – *The Field Guide to Extraterrestrials*

The Gray or Grey may well be the most familiar form of alleged alien visitor on record, but it is by no means the only one. As revealed in Patrick Huyghe's invaluable *Field Guide to Extraterrestrials* (1997) and elsewhere, ETs apparently come in all shapes and sizes, yet most receive far less attention from researchers than the Gray. Consequently, this chapter surveys some of the more exotic examples on file, all of which are truly alien - in every sense of the word!

THE TRIPODAL TERROR OF WABASH RIVER VALLEY

It was roughly 9.00 pm on 25 April 1972 near Enfield, in Indiana's Wabash River Valley - the scene of several reports around that time concerning strange lights observed in the sky, and, more recently, of some very peculiar tracks discovered by local war veteran Henry McDaniel in a nearby wood. These tracks each measured 3-5 in across, and contained six toe impressions, with a small hoof-like mark in the centre.

Consequently, when McDaniel's family heard a strange scratching sound on the outside of their home's back door, he cautiously looked out of a window, to find out what manner of animal was responsible. Even so, he did not really expect to see anything more remarkable than a stray dog, a neighbour's cat, or perhaps an inquisitive raccoon. In reality, however, what he did see was so incredible, and alarming, that he immediately grabbed hold of his shotgun and fired four rounds directly at it.

There, no more than 3 ft away from him, was a surrealistic entity, 4-5 ft tall, with a hairy dirty-grey body and disproportionately large head, staring at him through two pink reflective eyes, and standing upright like a human, but on three legs!

Although one of McDaniel's shots hit this tripodal terror, it merely hissed and leapt away along a railway track close by, clearing 75 ft in three huge bounds - but it would be back.

A few days later, at around 3.00 am, McDaniel was woken up by his dogs, who were barking uncontrollably. Carefully opening his door, he looked out, and spied this intergalactic Jake the Peg standing by the railway line, looking at him. Now, however, it made no attempt to approach, and was not seen by McDaniel (or anyone else) again.

THE DISEMBODIED BRAINS AT PALOS VERDES

Bodiless brains are not the most appetising of sights at the best of times, and certainly not when they float towards you and begin communicating telepathically as you sit in your car on a lonely Californian road at 2 am in the morning! Nevertheless, this bizarre scenario was supposedly experienced by two men in their early 20s, known pseudonymously as John Hodges and Peter Rodriguez, sometime in August 1971 at Palos Verdes Estates, California.

While walking towards their car, they had seen a diffuse white beam close by, and once inside the car their headlights had revealed two grotesque brain-resembling entities hovering in the air outside. Both were blue in colour, but whereas one was only the size of a softball, the other was about 18 in high, and bore a large bright red spot on its surface.

The two men lost no time in driving away, and Hodges took Rodriguez home, but in the classic abduction tradition, after reaching his own home Hodges discovered that he had mysteriously 'lost' two hours.

Later, he recalled experiencing a dream-like state while in his car, during which he seemed to be in a room containing not only the larger of the two levitating brains but also several tall, grey, six-fingered, humanoid (yet non-human) entities. The brain informed him that Earth was being monitored because of its nuclear power; the humanoid entities claimed that they were from Zeta Reticulii, and that the brains were translators. They also voiced a number of prophecies, but these proved to be inaccurate.

THE CRAZY CRITTER OF BALD MOUNTAIN

This was the name applied by the local media to the incredible beast confronted by several shocked motorists during the evening of 17 November 1974 on Bald Mountain, situated approximately 20 miles east of Chehalis, in Washington State, USA. Three nights earlier, and only about 5 miles away, a UFO of the unidentified fiery object kind plummeted to earth, but this had attracted little publicity - until Seattle grocer Ernest Smith saw the 'crazy critter'.

According to Smith's description, cited in Jim Brandon's book *Weird America* (1978):

> ...it was horse-sized, covered with scales and standing on four rubbery legs with suckers like octopus tentacles. Its head was football-shaped with an antenna sticking up...The thing gave off this green, iridescent light.

That glow was also spied by Mr and Mrs Roger Ramsbaugh from Tacoma, as they were driving by, and when they went closer to investigate, they were confronted with the same weird wonder, complete with antenna and suckered legs, that Smith had seen earlier.

Such reports as these soon attracted the attention of the local authorities, headed by Lewis County Sheriff William Wister, but some accounts claim that he was instructed by airforce and NASA officials not to continue his investigations, and his own team of county officials was replaced by what Brandon refers to as "a special NASA team, including a heavily armed military unit wearing uniforms with no insignia". Sounds familiar?

The Crazy Critter, illustrated by Tim Morris

As for the crazy critter itself: this episode reminds me of a storyline greatly favoured by newspaper cartoonists (and as also recalled by Malcolm Smith in *Bunyips and Bigfoots*, 1996), in which the captain of a flying saucer is angrily reprimanding one of his crew members: "You fool! You know darned well that you should always keep the ship's mascot on a leash when you take it for a walk!".

THE FLYING SAUCER FAIRIES

Some researchers deem ETs to be fairies or other Little People who, keeping abreast of modern times, have traded in their traditional image of dancing merrily in elfin glades and hollow hills in favour of a more technologically-compatible lifestyle, flitting through the skies aboard flying saucers sporting a nifty turn of speed. Judging from the following incident, documented by Alfred Budden (*Fortean Times*, summer 1988), they may well have a point!

Standing in her garden at Rowley Regis, near Birmingham in the West Midlands, on the morning of 4 January 1979 after waving her husband off to work, Jean Hingley saw a large orange sphere hovering close to their garage's roof and emanating an appreciable amount of heat. As she watched, it turned white and floated nearer, hovering over her back garden.

Suddenly, she was shocked to see that her pet dog, Hobo, standing by her, had become paralysed (happily, he later recovered) - and was even more shocked when three small fairy-like beings quite literally, and audibly, 'buzzed' by her and into her house through the doorway. According to Budden's account, Hingley claimed that her unwonted (and unwanted!) visitors were each:

> ...about 3.5 ft tall, and dressed in a silvery tunic with six silver buttons down the front. They had large eyes like 'black diamonds' with a glittering lustre, set into wide white faces with no nose to speak of and a simple line for the mouth. Their heads were covered by transparent helmets like 'goldfish bowls', surmounted by small lights. Their limbs were silvery-green, ending in simple tapering points with no apparent hands or feet. They had large oval 'wings' which looked as if they were made of thin, transparent paper covered with dozens of glittering multi-coloured dots, like 'braille dots'. Each 'being' was surrounded by a halo, and numerous very thin streamers hung down from their shoulders. They hovered and flew about the room with their 'arms' clasped in front of their chests, while their 'legs' hung down stiffly. Their wings didn't flap like those of birds, but seemed to be for display and merely fluttered gently or occasionally folded inwards like a concertina. Their expression - "like a dead person's face" - never changed during the encounter, which lasted for about an hour.

During that hour, these 'space-age fairies' occasionally spoke to Hingley, but always in unison, and via a guttural, masculine voice. Often, however, when she attempted to speak with them, they would emit a very thin laser-like beam from their helmets' lights, directed onto her brow, which dazzled and paralysed her, and produced an intense burning sensation at first. They behaved in a mischievous manner, shaking her Christmas tree and jumping up and down on the sofa, but showed great curiosity about her newspapers, cassettes, and a picture of Jesus, and they often picked up small objects or simply touched them. Oddly, however, whenever they found themselves unable to do something, such as drinking water from a glass or answering certain questions, they chose to disable Hingley with their beam.

Eventually, a very loud electronic beeping noise resounded from her back garden, and when Hingley looked she could see an orange-glowing craft there, oval in shape and about 8 ft long, with two luminescent portholes and an external antenna bearing a series of spokes arranged in a wheel-like shape. The space-fairies floated out of her house, each carrying a mince pie(!), and into their craft, which promptly took off, emitting as it did so a blue light from its antenna. As soon as they had gone, however, Hingley became convulsed with pain, and remained in a very distressed state for a number of hours, before becoming well enough to call her husband, a neighbour, and the police. Moreover, the burning mark left upon on her forehead by the entities' beam remained there for several months.

It is possible, as speculated by Budden, that this exceedingly bizarre encounter was really a hallucination, triggered somehow by the UFO, but the UFO was no hallucination - because her human visitors could clearly see a peculiar impression in the snow covering her back garden's lawn that closely resembled a tank's caterpillar track, about 8 ft long and symmetrical. In addition, several electrical gadgets in her house had stopped working, as if affected by an intense magnetic field, and the cassettes handled by the entities were now useless.

THE FLATWOODS MONSTER

For inducing sheer terror in its eyewitnesses, few ETs have matched the fearsome reputation of the

'Flatwoods monster', famously encountered by a party of children and adults while investigating a UFO report on 12 September 1952 at Flatwoods, West Virginia.

Likened by some of its eyewitnesses, a group of youngsters, to a meteor, the UFO had landed on top of a hill near Flatwoods, and some of them set off in its direction, to see if they could find it. Along the way, they were joined by a matron called Kathleen May (or Hill, in some accounts), her two sons, and a teenage National Guardsman, Gene Lemon.

As this amassed company drew nearer, they spied a huge pulsating globe or sphere, about 20 ft in diameter, and one of the eyewitnesses also noticed what he thought to be a pair of animal eyes, staring down at them from the branches of a tree close by. Shining his torch in the direction of these eyes, he and everyone else in the company were horrified to see an enormous figure, standing just beneath the tree's lower branches.

This macabre entity was 10-15 ft tall, and according to Mrs May it seemed to be dressed in a long cloak-like garb with a pointed hood, thus resembling the habit of a monk - but the face that stared out at them from inside the hood was certainly not that of any monk!

**The Flatwoods Monster
as drawn by an eyewitness *(FPL)***

Instead, it was round in shape and blood-red in colour, with a pair of bulbous eyes that glowed with an eerie greenish-orange hue. And as the terrified group of UFO-seekers gazed at it, this weird apparition began to float slowly down the hill towards them, hissing!

Needless to say, everyone fled at once, and such was the horror of this experience that a number of the eyewitnesses were hysterical and violently sick for several hours afterwards. During the following day, the local newspaper's editor and a team of other investigators scoured the area where the 'monster' and the giant globe had been encountered, but both had disappeared. However, they did find some odd tracks on the ground, a patch of flattened grass, and a peculiar, irritating odour persisting just above ground level.

MUMMIES FROM SPACE!

Imagine a 5-ft-tall, neck-less, humanoid figure resembling an Egyptian mummy, with grey wrinkled skin of elephantine appearance and texture, a pair of disproportionately long arms whose hands resembled mittens (i.e. with a single thumb but no differentiated fingers), a pair of legs held together like a pedestal and terminating in two bulbous elephant-like feet, and a seemingly eyeless face sporting a pair of pointed retractile ears, a conical nose-like structure, a short slit of a mouth that never opens, and a terrifyingly blank expression.

Multiply this grim-looking apparition by three, then add the domed football-shaped buzzing craft, with two windows, two blue lights, an invisible door, and a highly-illuminated interior, from which they emerged - and the result is the living nightmare experienced by two terrified Mississippi fishermen,

Charles Hickson and Calvin Parker, at the Pascagoula River, on the night of 11 October 1973. And their sighting of these sinister entities was only the beginning of their ordeal. For as the two men stood there, prisoners of the combined fear and fascination aroused by their incredible encounter, the 'mummies' floated up to them, seized hold of them, and physically immobilised them before carrying them into their craft.

Once inside, the mummies subjected their human captives to a close physical examination, after which they returned them to the spot from where they had abducted them, then soared away through the sky inside their blue-lit vessel.

LOOK OUT - IT'S VEGETABLE MAN!

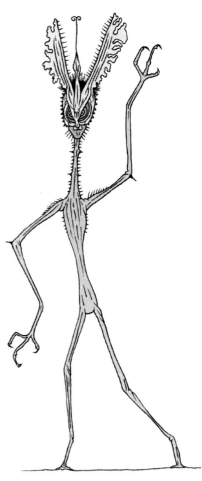

Never trust anyone whose body resembles the green, slender stalk of a plant, and who sucks out your blood through three 7-in-long fingers with needle-like tips and suction cups - that's what I always say! And Jennings Frederick would certainly agree, because he allegedly met just such a being while hunting one day in a West Virginia woodland during July 1968.

According to American paranormal researcher Brad Steiger, Frederick suddenly became aware of what he later described as "...a high-pitched jabbering, much like that of a recording running at exaggerated speed", yet which he could somehow understand, and which was informing him that it came in peace but needed medical assistance.

At the same moment, Frederick saw beside him the extraordinary quasi-botanical entity described above, with a semi-human face, long ears, yellow slanted eyes, and two stick-like arms. Before he had time to be surprised, however, he felt a pricking sensation in one of his hands, as if it had become entangled in some thorns - but when he looked, he discovered to his horror that the entity was draining blood from it, through its own fingers. Moreover, its eyes suddenly changed colour, becoming bright red and yielding a rotating, hypnotic effect that rendered its blood-sucking operation painless.

A minute later, Frederick's enforced transfusion was over, and his mesmerising recipient fled, bounding away rapidly up a hill, each leap covering a distance exceeding 25 ft. Unfortunately for Frederick, however, once the entity had disappeared, the pain in his hand reappeared. And as he set off back home, he heard a strange humming sound, which he believed to be the entity's craft, transporting it back from whence it had come.

West Virginia's Vegetable Man, illustrated by Tim Morris

Frederick was so disturbed by his grotesque experience that he did not speak about it for several months.

Some researchers have speculated that it may be a hoax, but those who have spoken to him seemed convinced that his account, albeit highly unusual, is genuine.

FLYING JELLY BAGS AHOY!

Not everyone can claim the dubious honour of having been abducted by a trio of flying jelly bags - an honour that Stig Rydberg and Hans Gustaffson would be more than happy to relinquish if they could somehow relive 20 December 1958 and arrange to be somewhere far away from the Swedish forest that they were driving through on that fateful day.

It was 3 am when they noticed a strange glow and then spotted a mysterious tripodal craft, over 12 ft long, resting nearby on the ground. Just in front of it were four extraordinary entities, each about 3 ft long, blue-grey in colour, and virtually amorphous, with no visible limbs, head, or any other recognisable features.

Likened to jelly bags, they were leaping around their craft at first, but when they somehow perceived the two men, three of these animate blobs swiftly approached them and attached their shapeless forms to them, yielding a powerful suction force as they strove to haul their frightened human hostages back towards their craft. In the ensuing struggle, the men could smell a vile stench emanating from their eerie antagonists, combining the overpowering odour of ether with the nauseating stink of burned sausage.

During Rydberg's frantic attempts to escape, one of his arms forced itself deep within the body of a blob, yet with no detrimental effect, either to him or to the blob. Nonetheless, his struggle eventually succeeded, and he raced back to the car, where he sounded the horn loudly, which so startled the blobs that they released Gustafsson, and fled back into their vessel, which duly rose up into the sky with a high-pitched noise and sped away. The most bizarre five minutes of the two men's lives were over.

UNIPODAL ALIEN ROBOTS?

One of the least humanoid categories of ETs ever reported must surely be the astonishing version reputedly encountered by Brazilian bus driver Antonio La Rubia at 2.20 am one morning in September 1977 while driving to work in Paciencia, Rio de Janeiro. After spying a huge hat-shaped vessel exceeding 200 ft in diameter hovering above a football field, La Rubia was abruptly immobilised by a beam of blue light, and then saw three decidedly futuristic robotic entities, who transported him inside the vessel. Here, he was subjected not only to the customary alien examination but also to a slide show of sorts, before being somewhat rudely ejected back into the street again, after which the craft disappeared.

This scenario may be a common one to ufologists, but La Rubia's abductors were decidedly uncommon in appearance. Each was about 4 ft tall, excluding the vertical antenna on top of a featureless head shaped like a rugby ball standing erect on its tip, and equipped with a horizontal belt of blue mirror-like structures encircling its diameter. Beneath its head, which had no neck, was a sturdy ovoid body covered in scales resembling dull aluminium. It also had a waist-high belt bearing a series of hooks holding syringe-like objects, plus two arm-like appendages that curved downwards but terminated in a point, lacking hands. Its body was borne upon a tall, slim, central pedestal, with a circular base at its tip.

THE VENEZUELAN LITTLEFOOTS

It is well known that a number of ufological reports on file concern sightings of shaggy bipedal entities resembling North America's famous mystery man-beast, the bigfoot or sasquatch, in association with alleged UFOs and landed extraterrestrial spacecraft. However, the seemingly invulnerable troll-like category of alien encountered by Gustavo Gonzales and José Ponce while driving their truck from Caracas to Petare, Venezuela, during the early morning of 28 November 1954 could more aptly be referred to as a littlefoot.

Before reaching Petare, they encountered a huge glowing globe, hovering about 6 ft above the ground, which was virtually blocking the entire road ahead. Consequently, the two men got out of their truck to investigate, and as they approached the globe a furry bipedal being appeared, and began to approach them. Standing no more than 3 ft high, it was covered with stiff, bristly hair, and had large clawed hands and feet. Gonzales seized hold of this hirsute 'littlefoot', in order to take it to the police, and was surprised to find that the creature was exceedingly light. He was even more surprised, however, to discover how powerful it was - for with a single push from one of its paws, it effortlessly propelled Gonzales through the air, sending him sprawling onto the ground about 5 yards away.

By now, Ponce was running back down the road, towards the local police station, and as he did so he spied another two of these littlefoots, gathering rocks and carrying them aboard the sphere. The first littlefoot, however, angered by Gonzales's action, began savagely clawing him, but when Gonzales tried to defend himself by stabbing this creature with his knife, the blade made no impression on its body. Suddenly, a fourth littlefoot appeared, emerging from the sphere, and stunned Gonzales with a beam of light, enabling the others to go aboard and depart.

When the police had established that neither Ponce nor Gonzales were drunk, they gave them sedatives, and also confirmed that Gonzales bore a long red scratch on his side. Furthermore, several days later, a medical doctor came forward to announce that he had actually witnessed from a distance the attack upon Gonzales by the littlefoots, but had not intervened because he did not want to be the focus of publicity.

How can such an amazing diversity of alien forms as these – and the many others also on record – be explained? The initial assumption is that if any or all of them are indeed real (a big assumption in itself, of course), they clearly originate from totally different worlds, whether those worlds be planets or dimensions.

However, as Huyghe and other researchers have pointed out, these entities' morphological differences may owe more to psychology than anatomy. Could it be, for instance, that the image that an alien eyewitness sees is not the true form of the alien in question but rather a false image placed in the eyewitness's mind by the alien, thereby concealing the latter's true self? An equally thought-provoking, obverse explanation is that the eyewitness is not seeing the alien as it actually is but rather as the eyewitness subconsciously chooses it to be.

It is often said that beauty is in the eye of the beholder, but who knows, perhaps the same is also true of alien morphology.

Chapter 5
FOLK WITH FINS
Fishing for the Truth about Merbeings

Every evening the young Fisherman went out upon the sea, and called to the Mermaid, and she rose out of the water and sang to him. Round and round her swam the dolphins, and the wild gulls wheeled above her head.

And she sang a marvellous song. For she sang of the Sea-folk who drive their flocks from cave to cave, and carry the little calves on their shoulders; of the Tritons who have long green beards, and hairy breasts, and blow through twisted conchs when the King passes by; of the palace of the King which is all of amber, with a roof of clear emerald, and a pavement of bright pearl; and of the gardens of the sea where the great filigrane fans of coral wave all day long, and the fish dart about like silver birds, and the anemones cling to the rocks, and the pinks bourgeon in the ribbed yellow sand. She sang of the big whales that come down from the north seas and have sharp icicles hanging to their fins; of the Sirens who tell of such wonderful things that the merchants have to stop their ears with wax lest they should hear them, and leap into the water and be drowned; of the sunken galleys with their tall masts, and the frozen sailors clinging to the rigging, and the mackerel swimming in and out of the open portholes; of the little barnacles who are great travellers, and cling to the keels of the ships and go round and round the world; and of the cuttlefish who live in the sides of the cliffs and stretch out their long black arms, and can make night come when they will it. She sang of the nautilus who has a boat of her own that is carved out of an opal and steered with a silken sail; of the happy Mermen who play upon harps and can charm the great Kraken to sleep; of the little children who catch hold of the slippery porpoises and ride laughing upon their backs; of the Mermaids who lie in the white foam and hold out their arms to the mariners; and of the sea-lions with their curved tusks, and the sea-horses with their floating manes.

Oscar Wilde – *'The Fisherman and His Soul'*

Since the very earliest times, stories of merfolk or merbeings - sea-dwelling entities with the upper bodies of humans and the tails of fishes - have been reported by maritime travellers. Such accounts have been traditionally dismissed as nothing more than hoaxes (with many so-called stuffed mermaids, skilfully constructed from the bodies of dead monkeys fused to the tails of large fishes, as silent testimony to this), or quaint folklore.

Fake stuffed mermaid photographed by Czech cryptozoologist Ivan Mackerle

The blue men of Minh (=Minch) constitute a prime example from the latter category. Named after their entirely blue skin, these are said to be malevolent mermen who inhabit the Minh Strait, separating the Hebridean islands of Long and Shiant. Nevertheless, they pose no problem to seagoers with a little knowledge of how to deal with such matters - simply address them solely in rhyme, and they will swiftly depart elsewhere!

In stark contrast to such a whimsical tale as that, however, there are numerous, decidedly sober accounts on file that seem to report factual sightings of bona fide merbeings, unembellished by fancy, fraud, or folklore – as revealed in the following selection.

FUNERAL FOR A MERMAID ON BENBECULA.

The mourners stood silently, a solemn congregation met in sadness beside a small coffin about to be buried - but this is where any resemblance to a typical funeral ended. For the setting was a bleak, remote seashore; and if the eyewitnesses were truthful, the deceased was not human - instead, the corpse inside the coffin was that of a murdered mermaid.

The history of this astonishing incident takes us back over 150 years, and to the island of Benbecula in the Outer Hebrides. According to Alexander Carmichael's *Carmina Gadelica*, vol. 2 (1900), one day in or around 1830 some people were cutting seaweed on Benbecula, at Sgeir na Duchadh in Grimnis, when one of them spied a small woman-like creature splashing in the sea just a few feet away. Soon, all of the people had run to the spot to catch sight of this amazing entity, cavorting playfully in the water. Greatly intrigued, the men in the group tried to catch the creature, but all in vain as it readily eluded their attempts by swimming out of reach - but then some of the people's children, accompanying them, picked up a handful of stones and started throwing them at it. One of the stones struck the creature in the back, and it promptly disappeared beneath the waves.

Tragically, however, the blow had clearly been a fatal one, because a few days later the lifeless body of their unexpected visitor was washed ashore, at Cuile (=Culla) in Nunton, nearly 2 miles away from where it had been seen alive. In *Carmina Gadelica*, Carmichael records the following remarkable details concerning the nature of the corpse, and its ultimate fate:

> The upper portion of the creature was about the size of a well-fed child of three or four years of age, with an abnormally developed breast. The hair was long, dark, and glossy, while the skin was white, soft, and tender. The lower part of the body was like a salmon, but without scales. Crowds of people, some from long distances, came to see this strange animal, and all were unanimous in the opinion that they had gazed on the mermaid at last.

> Mr. Duncan Shaw, factor for Clanranald, baron-bailie and sheriff of the district, ordered a coffin and shroud to be made for the mermaid. This was done, and the body was buried in the presence of many people, a short distance above the shore where it was found. There are persons still living who saw and touched this curious creature, and who give graphic descriptions of its appearance.

In August 1994, archaeologist Adam Welfare visited Benbecula to examine what was thought to be a headstone in sand dunes at Cuile Bay (discovered there in March of that same year by Dr Shelagh Smith from the Royal Museum of Scotland during a visit) that possibly marked the burial site of the mermaid. After studying it, however, he concluded that it was not a headstone. The stone in question was on land tenanted by Mr A. MacPhee, who informed Welfare that according to traditional belief passed down through several generations of his family, the precise location where the mermaid's lifeless body had been washed ashore was a tiny rocky inlet known as Bogha mem Crann or 'Stinky Inlet' (on account of the foul-smelling seaweed that grows there), lying immediately south of Cuile Bay.

Moreover, Scottish writer R. MacDonald Robertson claimed that he had actually been shown the mermaid's grave – sited somewhere in the burial ground at Nunton. This burial ground is the graveyard of the Chapel of St Mary, a small medieval church lying immediately to the east of Cuile Bay. Although the church is itself ruined, its graveyard is still used today as the principal burial ground for Benbecula's inhabitants, and is indeed, therefore, a likely location for the grave of so humanoid an entity as the mermaid.

During May-June 2008, American cryptozoologist Nick Sucik spent three weeks on Benbecula investigating this engrossing case. He informed me that while there he discovered a hitherto-undocumented grave-shaped mound in a field next to the bay where the creature's body was seen, and which many locals believed to be its burial site. He did not have time to initiate any excavations here, however, but he plans to continue his investigations during a second trip to the island.

Meanwhile, if this creature were indeed real, what could it have been? Is it possible that it was a malformed child (there is, for instance, a rare congenital condition called sirenomelia, in which the legs are fused together to yield a mermaid-like single limb with feet at the end) around whose premature death a fictional story of a mermaid subsequently arose? If not, the most popular scientific identities for merfolk are misidentified seals, and, in particular, those herbivorous aquatic mammals the sirenians or sea-cows - comprising the dugong and manatees. Moreover, there is no doubt that such explanations are perfectly reasonable for quite a number of cases.

RIDDLE OF THE RI

One of the most significant of those cases featured a mysterious water-beast called the ri - which came to the attention of Virginia University anthropologist Dr Roy Wagner while working on the Papuan island of New Ireland during the late 1970s. According to descriptions given by the Barok tribespeople inhabiting the centre of this island, the ri was a bona fide mermaid - for it allegedly combined the head and torso of a human with the legless lower trunk and tail flukes of a fish, and even uttered plaintive near-human cries. Further south, this creature was also known to the the Susurunga tribe, who referred to it as the ilkai, and saw specimens on a very regular basis. They informed Wagner that it had human-like arms fused to its flanks, eyes set toward the front of its head, a small but protruding, odd-looking mouth, and a fish-like posterior body region that was smooth and scaleless.

Fascinated by what seemed likely to comprise a major cryptozoological discovery if unmasked, Wagner instigated searches for this mystifying creature, and success was finally achieved on 11 February 1985, when a team of American researchers photographed an animal that was positively identified by the natives as a ri. Sadly, however, cryptozoologists' hopes were swiftly dashed - for the ri proved to be nothing more novel than a dugong *Dugong dugon*.

MERFOLK - IN THE EYE OF THE BEHOLDER?

It has also been argued that some merfolk sightings might simply be optical illusions. Thus, Canadian researchers Drs Waldemar Lehn and I. Schroeder (*Nature*, 29 January 1981) postulated that medieval Norse writings describing mermen were, in reality, accurate descriptions of a natural meteorological phenomenon - whereby images of common sea mammals such as pinnipeds and whales are severely distorted vertically by strong, non-uniform atmospheric refraction, i.e. temperature inversions that occur when a mass of warm air moves over cold air, which bends the light so that objects are distorted beyond the horizon.

Early engraving of a merman and mermaid

None of the above explanations, however, seems very applicable to the Benbecula mermaid. After all, except for a handful of unconfirmed straggler records, sirenians do not occur in the waters around Great Britain - which leaves only seals, dolphins, whales, or similar forms as serious alternatives among known species for the identity of mermaids in this area of the world. In the case of the Benbecula mermaid, moreover, it surely defies belief that seafaring people like those of the Outer Hebrides, well-acquainted with seals and other marine creatures, could be fooled by the dead body of any such animal into believing that they were gazing upon the corpse of a mermaid.

The author alongside a life-sized manatee statue at Sea World in
San Diego, California *(Dr Karl Shuker)*

And whereas animals spied at a distance and/or in conditions of poor visibility may well appear strange or unfamiliar, they could hardly be mistaken for anything other than their true identity when observed as closely as the body of the Benbecula mermaid. It must have been decidedly humanoid to have elicited the decision to lay it to rest ceremoniously in a coffin and shroud, with a sizeable throng of persons in attendance - reverences not likely to have been accorded to many dead seals or whales, I would have thought.

MUNRO'S MERMAID

A number of other striking reports are on file for which it is equally difficult to offer, with any degree of confidence, any of the 'orthodox' explanations discussed above. Take, for instance, the being encountered one summer day in or around 1797 by Scottish schoolmaster William Munro while strolling towards Sandside Head on the Highlands' northern coast, and later reported by him on 8 September 1809 in no less august a publication than *The Times*:

> My attention was arrested by the appearance of a figure resembling an unclothed human female, sitting upon a rock extending into the sea, and apparently in the action of combing its hair, which flowed around its shoulders, and was of a light brown colour. The forehead was round, the face plump, the cheeks ruddy, the eyes blue, the mouth and lips of a natural form, resembling those of a man; the teeth I could not discover, as the mouth was shut; the breasts and abdomen, the arms and fingers of the size of a full grown body of the human species; the fingers, from the action in which the hands were employed, did not appear to be webbed, but as to this I am not positive. It remained on the rock 3 or 4 minutes after I observed it, and was exercised during that period in combing its hair, which was long and thick, and of which it appeared proud, and then dropped into the sea, from whence it did not reappear to me. I had a distinct view of its features, being at no great distance on an eminence above the rock on which it was sitting, and the sun brightly shining.

A naked bather, an early naturist, a blatant hoax? Whatever it was, it certainly was not a seal, a whale, or a temperature inversion, that's for sure!

And how can we explain the detailed description of a mermaid given by two crew members aboard a ship captained by the renowned maritime navigator Henry Hudson?

According to Thomas Hilles and Robert Raynar, who spied the supposed mermaid on 15 June 1608 while sailing near the coast of Novaya Zemlya (an island cluster off northern Russia), she had very pale skin and long black hair flowing down over her back, the breasts and back of a woman but a body as big as a man's, and a speckled tail shaped like that of a porpoise.

SWALLOWED BY A SHARK

If merfolk exist, then they must surely fall prey at times to some of the seas' greatest predators - the sharks. One little-known case on file not only supports such a possibility but also provides an added in-sight into merfolk morphology. The creature in question was discovered inside the stomach of a shark caught on the northwest coast of Iceland, and was closely observed by the priest of Ottrardale. His de-scription was quoted in a 19th-Century book by Sabine Baring-Gould called *Iceland, Its Scenes and Sagas*, and reads as follows:

> The lower part of the animal was entirely eaten away, whilst the upper part, from the epigastric and hypogas-tric region, was in some places partially eaten, in others completely devoured. The sternum, or breast-bone, was perfect. This animal appeared to be about the size of a boy eight or nine years old, and its head was formed like that of a man. The anterior surface of the occiput was very protuberant, and the nape of the neck had a considerable indentation or sinking. The alae [lobes] of the ears were very large, and extended a good way back. It had front teeth, which were long and pointed, as were also the larger teeth.
>
> The eyes were lustreless, and resembled those of a codfish. It had on its head long black, coarse hair, very similar to the fucus filiformis [filamentous seaweed]; this hair hung over the shoulders. Its forehead was large and round. The skin above the eyelids was much wrinkled, scanty, and of a bright olive colour, which was indeed the hue of the whole body. The chin was cloven, the shoulders were high, and the neck uncommonly short. The arms were of their natural size, and each hand had a thumb and four fingers covered with flesh. Its breast was formed exactly like that of a man, and there was also to be seen something like nipples; the back was also like that of a man. It had very cartilaginous ribs; and in parts where the skin had been rubbed off, a black, coarse flesh was perceptible, very similar to that of the seal. This animal, after having been ex-posed about a week on the shore, was again thrown into the sea.

STELLER'S SEA-APE

If some merfolk are neither poorly-spied known creatures nor the product of other unusual optical phenomena, then we must consider the exciting prospect that they could comprise a dramatically new, still-undiscovered animal species - a prospect that certainly seems worthy of contemplation with regard to the mystifying sea-ape. This is the name given to a remarkable unidentified animal observed for no less than 2 hours by naturalist Georg Steller on 10 August 1741 while aboard a Russian exploration vessel near the Gulf of Alaska's Shumagin Islands. Said to resemble an aquatic monkey, it was 5 ft long, lacked front limbs, and possessed a very distinctive heterocercal tail - as with that of sharks, a tail whose upper fin is much larger than its lower fin.

Despite attempts made by some modern-day researchers to identify it as a vagrant Hawaiian monk seal *Monachus schauinslandi*, no species directly comparable with Steller's sea-ape is known either from our modern-day fauna or from the fossil record. (See also my detailed account of this cryptid in my book *Mysteries of Planet Earth*, 2000.)

THE MER-MONKEY OF YELL

As already shown with the shark-ingested specimen from Iceland, Steller's sighting is not the only record of a strange unidentified marine mammal resembling a sea-ape or sea-monkey.

During the 1800s, six fishermen acquainted with naturalist Dr Robert Hamilton were fishing off the island of Yell, one of the Shetland Islands north of Scotland, when they captured a decidedly monkey-like mermaid, which had become entangled in their lines:

> The animal was about three feet long, the upper part of the body resembling the human, with protuberant mammae, like a woman; the face, the forehead, and neck were short, and resembling those of a monkey; the arms, which were small, were kept folded across the breast; the fingers were distinct, not webbed; a few stiff, long bristles were on the top of the head, extending down to the shoulders, and these it could erect and depress at pleasure, something like a crest. The inferior part of the body was like a fish. The skin was smooth, and of a grey colour. It offered no resistance, nor attempted to bite, but uttered a low, plaintive sound. The crew, six in number, took it within their boat; but superstition getting the better of curiosity, they carefully disentangled it from the lines and from a hook which had accidentally fastened in its body, and returned it to its native element. It instantly dived, descending in a perpendicular direction.

RUNAN-SHAH – THE MASTER OF THE SEA AND RIVERS

In 2005, news emerged from Iranian media sources that for the previous two years, residents of coastal areas around the southern and southwestern Caspian Sea had been reporting a strange merman-like entity, which had been dubbed the runan-shah ('master of the sea and rivers') by Iranian sailors who had seen it. According to a detailed account published by the Russian newspaper *Pravda* on 25 March 2005:

> All the eyewitness accounts provide a similar description of the marine humanoid. His height is 165-168 cm, he has a strong build, a protruding ctenoid stomach, his feet are pinniped and he has four webbed fingers on either of his hands. His skin is of moonlight colour.
>
> The hair on his head looks black and green. His arms and legs are shorter and heavier than those of a medium-built person. Apart from his fingernails, he has nails growing on the tip of his aquiline nose that look like a dolphin's beak. No information as to his ears. His eyes are large and orbicular. The mouth of the crea-

ture is fairly large, his upper jaw is prognathic and his lower lip flows smoothly into the neck, his chin is missing.

Moreover, the same creature was also allegedly seen by various Azerbaijani fishermen in May 2004 in waters between the cities of Astara and Lenkoran. How reliable any of these reports are, however, is difficult to say.

SOME LITTLE-PUBLICISED MERFOLK REPORTS

The following reports have previously received little if any mainstream attention, so I am especially indebted to American cryptozoological author Michael Newton for passing them on to me.

The first of these was published by the *Daily Kennebec Journal* newspaper of Augusta, Maine, on 24 June 1873. The most pertinent section reads as follows:

> About the same time [c.1737] a story came from Virgo, in Spain, to the effect that some fishermen on that coast had caught a sort of a merman, five feet and a half from head to foot. The head was like that of a goat, with a long beard and moustache, a black skin, somewhat hairy, a very long neck, short arms, hands longer than they ought to be in proportion, and long fingers, with nails like claws; webbed toes, and a fin at the lower part of the back.

Interestingly, an engraving of a weird goat-headed merbeing, variously dubbed a sea-Pan, sea-satyr, sea-devil, or ichthyocentaur, based upon a skeleton and some mummified exhibits, appeared in the second edition of Conrad Gesner's *Historiae Animalium Liber IV. Qui est de Piscium et Aquatilium Animantium Natura* (1604). This must surely have been another cleverly-constructed fraud.

Continuing the *Daily Kennebec Journal* article:

> The magazines for 1775 gave an account of a mermaid which was captured in the Levant and brought to London. One of the learned periodicals gravely told its readers that the mermaid had the complexion and features of a European, like those of a young woman; that the eyes were light blue, the nose small and elegantly formed, the mouth small, the lips thin, "but the edges round like those of a codfish; that the teeth were small, regular and white; that the neck was well rounded, and that the ears were like those of an eel, but placed like those on the human specie [sic], with gills for respiration, which appear like cork." There was no hair on the head, but "rolls..." There was a fin rising pyramidally from the temples, "forming a foretop, like that of a lady's headdress." [']The bust was nearly like that of a young damsel, a proper orthodox mermaiden, but, alas! all below the waste was exactly like a fish! Three sets of fins below the waist, one above the other, enabled her to swim. Finally it is said to have an enchanting voice, which it never exerts except before a storm." The writer in the *Annual Register* probably did not see this mermaid, which the *Gentleman's Magazine* described as being only three feet high. It was afterward proved to be a cheat, made from the skull of the angle [angel] shark

> A Welsh farmer, named Reynolds, living at Pen-y-hold in 1782, saw a something which he appears to have believed to be a mermaid; he told the story to Dr. George Phillips, who told it to Mrs. Moore, who told it to a young lady pupil of hers, who wrote out an account of it to Mrs. Morgan, who inserted it in her "Tour to Milford Haven." How much...[it] gained on its travels - like the Three Black Crows or the parlor game of Russian Scandal - we are left to find out for ourselves; but its ultimate form was nearly as follows: One morning, just outside the cliff, Reynolds saw what seemed to him to be a person bathing in the sea, with the upper part of the body out of the water. On nearer view, it looked like the upper part of a person in a tub, a youth, say of sixteen or eighteen years of age, with nice white skin!, a sort of brownish body, and a tail was under the water. The head and body were human in form, but the arms and hands thick in proportion to

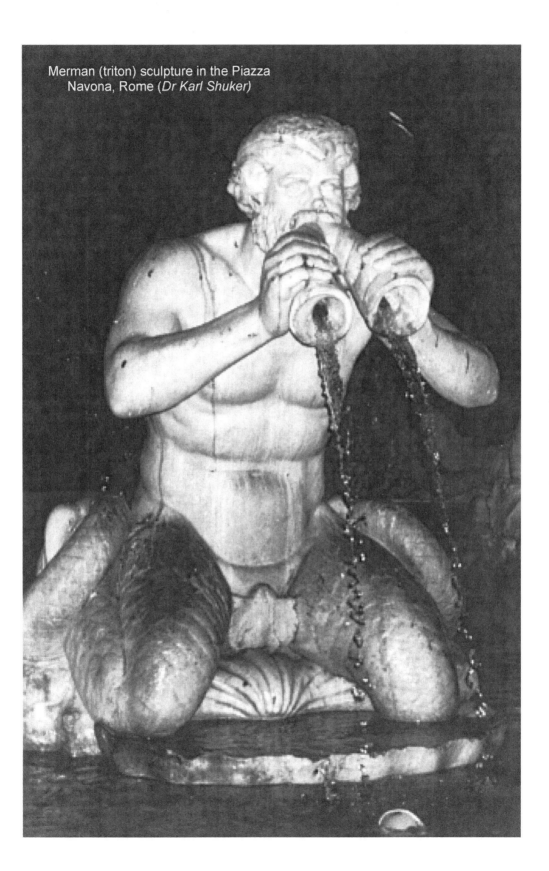

Merman (triton) sculpture in the Piazza Navona, Rome (*Dr Karl Shuker*)

ABOVE: Gesner's goat-headed merbeing **BELOW:** (Left) A sea-monk merbeing reputedly captured off the Norwegian coast during the 16th Century; (Right) A sea-bishop merbeing supposedly spied off the Polish coast in 1531; both were probably monkfishes

length, while the nose, seeming up high between the eyes, terminated rather sharply. The mysterious being looked attentively at Reynolds, and at the cliffs, and at the birds flying in the air, with a wild gaze, but uttered no...[sound]. Reynolds went to bring some companions to see the merman or mermaid, but when he returned it had disappeared. If we like to suppose that Reynolds had seen some kind of seal, and that the narration had grown to something else by repeating freely mouth to mouth, perhaps we shall not be very far wrong.

Also of interest is the following report, published by the *Cape Brooklyn Eagle* newspaper on 22 August 1886:

The fishermen of Gabarus, Cape Breton [an island off the coast of Nova Scotia, Canada], have been excited over the appearances of a mermaid, seen in the waters by some fishermen a few days ago. While Mr. Bagnall, accompanied by several fishermen, was out in a boat, they observed floating on the surface of the water a few yards from the boat what they supposed to be a corpse. Approaching it for the purpose of taking it ashore, they observed it to move, when to their great surprise, it turned around in a sitting position and looked at them and disappeared. A few moments after[,] it appeared on the surface and again looked toward them, after which it disappeared altogether. The face, head, shoulders and arms resembled those of a human being, but the lower extremities had the appearance of a fish. The back of its head was covered with long, dark hair resembling a horse's mane. The arms were shaped like a human being's, except that the fingers of one hand were very long. The color of the skin was not unlike that of a human being. There is no doubt, that the mysterious stranger is what is known as a mermaid, and the first one ever seen in Cape Breton waters.

Were it not for the mention of fingers, I would be inclined to identify this particularly hirsute merbeing as a seal, quite probably a fur seal. Could it be that there is an unknown species of seal with a foreshortened muzzle, rendering its face superficially humanoid, and with flippers in which the digits are more prominently revealed than in other seals? Such a species would correspond closely with a number of noteworthy merbeing reports.

On 3 November 1896, the *Cape Brooklyn Eagle* published another noteworthy merbeing report:

Seattle, Wash., November 3 - A party of Englishmen who have been porpoise fishing in the Pacific discovered and killed a monster that resembled a merman. The party was off the island of Watmoff on the hunting boat and Lord Devonshire, one of the fishers, had just shot a porpoise, when some one called out, "Look there!" pointing to a frightful looking monster about a cable's length away. Hastily raising his weapon his lordship fired and hit the creature between the eyes. The shot, though it did not kill it, so stunned the animal that it lay perfectly still on the surface of the sea.

It showed fight when hauled into the boat and had to be killed to prevent it from swamping the craft. The monster is said to be one of the strangest freaks ever put together. It measures 10 feet from its nose to the end of its fluke shaped tail and the girth of its human shaped body was just six feet. It would weigh close to 500 pounds. From about the breast bone to a point at the base of the stomach it looked like a man. Its arms, quite human in shape and form, are very long and covered completely with long, coarse, dark reddish hair, as is the whole body.

It had, or did have, at one time four fingers and a thumb on each hand, almost human in shape, except that in place of finger nails there were long, slender claws. But in days probably long since gone by, it had evidently fought some monster that had got the best of it, for the forefinger of the right hand, the little finger of the left and the left thumb are missing entirely. Immediately under the right breast is a broad, ugly looking scar which looked as if sometime in the past it had been inflicted by a swordfish. The creature is now being preserved in ice at Seattle and will be shipped to the British museum.

As I feel sure that the scientific world would have heard more about this entity had it indeed reached the

British Museum, I am assuming that it was never shipped there after all. In June 2008, I sent details concerning this fascinating case to Mandy Holloway of the British Museum (Natural History)'s Department of Zoology, who very kindly promised to pursue the matter for me through the museum's archives – just in case! However, no trace of any merman was found. Ah well, at least we tried!

AN UNDISCOVERED AQUATIC PRIMATE?

One further possibility exists relative to the likely nature of merfolk - the most radical but most exciting possibility of all. In July 1960, eminent British scientist Professor Sir Alister Hardy FRS published a highly thought-provoking hypothesis in *New Scientist* - speculating that instead of evolving from upright plains-dwelling ancestors, man learned to stand erect by having descended from aquatic ape-like forms. This startling idea has since been pursued in great detail by researcher Elaine Morgan in *The Aquatic Ape* (1982) and later books, and has provoked much dissension among scientists. I shall also be assessing this notion in detail within a future book, currently in preparation. For now, therefore, let me just say that if it is indeed valid, it has great bearing upon the question of mermaids and their kind.

After all, if, while some of these water-dwelling ape-like primates did indeed become terrestrial and evolve into modern-day *Homo sapiens*, others remained in the sea - continuing to evolve and surviving into modern times but retaining their fully aquatic form - what would they be like today? Perhaps we already know the answer - and perhaps that answer already has a name: merfolk.

Chapter 6
HOW GREEN WERE MY CHILDREN?
Unmasking the Weirdlings of Woolpit

I suddenly, quite without noticing how, found myself on this other earth, in the bright light of a sunny day, fair as paradise. I believe I was standing on one of the islands that make up on our globe the Greek archipelago, or on the coast of the mainland facing that archipelago. Oh, everything was exactly as it is with us, only everything seemed to have a festive radiance, the splendour of some great, holy triumph attained at last. The caressing sea, green as emerald, splashed softly upon the shore and kissed it with manifest, almost conscious love. The tall, lovely trees stood in all the glory of their blossom, and their innumerable leaves greeted me, I am certain, with their soft, caressing rustle and seemed to articulate words of love. The grass glowed with bright and fragrant flowers. Birds were flying in flocks in the air, and perched fearlessly on my shoulders and arms and joyfully struck me with their darling, fluttering wings. And at last I saw and knew the people of this happy land. They came to me of themselves, they surrounded me, they kissed me. The children of the sun, the children of their sun - oh, how beautiful they were!

Never had I seen on our own earth such beauty in mankind. Only perhaps in our children, in their earliest years, one might find, some remote faint reflection of this beauty. The eyes of these happy people shone with a clear brightness. Their faces were radiant with the light of reason and fullness of a serenity that comes of perfect understanding, but those faces were gay; in their words and voices there was a note of childlike joy. Oh, from the first moment, from the first glance at them, I understood it all! It was the earth untarnished by the Fall; on it lived people who had not sinned. They lived just in such a paradise as that in which, according to all the legends of mankind, our first parents lived before they sinned; the only difference was that all this earth was the same paradise. These people, laughing joyfully, thronged round me and caressed me; they took me home with them, and each of them tried to reassure me. Oh, they asked me no questions, but they seemed, I fancied, to know everything without asking, and they wanted to make haste to smooth away the signs of suffering from my face.

Fyodor Dostoyevsky – *The Dream of a Ridiculous Man*

It was then that he noticed a peculiarity in her flesh, which explained her strange pallor. The skin was not white, but a faint green shade, the colour of a duck's egg. It was, moreover, an unusually transparent tegument, and through its pallor the branches of her veins and arteries spread, not blue and scarlet, but vivid green and golden. The nails were pale blue, very like a blackbird's eggshell. The faint emanation of odour from her flesh was sweet and a little heavy, like the scent of violets.

Olivero looked up at the man, who stood glowering against the wall. 'It is the Green Child!' he cried. The man merely stared fixedly, but Olivero knew that his guess was right.

Herbert Read – *The Green Child*

T he two children that crawled out of a deep pit in the Suffolk village of Woolpit one sunny autumn day around 830-850 years ago shocked the local inhabitants for a number of reasons. They wore strange, unfamiliar clothes, they spoke in an unintelligible language – and their skin was green! Little wonder, then, that some researchers have speculated that perhaps they had inadvertently crossed into our world from some far-removed, alternative version, recalling Dostoyevsky's visionary dream quoted above. They also inspired Herbert Read's strange, yet hauntingly beautiful, only novel, *The Green Child* (1935), from which the quote above is excerpted.

More than eight centuries have passed since the emergence of these bizarre entities, but the mystery of their origin and identity continues to intrigue and challenge their numerous investigators. Not so long ago, I decided to visit Woolpit myself – until then, it had been just another name on a map to me - and pursue the tortuous trail of its most famous former denizens.

THE COMING OF THE GREEN CHILDREN

Today, Woolpit is a small, picturesque village, dominated not only by the beautiful presence of St Mary's Church, whose high spire features prominently in the local skyline, but also by the memory of its two mystifying visitors.

There is, for instance, a tall, elegant village name sign standing not far from the church that depicts the two children, the church, and a wolf, for Woolpit earns its name from the series of deep trenches, popularly termed pits, in which wolves were formerly captured at the perimeter of the village – 'Woolpit' is a corruption of 'Wolf Pit' – but most of these ancient hollows are now gone. The ornate, multi-coloured church banner also portrays the children and a wolf, and the church displays a long, handwritten scroll documenting their history.

Although small, Woolpit is very accessible to the outside world, and can be reached directly from my home county of the West Midlands (just follow the M6 south and then the A14 all the way there). Back when the green children appeared, conversely, it was very different. In those long-departed times, villages were much more insular; many of their inhabitants spent their entire lives within their confines, never venturing beyond, and neighbouring villages just a few miles away seemed as far-distant as foreign countries. Little wonder, therefore, that the coming of the green children was such a sensation for the Woolpit villagers.

King Henry II reigned in England from 1154 to 1189 AD, succeeding King Stephen (1135-1154), and chroniclers of the green children episode place it variously within one or other of these periods, with most favouring sometime towards the beginning or the middle of Henry II's reign. Down through the centuries, several chroniclers have penned a version of the green children's history, but perhaps the most

The author alongside Woolpit's village name sign, depicting the green children *(Dr Karl Shuker)*

reliable, and certainly the least sensational, of these is the account authored by an Augustinian monk called William of Newburgh (1136-98). He spent his entire life at his Yorkshire-based priory, but corresponded with many outsiders and visitors during his preparation of a history of England. This included his account of the Woolpit weirdlings (a copy of which is written on the scroll maintained today inside St Mary's Church).

According to that account, one day while the villagers were harvesting their crops in the village fields two children emerged from one of the pits, a girl and a younger boy, both with entirely green skin, and dressed in clothes of unusual material and colour. When they began wandering through the fields, they were captured by some of the villagers, who brought them back to Woolpit, where they attracted considerable attention. Attempts were made to feed them, but the uncanny pair refused everything offered, even though they were clearly starving, until some beanstalks were brought and the beans removed from the pods for them, which the two children ate. Thereafter they lived exclusively upon a diet of beans for some months, until eventually they became accustomed to bread. Gradually, moreover, their skin colour transformed to a normal healthy pink, and they learnt to communicate with the villagers. Later, they were baptised, but soon afterwards the boy died. His sister, however, thrived, no longer looked much different from any other woman, and, it was said, eventually married a man from Kings Lynn (in some later accounts, she then became known as Agnes Barre), in southern Norfolk.

The banner of St Mary's Church, depicting the green children *(Dr Karl Shuker)*

Once the children had learnt the villagers' language, they were asked where they had come from. In reply, they claimed that they were from the land of St Martin, where there were Christian churches, but where the sun was faint, not rising, so that it was perpetually twilight. However, they could see another land on the other side of a great river (described in some translations of William of Newburgh's original Latin account as a land of light or luminous land), and one day, while tending their father's flocks in a field, they had heard a great noise, just like the villagers of Woolpit hear the bells from the nearby monastery of St Edmund, and that somehow the noise had led them to Woolpit's fields.

Another major account from this same period was penned by Cistercian monk Ralph of Coggeshall, at the Abbey of Coggeshall in Essex, who travelled widely in East Anglia during the late 1100s and early 1200s while gathering information for the abbey's chronicle of England. His version of the green children's arrival in Woolpit tallies closely with that of William of Newburgh, but also contains some interesting additional information – not least of which is that after the children had been captured by the villagers, they were brought to the house of a local landowner, Sir Richard de Calne, who lived north of Woolpit, near Bardwell. He took the two in and cared for them from then on. Moreover, Abbot Ralph's source for this information is beyond reproach – for it was none other than Sir Richard himself, who related the entire history of the green children to the abbot. The green shade of their skin eventually faded away, but the boy died after a time.

However, his sister survived, grew up, and stayed in the knight's service for many years. According to her, all of the inhabitants of her native land, and even all of their possessions there, were green, and she also claimed that she and her brother had entered a cavern after following their flocks for a while, whereupon they heard the sweetest sound of bells, which so entranced them that they pursued it for a time through the cavern until they came to its mouth, emerged, and were stunned by the brightness and warmth of the sun. Terrified, and unable to relocate the cavern's entrance, they were finally captured by the Woolpit villagers.

It is interesting to note that Ralph of Coggeshall's account has a more paranormal edge to it than William of Newburgh's, which might be because the former author had a passion for supernatural tales, so it may be wise to treat his more extraordinary claims, such as everything being green in the children's original land, with caution. However, both authors, and Sir Richard de Calne, were adamant about one fundamental aspect of this curious case – when the children were discovered by the Woolpit villagers, their skin was definitely green. So how can we explain them? Over the years, a wide range of identities and solutions to this longstanding mystery have been offered.

BABES IN THE WOOD, OR WOLF CHILDREN?

There is a centuries-old East Anglian legend that tells of two young Norfolk children who were heirs to the estate of their dead parents, but were being reared by an evil guardian who desired their fortune for himself. Consequently, to clear the path to inheriting it, he slowly poisoned the children with doses of arsenic, but, impatient at the length of time that this method was taking, he finally told one of his servants to take them into the depths of the forest beyond his own estate and kill them there, where their bodies would soon be eaten by wild animals. However, the servant could not bring himself to do this, and merely abandoned the children there instead, where they later died. Some Woolpit inhabitants believe that these babes in the wood – upon whose tragic tale the popular pantomime of the same name is based – were real, not legendary, and did not die, but rather found their way to Woolpit, and it is noteworthy that one of the symptoms of arsenic poisoning is a greenish tinge to the skin. Another interesting explanation proposed for the green children is that they had been abandoned or orphaned as youngsters and thereafter reared by wolves. Because these feral children would have lived in caves with the wolves away from sunlight and would probably have had a very poor diet, they may have suffered from chlorosis, turning their skin green. Moreover, in a *Daily Mail* letter of 2 July 1997 discussing this theory, Laraine Bates of Brome, Suffolk, stated that after appearing at Woolpit both children were said to howl at a full moon and were sometimes seen running on all fours.

FAERIE FOLK?

Eminent British folklorist Dr Katharine Briggs has emphasised various parallels between the green children's history and traditional storylines and motifs associated with fairies and the land of Faerie, which in turn are closely entwined with the Celtic realm of the dead. Green is the colour of Faerie and the colour of death in Celtic belief, both of which are reflected in the medieval tale of the enchanted green knight confronted by Sir Gawain. Fairies are also associated with subterranean abodes and twilight, as is the Celtic land of the dead, beans are the sole diet of the dead in Celtic legend, and St Martin is a saint linked with the dead.

So could the green children episode have originated as nothing more than a simple fairy story, of elves or fairies visiting the land of humans from their own hidden, twilit domain, which was gradually elaborated as it was passed down numerous successive generations via countless retellings until it was eventually but mistakenly assumed to be based upon truth? Indeed, is it possible that the green children are nothing more than variations upon the popular Green Man or Jack-in-the-Green theme, i.e. personifications of nature and fertility?

In reality, the facts that this episode's two main chroniclers actually lived at the time when it took place (rather than centuries later, with no direct, contemporary knowledge of it), and one of the main sources for it, Sir Richard de Calne, was undeniably real and featured centrally within it, argue persuasively against this notion.

BEINGS FROM ANOTHER WORLD?

Several different options can be considered within this broad category. For example: supporters of the Hollow Earth theory, which alleges that our planet is hollow, and conceals a secret, superior civilisation within its vast interior, have proposed that the green children must have absconded from this underground realm and found their way to the surface through a network of tunnels and passages, lured by the sound of church bells. Ironically, however, there are no caves or even any mines in the vicinity of Woolpit, though there are plenty of flint mines in the centre of Thetford Forest to the northwest and also many associated tunnels there.

Another possibility is that they crossed over into our world from a parallel universe, perhaps even containing a parallel Earth – thus recalling, for instance, Fyodor' Dostoyevsky's *The Dream of a Ridiculous Man* (quoted from at the beginning of this chapter) - emerging through some interdimensional window that had opened at Woolpit. Such windows have been postulated by paranormal researchers to explain the occurrence of all manner of uncanny entities, ranging from phantom black dogs, America's mothman, and Cornwall's owlman to 'grey' aliens, fairies, and the big grey man of Ben MacDhui. However, with no evidence to hand that such windows actually exist, it is difficult to progress with this theory as an explanation for the green children.

A third prospect is that the green children are extraterrestrials. As long ago as 1651, Robert Burton opined in his tome *Anatomy of Melancholy* that they may have come from Venus or Mars. Much more recently, the extraterrestrial hypothesis has been pursued enthusiastically by astronomer Duncan Lunan, assistant curator at Scotland's Airdrie Observatory. Based upon the children's description of their twilit St Martin's land, and the great river separating it from a luminous land beyond, Lunan has speculated that they may have originated from a planet whose one side permanently faces the sun and whose other is permanently cloaked in darkness with a twilit zone sandwiched between them. As for the great river, Lunan has postulated that this is actually a huge canal that encircles the entire planet and is used for planet-wide thermoregulatory purposes. He believes that they must have reached Earth by teleportation, and has suggested that this was accompanied by a bright auroral display, thereby interpreting the children's description of a sweet sound of bells as a visual rather than an aural stimulus.

Bearing in mind, however, the claim by both contemporary chroniclers of the green children episode that once the two began eating normal food their green skin colour slowly vanished, and that the girl grew up into a typical-looking woman and married locally (there are even claims that some modern-day descendants of her lineage exist today, including one branch in the USA), it seems unlikely that they belonged to some alien species.

A CATALONIAN COINCIDENCE?

One of the most peculiar aspects of this already-strange story is that an almost identical episode to that of the Woolpit green children allegedly took place in Spain, several centuries later, at the Catalonian village of Banjos. Apparently, one day in August 1887, the Banjos villagers were amazed to see two green-skinned children, a girl and a younger boy, emerge from a cave, wearing odd-looking clothes and speaking a language that no-one could understand. They were taken to a local landowner, and numerous attempts were made to feed them, but they refused everything until they were offered beans, which they lived on from then on. The boy died not long after the two had first appeared, but the girl survived for five years before she too died. By then, however, she had already learnt enough Spanish to be able to tell the Banjos villagers that she and her brother had come from a land where it was always twilight, and which was separated by a great river from a much brighter sunnier land.

And as if all of this (with the sole exception of the girl's death in the Spanish account) were not similar enough to the much earlier Woolpit version, the name of the Spanish landowner who had looked after them was none other than Señor Ricardo da Calno (an exact Spanish translation of Woolpit's Sir Richard de Calne!). Not surprisingly, many researchers have viewed the Spanish tale with great scepticism, and Sussex-based researcher Frank Preston has conducted detailed enquiries in an attempt to uncover any evidence for its veracity. However, none could be found – no Spanish library or museum archive with even a single reference to the incident was located, and the village of Banjos has apparently never existed. We can only assume, therefore, that someone had read about the Woolpit episode and decided to initiate a Spanish spoof version, which in due course became accepted as true.

FLEEING FLEMISH?

By far the most conservative but also without doubt the most persuasive and comprehensive explanation on offer for the whole Woolpit green children scenario (though as noted in Wikipedia's green children entry, even this theory is not without some shortcomings) is that of veteran green children investigator Paul Harris, who has published extensively on this subject, and with whom I have corresponded concerning it. Eschewing origins from beneath or beyond our world, Paul believes that the answer to the green children riddle can be readily uncovered by closely analysing the local history and geography of Woolpit's environs at the time of this mystifying episode.

Prior to the reign of Henry II, there had been a significant influx of Flemish weavers and merchants into Eastern England, but these were severely persecuted by Henry, culminating in a massacre of the Flemish at a battle in 1173 near Bury St Edmunds. Just 8 miles northwest of Woolpit and only a mile north of Bury St Edmunds is the village of Fornham St Martin. Paul deems it very plausible that the green children were Flemish children from the latter village (explaining why they called their homeland St Martin's Land) whose parents had been killed, and who had fled away northward into the dense woodland terrain of Thetford Forest, whose dark shadowed interior would have reminded them of twilight. With little to eat, the two frightened children would have gradually weakened to a state of near-starvation, which can yield a distinctive greenish tinge to the skin, a condition of severe malnutrition known as chlorosis. Tellingly, however, once normal feeding resumes and malnutrition recedes, so too does this chlorotic skin colouration – thus corresponding precisely with the history of the green children.

Looking out across the River Lark to the south, they would have seen the much more open, sunnier town of Bury St Edmunds, and perhaps lured one day by the sound of pealing bells from its church, they may well have found their way across to the other side of the Lark (possibly via various mining tunnels leading away from the forest), and thence to Woolpit, where they may have fallen into or taken shelter inside one of its wolf pits. Crawling out again when the sun came up and all around was brightly lit, they would have been dazzled and frightened, especially after having spent so much time previously concealed amid the gloom of Thetford Forest.

While in this bewildered, disorientated state, they were encountered by the Woolpit villagers, who would have been perplexed by these children's strange Flemish costumes and equally unfamiliar Flemish dialect, not to mention the eerie green pallor of their skin, caused by their severe malnutrition. And in their later recollection of their origin, it is not unreasonable to suppose that two young, orphaned children's memories of their long-lost original home, Fornham St Martin, and their enforced shadowy hideaway in Thetford Forest, might well have become somewhat confused and intertwined, yielding the composite, crepuscular St Martin's Land (or perhaps their grasp of the Woolpit dialect was not sufficient for them to differentiate verbally the two localities to the villagers and Sir Richard). Suddenly, many of the odd and seemingly opaque aspects of this enigmatic history are elucidated and plausibly resolved. Of course, far too long a time has elapsed for this theory ever to be unequivocally confirmed. Nevertheless, it ties in so well with such a diverse array of anomalies relating to this case that there seems little doubt it provides the most lucid appraisal yet offered regarding the true nature of the hitherto-inexplicable history of Woolpit's weird (albeit less than truly wonderful) green children.

THE GREEN CHILDREN OF OZ?

Finally: Devotees of L. Frank Baum's enchanting 'Wizard of Oz' series of children's books will no doubt have observed a strong similarity between the Woolpit green children's claim (at least according to Abbot Ralph's account) that everything in their St Martin's Land was green, and Baum's description of the Emerald City of Oz. Who knows – perhaps Baum had even come across a telling of the Woolpit weirdlings' history that inspired his imaginary viridescent city? Walking through St Mary's Church, moreover, I was surprised to find a second unexpected correspondence between Oz and Woolpit.

The church contains numerous carvings of animals, some real and others mythological, but one of the most startling of these, perched at the end of a pew, is an extraordinary composite beast that looks remarkably like a flying monkey! In *The Wizard of Oz*, Dorothy and her friends were, of course, pursued and harried by a flock of flying monkeys sent by the Wicked Witch of the West.

Meticulously carved, with every feather beautifully delineated, this mini-masterpiece may be an opinicus – a griffin-related hybrid, sometimes combining a simian face with a lion's body and the plumed wings of an eagle – yet another bizarre being finding shelter in the magical village of Woolpit.

The winged monkey carved on a pew inside St Mary's Church, Woolpit
(Dr Karl Shuker)

Chapter 7

M.I.S.

On the Trail of Men In Scales

And at last mankind conquered, so long ago that naught but dim legends come to us through the ages. The snake-people were the last to go, yet at last men conquered even them and drove them forth into the waste lands of the world, there to mate with true snakes until some day, say the sages, the horrid breed shall vanish utterly. Yet the Things returned in crafty guise as men grew soft and degenerate, forgetting ancient wars. Ah, that was a grim and secret war! Among the men of the Younger Earth stole the frightful monsters of the Elder Planet, safeguarded by their horrid wisdom and mysticisms, taking all forms and shapes, doing deeds of horror secretly. No man knew who was true man and who false. No man could trust any man. Yet by means of their own craft they formed ways by which the false might be known from the true. Men took for a sign and a standard the figure of the flying dragon, the winged dinosaur, a monster of past ages, which was the greatest foe of the serpent. And men used those words which I spoke to you as a sign and symbol, for as I said, none but a true man can repeat them. So mankind triumphed. Yet again the fiends came after the years of forgetfulness had gone by--for man is still an ape in that he forgets what is not ever before his eyes. As priests they came; and for that men in their luxury and might had by then lost faith in the old religions and worships, the snake-men, in the guise of teachers of a new and truer cult, built a monstrous religion about the worship of the serpent god. Such is their power that it is now death to repeat the old legends of the snake-people, and people bow again to the serpent god in new form; and blind fools that they are, the great hosts of men see no connection between this power and the power men overthrew eons ago.

Robert E. Howard – *The Shadow Kingdom*

One morning in July 1983, Ron and Paula Watson of Mount Vernon, Missouri, spied in a pasture adjacent to their farmhouse a pair of strange silver-suited humanoid beings, examining a black cow, lying motionless on the ground. Watching this unexpected scene through binoculars, the Watsons were amazed to see the cow and its mysterious investigators abruptly float through the air towards a large conical craft standing by some trees. They were even more amazed, however, when they spied two additional bipedal entities standing near the craft. One resembled a tall big-foot-like beast, but the other was a veritable lizard-man! Approximately 6 ft tall, it sported green reptilian skin, webbed hands and feet, and glowing cat-like eyes with vertical pupils. As the Watsons continued to watch, all four beings, together with the immobile black cow, entered the craft, which duly departed. Later, however, during a session of hypnotic regression, Paula Watson claimed that a few days before this, she had been abducted by these entities.

CLOSE ENCOUNTERS OF THE SCALY KIND

Although the 'Gray', with its huge wrap-around eyes, smooth skin, and slit-like nostrils, is by far the most familiar category of supposed extraterrestrial entity, it is not the only one that has been reported. Less familiar, but by no means less fascinating, is the reptilian alien or reptoid - as epitomised in fiction by the popular TV series 'V (The Visitors)', and in fact by the above case, from Patrick Huyghe's excellent *Field Guide to Extraterrestrials* (1996).

Another reptilian alien case documented by Huyghe featured eyewitness Fortunato Zanfretta, a 26-year-old night watchman, and took place on 6 December 1978, behind a house at Marzano in Genoa, Italy. This was where Zanfretta spied four bright moving lights, but when he investigated he was horrified to confront a 10-ft-tall bipedal reptoid, with horn-like projections above its luminous yellow triangular eyes, pointed spines on either side of its head, and a burly dark-green body whose skin sported a series of horizontal folds or ribbing.

This weird entity pushed Zanfretta to the ground, then vanished, but when Zanfretta stood up again and ran in panic to his car, he heard a loud whistling sound, felt a sudden heat, and saw a huge triangular craft rise up through the sky. He radioed his colleagues, and when they arrived they discovered a 24-ft-wide impression on the ground, resembling a giant horseshoe. Furthermore, as in the Watson incident, under hypnosis Zanfretta claimed to have been abducted by the reptoid and taken on board its craft for a time.

One person who is particularly interested in these 'Men in Scales' is veteran UFO investigator John Carpenter from Springfield, Missouri; and in a *MUFON UFO Journal* account from April 1993, he provided the following 'identikit' for reptilian aliens:

1. Height: 6-8 feet tall, upright always.
2. Features: lizard-like scales, smooth texture.
3. Colour: greenish to brownish.
4. Hands: four-finger claw with brown webbing.
5. Face: cross between human and snake.
6. Head: central ridge coming down from top of head to snout.
7. Chest: external "ribbing" may be visible.
8. Eyes: cat-like with vertical slit, gold iris.
9. Effect: repulsive, grotesque, disgusting.
10. Manner: intrusive, forceful, insensitive.
11. Behaviour: intrudes and rapes.
12. Physical effect: large claw marks photographed.
13. Communication: none reported!

Nor is Carpenter alone in his curiosity regarding this little-publicised category of ET. Other leading ufological researchers, including Budd Hopkins, Linda Moulton Howe (who first documented the Wat-sons' encounter), and Yvonne Smith, have collected similar data from eyewitnesses. Such is the consis-

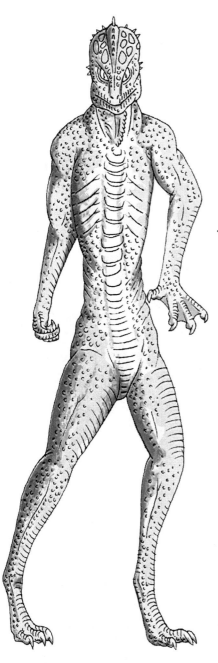

Reconstruction of alleged reptoid form by Tim Morris.

tency of descriptions, moreover, that when speaking in July 1995 at a MUFON symposium in Seattle, Washington, researcher Dan Wright readily conceded: "When a so-called reptilian is repeatedly described as having the same scaly skin, tone, claws for fingers, and an extreme interest in sexuality, one must pay attention".

One of the greatest controversies regarding alien visitations is whether the aliens really are extraterrestrial in origin, or whether they could actually be interdimensional. For example, some UFO researchers have speculated that the Gray may constitute a future stage in the evolution of our own species, which travels not through space but back through time to visit us. Similarly, as revealed here, the reptoid has also been the basis of some highly thought-provoking theories and scenarios intimately linking it to humanity - but from a very different perspective indeed.

REPTOIDS IN THE GARDEN OF EDEN?

Human origin has always incited considerable controversy among scholars. On one hand, DNA analyses reveal that humans share practically all of their DNA with higher apes such as the chimpanzee, thereby indicating a very close taxonomic relationship. Conversely, it has also been claimed that the theory of evolution cannot satisfactorily explain the profound intellectual hiatus separating *Homo sapiens* from other species. This in turn has led to much speculation concerning the possibility that human evolution was somehow accelerated artificially, by higher beings visiting our planet from Outer Space - and that accounts alluding to this significant event have been preserved in certain ancient documents.

American ufologist Dr Joe Lewels has been researching reptilian aliens for many years, and has published several articles, in which he presents some eye-opening clues to what he believes to be the true nature and reality of reptoids, as yielded by the Bible and other early religious works.

Noting that there are myths and religions worldwide telling of an ancient race of scaly super-beings that descend from the sky to assist in creating and teaching humanity, Lewels draws particular attention to certain alternative versions of the Biblical Creation, which provide some significant details not alluded to by the familiar version contained within the Book of Genesis. For example: in detailing the fateful consequences experienced by Adam and Eve after eating the forbidden fruit from the Tree of Knowledge, an ancient Jewish document called the Haggadah implies that they were originally very different outwardly from humans today:

> The first result was that Adam and Eve became naked. Before, their bodies had been overlaid with a horny skin and enveloped with the cloud of glory. No sooner had they violated the command given them than the cloud of glory and the horny skin dropped from them, and they stood there in their nakedness and ashamed.

This suggests that the first two humans were initially covered in a shining scaly skin, and part of their punishment for consuming the forbidden fruit was to lose their scales.

Continuing that line of speculation: if Adam were indeed created in the image of his maker, his maker must have been reptilian. This radical, revolutionary supposition is substantiated by a passage in another early document - the Nag Hammadi texts, a series of ancient scrolls found inside a clay jar in a small Egyptian town of the same name in 1945. The passage reads:

> Now Eve believed the words of the serpent. She looked at the tree. She took some of its fruit and ate, and she gave to her husband also, and he ate too. Then their mind opened. For when they ate, the light of knowledge shone for them; they knew that they were naked with regard to knowledge. When they saw their makers, they loathed them since they were beastly forms. They understood very much.

The concept of reptilian creators as indicated in early texts such as these has also been extensively explored by American researcher R.A. Boulay in his book *Flying Serpents and Dragons: The Story of Mankind's Reptilian Past* (revised 1997). Based upon these texts' testimony, Boulay concludes:

> "The sad fact is that in the West we have created God in our image and not the other way around. In this way we have hidden the true identity of our creators".

This is in stark contrast to Eastern traditions, in which humanity's supposed descent from reptilian ancestors is openly lauded.

Early Chinese emperors, for instance, widely proclaimed their lineage from dragons, and noble Indian families even today claim descent from the revered Indian serpent deities or nagas.

Eastern dragon carving in a Malaysian temple near Kuala Lumpur visited by the author during 2005 *(Dr Karl Shuker)*

WATCH OUT FOR THE WATCHERS

According to Lewels and Boulay, ancient documents testify to humanity's deliberate creation by scaly sky visitors - reptoids. But after they had inflicted their punishment upon Adam and Eve - transformation from reptilian into mammalian form, and banishment from Eden - what happened to these entities? Did they leave Earth, never to return? Perhaps not.

Lewels has noted that the Dead Sea Scrolls contain portions of the ancient books of Enoch which speak of the good and bad Watchers - angels who come down to Earth to teach good but also evil to humanity. Some descriptions of bad Watchers refer to them as serpent- or viper-like. Moreover, some modern-day alien abductees, such as Betty Andreasson Luca from Massachusetts, have claimed that their abductors have also informed them that there are good and bad Watchers - and that the bad Watchers will hurt and destroy humanity. Could these ominous figures be one and the same as the reptoids?

NOT SUCH A SNAKE IN THE GRASS?

Worthy of attention is the infamous serpent that tempted Eve in the Garden of Eden. Far removed from the limbless vermiform snakes of today, he is described in the Haggadah as:

> Among the animals, the serpent was notable. Of all of them, he had the most excellent qualities, in some of which he resembled man. Like man, he stood upright on two feet, and in height he was equal to the camel... His superior mental gifts caused him to become an infidel.

In other words, the serpent would have closely resembled a tall intelligent bipedal lizard - just like a modern-day reptoid, in fact. Only after he had lured Eve into temptation was he cursed by God to crawl ever afterwards upon his belly. Accordingly, some early artists depicted the serpent in his original bipedal form - one of the most famous works being a painting by 15th-Century Flemish artist Hugo van der Goes portraying Eve's reptilian inveigler as a human-headed web-footed lizard. Only later was this image superseded by renderings of the pre-cursed serpent as a normal, limbless snake. A detailed account of the pre-cursed serpent's appearance, illustrated by van der Goes's thought-provoking painting, can be found in an earlier book of mine, *Mysteries of Planet Earth* (1999).

Even more dramatic and radical is the 'serpent seed' belief, which in varying forms is central to the Two-Seeds-in-the-Spirit Predestinarian Baptists founded by Daniel Parker, and also to the Christian Identity movement, and includes faith-healing movement founder William M. Branham and Shepherd's Chapel pastor Arnold Murray among its proponents. According to this belief, Cain was not the son of Adam and Eve, but was a reptilian-human hybrid who resulted from sexual intercourse between the pre-cursed serpent and Eve.

HUMANOID REPTILES AND A PET DINOSAUR CALLED *HOMO SAPIENS*

According to the current fossil record, the earliest hominids (humans) date back only a few million years. Yet within that relatively short span of time (geologically-speaking), our own highly-advanced, super-intelligent species, *Homo sapiens*, has arisen. Surely, then, within the 150 million years that dinosaurs occupied our planet, a dinosaurian equivalent no less technologically advanced or intelligent than humans might also have arisen?

That such a dinosaur did indeed evolve is the central premise of David Barclay's book *Aliens: The Final Answer?* (1995). But that is not all. Barclay postulates that this humanoid reptile incited some global catastrophe, such as a massive nuclear war, at the close of the Cretaceous Period 65 million years ago, wiping out most of its kind, alongside more conventional dinosaur species.

A similar premise has also been presented by Mike Magee in his own book *Who Lies Sleeping?* (1993), in which he proposes that a humanoid dinosaur of advanced technological capabilities, dubbed by him *Anthroposaurus sapiens* ('wise human-lizard'), may indeed have arisen, eventually annihilating itself via a nuclear war, assisted by atmospheric pollution.

Barclay's theory, however, propels this concept much further. He suggests that a small contingent of the super-intelligent humanoid dinosaur survived, augmented by individuals that had been maintaining extraterrestrial outposts at the time of the nuclear devastation. Also surviving this devastation, according to Barclay, was the humanoid dinosaur's specially-bred 'pet' - itself a much-modified dinosaur, which evolved into...*Homo sapiens*!

During the next 65 million years leading up to the present day, the humanoid dinosaur continued to evolve too, ultimately transforming into the infamous Gray alien - thereby explaining why Grays seem so interested in humans. They are, in effect, continuing their studies and experiments with one of their own creations!

All in all, a memorable scenario, though unsubstantiated as yet by the fossil record, or indeed by human anatomy and molecular genetics. Even so, it does offer one tantalising thought upon which to reflect. If humanity ultimately wipes itself out via a nuclear holocaust or suchlike, but living organisms of some other kind survive, repopulate the planet, and are still alive here in vastly evolved, advanced form in 65 million years time, will there be sufficient clues persisting for them to deduce accurately what *Homo sapiens* and its civilisations were like? If not, who can state with absolute confidence that reptilian entities comparable to the humanoid dinosaurs postulated by Barclay and Magee are indeed wholly hypothetical?

DIMENSIONS AND DINOSAUROIDS

Drawing of the dinosaur-oid's head, by the author
(Dr Karl Shuker)

In an enthralling *Syllogeus* paper from 1982, Canadian palaeontologist Dr Dale A. Russell and taxidermist/model-maker Ron Séguin presented a detailed study of a bipedal 3-ft-tall theropod dinosaur from the Cretaceous Period called *Stenonychosaurus* (=*Troodon*) *inequalis*, which possessed a highly advanced degree of encephalisation (brain development). They concluded their paper by examining the fascinating prospect of how this dinosaur's evolution may have progressed if, instead of dying out over 65 million years ago like all other dinosaurs, it had survived and, via continuing evolution, had attained in the present day a level of encephalisation similar to our own. Their absorbing speculations produced an extremely interesting, albeit entirely hypothetical, reconstructed entity that they referred to as a dinosauroid.

Tailless, approximately 4.5 ft tall, walking upright in fully bipedal stance, with stereoscopic vision, opposable digits on its hands, and as intelligent as we are, their dinosauroid is undeniably the reptile world's answer to a human, but with some noticeable discrepancies. Its eyes are very much larger than ours and have reptilian vertical pupils;

its ears, nostrils, and lipless mouth are little more than slits; its hands each possess three claw-tipped fingers; and its feet each possess three principal toes, also claw-tipped.

Certainly a very distinctive figure - but also, for reptoid researchers, an uncomfortably familiar one. And for good reason - after all, this purportedly imaginary entity bears a striking similarity to eyewitness accounts of reptoid aliens, reptoids from the Black Lagoon (see later in this chapter), and even some reports of Gray aliens. Just a coincidence - or, as a few bold investigators have suggested, evidence for believing that some modern-day reptoids are visitors from parallel universes - in which the dinosaurs did not die out on planet Earth, but survived, continuing to evolve right up to the present day?

Conversely, as noted by Dr Darren Naish (*Fortean Times*, August 2008), some dinosaur experts have denounced the dinosauroid specifically for being *too* humanoid in form, claiming instead that an evolved troodontid dinosaur would more probably have retained a horizontally-aligned body (rather than acquiring a vertical human stance), as well as a long tail. Moreover, it might also have retained the feathery covering that troodontids are now known to have sported.

RHODES AND THE REPTOIDS

Among the most detailed sources of reptoid data currently accessible on the Internet is leading reptoid investigator John Rhodes's 'Reptoids Research Center' website (at http://www.reptoids.com/). Hailing from Nevada, Rhodes has appeared on many radio and TV programmes (including 'Unsolved Mysteries' and 'Strange Universe'), and has lectured on reptoids at various national and international UFO conferences.

On 24 August 1995, speaking at Mexico City's First International UFO Congress, he announced that the wave of UFO activity over Mexico City in the 1990s could be related to the reptoids, claiming: "...they may be preparing the way for the prophesied return of the Feathered Serpent god of ancient Mexico - Quetzalcoatl".

Rhodes's website contains several lengthy excerpts from his reptoid book, *Dragons of the Apocalypse*, including a detailed 'identikit' description of reptoid morphology. Based upon eyewitness accounts, it corresponds well with John Carpenter's version, except for one notable difference. Rhodes delineates from the typical reptoid form a morphologically distinct 'royal elite' of reptoid, which he terms the Draco.

Less frequently seen, they can be readily distinguished by their greater height (up to 12 ft tall), more athletic build, cranial horns, and leathery wings. Indeed, Rhodes considers that the mysterious airborne Mothman reported from West Virginia during the 1960s may have been a Draco.

Like certain other researchers, Rhodes believes that reptoids are not extraterrestrial in origin but are native to planet Earth. According to his Terrestrial Reptoid Hypothesis, they have descended from an intelligent race of

Mexican statuette of Quetzalcoatl
(Dr Karl Shuker)

surviving dinosaur, and are now living in underground cities in remote locations worldwide, from where they secretly communicate with civilians and the military. In addition, they host certain off-planet and even alternate-dimension stations.

However, Rhodes also devotes space to the extraterrestrial origin theory favoured by various other reptoid investigators, some of whom claim that these reptilian entities' home is a dying planet in (aptly) the star system Draco ('dragon'). In the past, it was also claimed that they plan to invade Earth during the 1990s in a planetoid spacecraft (though this did not happen, unless very covertly), and that they already use implants to control the Gray aliens who in turn use implants to control humanity.

THE DOGON OF MALI, AND REPTOIDS FROM SIRIUS?

Oannes

The Dogon are a shy, long-isolated tribe of hill- and cave-dwelling farming people inhabiting a barren plateau in Mali, West Africa - thus making it all the more remarkable that their ancient traditional lore embodies accurate knowledge concerning the star Sirius B, which they term po. For this tiny star, orbiting the dog star Sirius A in the Sirius system, is invisible to the naked eye, and was not even discovered by scientists until 1862 - so where did the Dogon derive their unexpected data concerning it?

Their answer is even more surprising than their fund of astronomical knowledge, because they claim to have received it long ago from a race of scaly amphibious extraterrestrials called the Nommo. According to the Dogon, the Nommo came from a planet orbiting a second tiny star in the Sirius system, which the Dogon term emme ya, but which is still unknown to scientists, who provisionally label it Sirius C.

The Dogon's amazing claims have been extensively investigated by well-respected scholar Robert K.G. Temple, who proposed in his bestselling book *The Sirius Mystery* (1976) that the Dogon's lore probably derived from pre-dynastic Egyptian belief, dating back well before 3000 BC. Several deities fitting the Nommo's description were revered in this region at that time, among the earliest of which were the Babylonian 'fish-people' or Annedoti, whose name translates as 'repulsive' - an adjective also incorporated in Carpenter's reptoid 'identikit'. The most eminent of these was Oannes, portrayed in ancient Babylonian depictions as a bizarre crossbreed of human and fish, with a bearded man's head beneath a fish's head, and a scaly fish's body borne upon the back of a man's body.

The Annedoti reputedly taught the ancient Babylonians the rudiments of civilisation; Oannes in particular would come onto land during the day to teach the people, then dive back into the Persian Gulf at night. Conceivably, therefore, Oannes was the original Nommo of the Dogon - but where did he come from? According to Dogon lore, the Nommo arrived on Earth in a huge fiery ark that generated a mighty whirlwind of swirling dust and thunderous noises as it landed, leaving behind a stationary star-like object in the sky. Such descriptions suggest that the latter may have been a mother spacecraft and the ark a shuttlecraft. But who were their occupants - an extraterrestrial race of scaly amphibious reptoids, perhaps?

REPTOIDS FROM THE BLACK LAGOON?

On the evening of Saturday 8 November 1958, Charles Wetzel was driving his 2-door green Buick Super along North Main Street, bordering the Santa Ana River near Riverside, California, when he came to a stretch of flooded road, caused by the river overflowing its banks, and began to slow down. For no apparent reason, his radio started to crackle, and Wetzel tried vainly to eliminate the problem by changing channels - but as he did so, something happened that totally dispelled all thoughts of radios from his mind.

Abruptly, an extraordinary figure leapt into view, and stood directly in front of his car, staring at him. Standing upright on its hind legs and at least 6 ft tall, this bizarre entity was covered in leaf-like scales, had a round pumpkin-like head that lacked a nose and ears but possessed a pair of bright fluorescent eyes and a beaky protuberant mouth, and waved its disproportionately long arms as its entire body swayed from side to side. Resembling some grotesque hybrid of lizard and human, its reptilian similarities were enhanced by Wetzel's description of its legs - for he claimed that instead of emerging from directly beneath its torso like a human's (and also like those of other mammals), they splayed out from the sides of its torso instead (as do those of lizards and other reptiles, interestingly enough).

The author with a model of an amphibious reptoid *(Dr Karl Shuker)*

Emitting a peculiar high-pitched gurgling scream, this saurian horror charged right up to the bonnet of Wetzel's car, reached across it, and began clawing the windscreen in a frenzied attempt to reach Wetzel - who grabbed his rifle but was scared to shoot, as it would destroy the windscreen, which was the only barrier, however fragile, separating him from his nightmarish attacker. Instead, in a fit of absolute terror, he promptly accelerated and ran the reptile-man down, feeling it scraping the undercarriage as the car sped over it and away.

This frightening episode was investigated by police, who failed to locate the scaly entity's body, even with the aid of bloodhounds. However, laboratory tests confirmed that something had certainly scraped the grease from the car's undercarriage, and also in need of explanation were the sweeping clawmarks visible for all to see on the car's windscreen. Years later, veteran Fortean investigator Loren Coleman interviewed Wetzel concerning his macabre encounter - just one of many such cases uncovered by Coleman that feature weird entities more akin to the category of alleged extraterrestrial reptoid than anything recognised by science to exist on planet Earth today.

Earlier in this chapter, cases and theories were examined which suggested that some modern-day reptoids may indeed have been extraterrestrials and were actually responsible in an earlier age for creating humanity. Also examined was the startling prospect that reptoids (and also Grays), even if extraterrestrial in abode today, might be unrecognised, highly-evolved descendants of ancient reptiles native to our planet yet which 'officially' became extinct millions of years ago. But what if some reptoids are not extraterrestrial, and instead still inhabit Earth? Could Wetzel's attacker be one such individual?

Loren Coleman refers to Wetzel's reptile-man and other similar beings researched by him over the years as 'Creatures From the Black Lagoon', on account of their surprising similarity to the amphibious scaly reptile-man in the classic sci-fi film of the same name from 1954. This term is especially applicable to the entity encountered by two sets of eyewitnesses at Thetis Lake, near Colwood in British Columbia, Canada, during August 1972, and subsequently documented by Coleman in his book *Curious Encounters* (1985).

The first sighting, occurring on 19 August, was an uncomfortably close encounter, featuring Gordon Pile, Robin Flewellyn, and a silver-scaled bipedal reptoid that emerged from Thetis Lake and chased the terrified youths, approaching close enough to cut Flewellyn's hand with the six razor-sharp spikes projecting from the top of its head in a central longitudinal ridge. Both youths were later interviewed by an officer from the Royal Canadian Mounted Police, who seemed satisfied by their apparent sincerity, and terror.

On the afternoon of 23 August, the Thetis reptoid was spied emerging from the lake again, but this time on the opposite shore, by Russell Van Nice and Michael Gold, who observed it in detail before it re-entered the water and vanished. They stated that its body was humanoid in shape, but silver in colour and covered in scales. Its face, however, was monstrous, with huge ears, and at least one large spike projecting from its head.

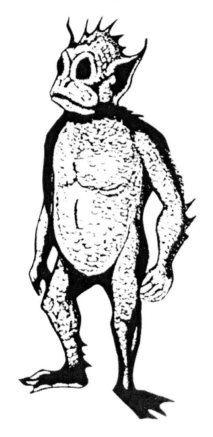

Reconstruction of the Thetis Lake reptoid's appearance
(Loren Coleman/FPL)

The following day, a sketch attempting to convey this entity's appearance, prepared by Walter McKinnon, accompanied an article documenting the episode in a British Columbia newspaper, the *Victoria Times*, and provided the creature with three-pronged flippers for feet, a fish-like mouth, and very large fish-like eyes. An enhanced version of this sketch has since been prepared by Loren Coleman, as featured here.

Remarkably, this bizarre entity was subsequently 'explained' by local police as nothing more mysterious than an escaped pet tegu lizard. Native to South America, tegus can grow up to 4 ft long, but are not bipedal or humanoid in shape, do not have crests, are not silver in colour, have typical five-toed lizard-like feet as opposed to three-toed flippers, do not have a fish-like mouth, but do have a very long, noticeable tail (unlike the Thetis Lake monster). Not surprisingly, given the considerable morphological discrepancies noted above, investigators of this case do not share the police's enthusiasm for accepting a tegu as the Thetis Lake monster's identity.

No less remarkable was the sober statement made in 1977 by respected New York State Conservation naturalist Alfred Hulstruck, who claimed: "a scaled, man-like creature...appears at dusk from the red, algae-ridden waters to forage among the fern and moss-covered uplands". This reptoid's reputed territory is New York State's Southern Tier region.

Nor are such reports only of recent origin. While perusing through the early archives of a Kentucky newspaper called the *Louisville Courier-Journal*, Coleman stumbled upon an extraordinary article from its 24 October 1878 issue, which told of an amazing entity allegedly captured alive in Tennessee and later exhibited in Louisville. According to this article, the entity stood 6 ft 5 in tall, was "covered with fish scales", and had eyes twice as large as normal. No clues to this apparent reptoid's final destiny have been uncovered.

Fascinating though they may be, however, all of these 'Reptoids From the Black Lagoon" pale into insignificance against the Lizard Man of South Carolina.

LOOK OUT - IT'S LIZARD MAN!

Reconstruction of Lizard Man's appearance by Richard Svenssen

One of the most surrealistic media stories of modern times began when 17-year-old Christopher Davis stopped to change a flat tyre. The time was approximately 2.00 am, the date 29 June 1988, when Davis pulled up near Scape Ore Swamp, just outside the backwater village of Bishopville in South Carolina's Lee County. After changing the tyre, Davis was placing the jack into the boot of his car when he saw something racing towards him, very rapidly, across a nearby field.

Just over 7 ft tall, it was running on its hind legs and had a humanoid outline, but when this unexpected figure drew closer, Davis was horrified to see that it resembled a giant bipedal lizard - with green wet scaly skin resembling a snake's, and slanted glowing red eyes. It only had three fingers on each hand, and three toes on each foot, but each of these digits sported a 4-in-long black claw.

Petrified with fear, Davis scrambled into his car and attempted to slam the door shut - but he was almost too late. By now, his scaly attacker, soon to be dubbed 'Lizard Man' by the marvelling media, had reached his vehicle, and grasped the door's mirror as it tried to wrench the door back open and thereby reach Davis. So determined was this frenzied reptoid that when Davis accelerated to escape its clutches, it leapt onto the car's roof and sought to hold on as the terrified driver drove frantically through the swampy wilderness at up to 40 mph. Fortunately for Davis, it evidently lost its grip when it reached down to pull at his car's windscreen wipers, because the reptoid promptly dropped off and was left far behind.

As might be expected, when Davis got home he was shaking with fright, and his car had several long deep scratches on its roof, as well as a severely twisted wing mirror. Speaking to the *Houston Chronicle* on 31 July, Davis's father, Tommy, said: "All I can tell you is that my son was terrified that night. He was hysterical, crying and trembling. It took awhile before he was calmed down enough to tell us what happened". Worthy of note is that eyewitness Davis insisted that what he saw was not simply a man in a suit; also, he was reluctant for his name to be reported to the media, because he was concerned that any resulting publicity might jeopardise his planned military career.

Lee County's Sheriff Liston Truesdale also accepted Davis's story, commenting: "We checked out his reputation, and he's a pretty clean-cut kid, no drinking or drugs. He's also agreed to take a polygraph test or go under hypnosis". In the weeks that followed, sightings describing a similar entity were recounted to the local police by many other eyewitnesses too - "and these are reputable people", to quote Truesdale's words.

Scape Ore Swamp, meanwhile, was itself swamped - with TV crews battling for filming sites, and a local radio station offered a $1 million reward to anyone capturing Lizard Man, after receiving over a thousand phonecalls a day concerning this scaly Scarlet Pimpernel. Naturalists claimed that a car said by Tom and Mary Waye to have been attacked by Lizard Man on 14 July had merely been clawed by a fox. And they discounted Davis's reptilian persecutor as a drunken hobo crawling out of a muddy ditch - though any 7-ft-tall hobo covered in green scales and with 4-in-long claws on each of its hand's (and foot's) three (not five) digits would surely have been every bit as extraordinary as a bona fide Lizard Man!

Several three-toed footprints, each measuring 14 in long, were later discovered in the swamp, and casts were made, but police dismissed these prints as most probably the work of a prankster, inspired by the extensive media coverage devoted to Lizard Man. In mid-August, just to add a spice of variety to the proceedings, a man from Florence, South Carolina, reported seeing a lizard woman; and a brown-scaled lizard man was reputedly spied by an Army colonel from Bishopville at the end of that same month, when it ran across the road along which he was driving - McDuffy Road, about 1.5 miles from Scape Ore Swamp.

Interestingly, some investigators suggested that Lizard Man may have ufological links - a line of speculation undeniably suggested by reports on file of similar reptoids encountered in association with strange aerial craft. However, this was played down by the media, who preferred to connect Lizard Man with bigfoot - despite (or possibly because of?) the fact that South Carolina had never previously been the scene of any alleged bigfoot sightings.

Even so, just like many other mysteries, Lizard Man eventually lost its grip on the headlines, especially when some later reports were exposed as confirmed hoaxes. The enigmatic entity itself was never identified - let alone captured - but its flurry of publicity has secured a place for it within the chronicles of unexplained phenomena.

But if Lizard Man and the various other Reptoids of the Black Lagoon variety reported here were, and are, truly real, what could they be? As speculated by Loren Coleman:

Are these beasts future time travelers lost in some time/space warp? Or infrequent visitors?

Or do you feel more comfortable with the idea there is a breeding population of scaly, manlike, upright creatures lingering along the edges of some of America's swamps?

Something is out there. That's for sure.

Nor are these entities confined to the New World. On 5 January 2003, a brief posting on the cz@yahoogroups.com online cryptozoology news group referred to humanoid lizard men having been reported from Italy's Rivers Po and Pijava. Accounts of these have been collected by chemist Sebastiano di Djenaro, who has even made plaster casts of handprints and footprints allegedly made by such entities. Similar beings have also been reported from large ponds in Poland.

FROG-MEN FROM AFAR?

Not all reptoids are scaly and saurian. Some seem more frog-like than lizard-like, with smoother amphibian-type skins. In his classic book *Passport To Magonia: From Folklore to Flying Saucers* (1969), long-standing UFO researcher Jacques Vallée referred to a remarkable encounter that took place during autumn 1938-9 at Juminda in Estonia. This was where two eyewitnesses spied a peculiar 3-ft-tall entity with greenish-brown skin. Although it was somewhat reminiscent of a giant frog, its eyes and mouth were slit-like, and even though it stood on its hind legs it seemed to have difficulty in walking on land. Even so, this did not prevent it from fleeing rapidly when pursued by its two astonished observers.

A direct ufological link featured in steelworker Harrison Bailey's disturbing encounter with a host of 18-in-high 'frog-men'. While walking through woodland at Orland Park, Illinois, on 24 September 1951, Bailey suddenly felt a burning, cramp sensation on his neck, and turned round to see a grey whirlwind-like object behind him. His next memory was of spying a silver-grey oval craft straddling a road close by, and two figures with green-tinted visors asking him where he was from and where he was going. At that time, Bailey felt paralysed, but afterwards he was able to walk away. Some men told him the next day that they had actually seen him come out of a flying saucer, but he did not believe them.

Nevertheless, Bailey was troubled by the memories of his odd experience for many years until, in 1975, he finally decided to undergo hypnosis. During one session, he stated that while walking in the woods he had been surrounded by a horde of 18-in-tall frog-like beings, running swiftly about on their hind legs, with large protruding eyes, a long slit-like mouth, smooth brown skin patterned with darker stripes, and only three toes on each foot. Curiously, while these beings were swarming all around Bailey, the grass on which they ran was bustling with strange tiny bug-like entities, only about an inch across.

Bailey ran away to try to escape from these frog-men, but they evidently caught up with him because his next recollection was of awakening inside their craft, where he saw the two visor-wearing beings who questioned him about his origin and destination. Bailey claimed that these beings informed him telepathically that they meant him no harm, desiring only to communicate with humanity and to utilise him as a spokesman for them.

These frog-like reptoids' ability to communicate with Bailey and their alleged amicable intent are in contrast to the sinister silence and antipathy sometimes displayed by their more lizard-like, scaly counterparts encountered by various other eyewitnesses.

One of the most mystifying reptoid cases falls within the frog-man category. Dating from 1972, it took place in Ohio, consists of two separate incidents, and features two well-respected eyewitnesses - as well as an incredible entity popularly termed the Loveland frog-man.

The eyewitnesses in question were a pair of police officers, Ray Schocke and Mark Matthews. The first incident occurred at 1 am on the morning of 3 March, when Schocke was driving towards Loveland along Riverside Road. His headlights lit up what he initially took to be a dog - until the 'dog' stood up on its hind legs, and revealed itself, according to Schocke's subsequent description, to be a 3-4-ft-tall entity with a frog-like or lizard-like face, textured leathery skin, and weighing about 60 lb. The 'frog-man' looked at him for a moment before turning away, leaping over a guard rail, and moving off down an embankment into the Little Miami River. Schocke drove off to the police station, and returned with fellow officer Mark Matthews to seek evidence of the entity's erstwhile presence, but all that they found were some scrape marks leading down to the river.

However, on or around 17 March, Matthews had his own encounter, when, while driving just outside Loveland alongside this same river, he saw what seemed to be a dead animal lying on the road up ahead. He stopped the car to pick up the carcase, but as he opened his car's boot the 'carcase' raised itself up into a crouching position, then moved to the guard rail without taking its eyes off Matthews, lifted its legs over it, and disappeared. Matthews shot at it with his gun, but missed.

The Loveland frog-man, illustrated by Richard Svenssen (FPL)

Other eyewitnesses had previously reported weird frog-like creatures in this region of Ohio, but in recent times some researchers have begun to wonder whether the two officers were mistaken in their sightings, perhaps subconsciously influenced by earlier stories.

REPTOIDS ON SCREEN

Our enduring fascination with terrifying reptiles does not seem entirely limited to dragons and dinosaurs. Although the Gray is the most commonly reported alien type in real life, on the big and small screens it is the reptoid that has proven most popular, judging from the frequency with which such aliens appear in films and TV shows.

The sci-fi movie 'Enemy Mine' (1985), starring Dennis Quaid and Louis Gosset Jnr, is a good example, featuring a human (Quaid) and a reptoid (Gosset) that start out as implacable enemies, but in order to survive are forced to become allies, and, ultimately, friends, when they are stranded together upon a barren, hostile planet.

Equally entertaining is 'Stargate' (1994) - a special-effects masterpiece, in which time-travel, ancient Egypt, and our human hero (played by Kurt Russell) are brought into conflict with an enigmatic god-alien being (Jaye Davidson), whose true reptilian nature is exposed in the film's final scene as its skin is peeled away.

Nor should we forget the granddaddy of all reptoid movies – 'The Creature From The Black Lagoon' (1954), in which an Amazonian scientific expedition encounters an aquatic scaly 'gill-man'. Although ostensibly a cold-blooded prehistoric survivor, this amphibious reptoid displays a decidedly hot-blooded interest in the character played by Julie Adams, especially when she decides to take a swim wearing her smooth, body-hugging bathing costume!

(Intriguingly, it has been claimed that this film was inspired by Mexican cinematographer Gabriel Figueroa mentioning a legend regarding a reptilian humanoid said to live somewhere in South America.)

Imagine a parallel planet Earth, in which the dinosaurs survived into the present, evolving into reptoid thugs every bit as aggressive and unpleasant as their human counterparts on our planet Earth. Then imagine what would happen if these reptoids found a way of reaching our Earth. Who could save us? Why, the Mario Brothers, of course! That, at least, is the plot of the hugely enjoyable 'Mario Brothers' film.
By far the most memorable reptoid-inspired screen creation, however, is the 1980s TV series 'V (The Visitors)'. Earth is visited by a race of ostensibly humanoid aliens, technologically superior to humanity but seemingly peaceful and willing to share some of their scientific advances with us. They befriend the authorities, and lull much of the human population into a sense of passive security - except for a small group of militia who fail to succumb to their saccharine charm. Suspicious of the Visitors' true motives, these non-believers begin secretly investigating them, and make the horrific discovery that the 'human' Visitors are actually disguised reptoids, planning - in best alien tradition - to take over the world! Happily, however, the Visitors do not have things all their own way, leading to some classic confrontations and storyline twists as the saga progresses.

REPTOIDS FROM INNER SPACE?

Why does humanity display such a profound, persistent fascination for dinosaurs and dragons? After all, one group is long-deceased, and the other never existed at all, so it is not as if we are ever likely to encounter them in the flesh. And why too do these two groups of creatures, bearing in mind their entirely distinct respective identities - 'extinct yet real' versus 'wholly imaginary' - share such marked morphological similarities? Both groups feature gigantic reptilians, whose respective carnivorous contingents, sharing remarkably alike body forms, would, if alive, terrorise and actively prey upon humanity.

In his book *The Dragons of Eden: Speculations On the Evolution of Human Intelligence* (1977), the late Professor Carl Sagan recalled the pioneering brain studies of Dr Paul MacLean, head of the Laboratory of Brain Evolution and Behavior of the USA's National Institute of Mental Health. MacLean's work led him to propose in his treatise *A Triune Concept of the Brain and Behaviour* (1973) that the forebrain in higher vertebrates (reptiles, birds, and mammals) is composed of three distinct regions, each acquired during a subsequent phase of evolution.

The most ancient of these three regions surrounds the midbrain and is known as the reptilian or R-complex, which probably evolved several hundred million years ago and is shared by reptiles, birds, and mammals. Surrounding it is the second region, called the limbic system, shared by all mammals but not, in its full elaboration, with the reptiles and birds, and probably evolving over a hundred million years ago. Lastly, encompassing this is the third and most recently-evolved region, known as the neocortex, which is most elaborately developed in humans and whales, and, to a lesser extent, in non-human primates and carnivoran mammals. Although it probably evolved several tens of millions of years ago, the neocortex's greatest flourish of development occurred just a few million years ago, when humanity first appeared.

In his own book, *Dreams of Dragons* (1986), biologist Dr Lyall Watson also recalled MacLean's views, noting that whereas humans think with the neocortex and experience emotions with the limbic system, when we are asleep (and especially when we dream) it would appear that it is our long-suppressed reptilian complex which takes control - held tightly in rein during our consciousness, but unchained like a veritable dragon as we slumber.

And what does this archaic portion of the human mind contain - memory remnants, perhaps, of long-distant ages when our bygone ancestors, the shrew-like non-entities constituting the first mammals, ran for cover beneath the mighty feet of the dinosaurian lineage more than 65 million years ago? Memories of what it was like in those terrifying times when giant reptiles ruled the earth would not be easily banished even by many millions of continuing evolution long after the dinosaurs had finally met their end. Nonetheless, just like orally-preserved traditions and folklore, they would inevitably become distorted, especially over millions, rather than merely hundreds or even thousands, of years.

Perhaps, therefore, as suggested by Sagan and Watson, this is the key to the intriguing morphological and behavioural similarities between real dinosaurs and unreal dragons, and also to our tenacious fascination for both of these formidable reptilian groups. Quite possibly they are one and the same - that humanity's worldwide myths and legends of dragons are preserved, distorted reflections of far-distant recollections of the last dinosaurs, passed down to us from the primitive mammals that shared their world and ultimately supplanted their reptilian dynasty with a mammalian one. But that is not all.

Just as speculation is increasing that some supposed meetings with Gray aliens and even fairies and other supernatural beings may not be real but rather hallucinations induced by abnormal electrical activity within the brain's temporal lobe, could it also be that some alleged encounters with reptoids are actually hallucinations incarnated by irregularities within the forebrain's reptilian component?

How extraordinary it would be if one day we were to confirm that at least some reptoids originate not from Outer Space but from inner space, periodically conjured forth by our own brains to dance before us like hazy wraiths called to life from their prehistoric graves.

Chapter 8
ALL IN THE MIND?
Tulpas, Egrigors, and Other Thought Forms

Perhaps fairies, dwarfs, elves, leprechauns, dragons, monsters, vampires, werewolves, ghosts, poltergeists and flying saucers all exist. And perhaps the cynics who say that it is all in the mind are also right, because all these things exist or are produced at the second or etheric level. The strange behaviour of all apparitions suggests that they obey laws not quite like those of conventional physics, and that they probably belong to a reality with slightly different space-time references. The fact that those who come closest to these phenomena, usually receive information structured to support their own beliefs or fears, suggests that these apparitions cannot be entirely independent of the minds of those involved. Taken together, these two suggestions provide the basis for a concept that could account for a great many mysteries.

Dr Lyall Watson – *The Romeo Error*

I s it possible for the human mind to create visible, animate entities merely via the power of thought? This exceedingly radical but fascinating concept has received serious attention over the years from a number of well-respected researchers in the field of the paranormal, including Dr Lyall Watson, who penned the above-quoted suggestion while writing on the subject of survival without a physical body.

Expanding this line of speculation in *Species Metapsychology, UFO Waves, and Cattle Mutilations* (1977), retired American nuclear scientist Thomas Bearden proposed:

...the collective species unconscious is vastly more powerful than a personal unconscious, and under appropriate conditions it can directly materialise a thought form, which may be of an object or even of a living being.

Bearden believed that UFOs, fairies, dragons, and even the Loch Ness monster have originated as thought-forms, and have become increasingly substantial due to the continuing interest that they have generated, i.e. a self-perpetuating process that will ultimately lead to these entities becoming as physically corporeal as any 'real' organism.

But what (if any) evidence is there for the existence of thought-forms? In fact, there is a good deal more than one might expect, as disclosed here.

TULPAS FROM TIBET

The most famous example of a thought-form to be reported is undoubtedly the Tibetan tulpa. Usually (but not invariably) assuming human form, tulpas are reputedly generated by adept Buddhist lamas following intense periods of concentration sometimes spanning several months. These animate thought-forms are generally created for the purpose of fulfilling a specific mission, and can therefore be noticeably independent and vigorous - so much so that they may not only be mistaken for real, corporeal humans, but can sometimes even create their own tulpas! These secondary, tulpa-engendered tulpas are referred to as yang-tuls, and will in turn occasionally engender their own tulpas - rare, tertiary emanations termed nying-tuls.

Such claims may well seem implausible and fanciful to westerners unfamiliar with Tibetan mysticism, but one westerner who was far from unfamiliar with this subject and who experienced tulpa creation at first-hand in Tibet was Alexandra David-Neel.

A notable French scholar and former opera singer, David-Neel spent 14 years in this bleak, mountainous land, devoting her entire sojourn here to the painstaking study of Tantric Buddhism, and eventually becoming Tibet's first-ever woman lama. During her studies, she learnt all about tulpas, and as recalled in her book *With Mystics and Magicians in Tibet* (1931), her own attempt to create one of these mentally-induced manifestations was certainly successful - indeed, it proved to be a little too successful:

> In order to avoid being influenced by the forms of the lamaist deities, which I saw daily around me in paintings and images, I chose for my experiment a most insignificant character: a monk, short and fat, of an innocent and jolly type.

> I shut myself in tsams [seclusion for meditation] and proceeded to perform the prescribed concentration of thought and other rites. After a few months the phantom monk was formed. His form grew gradually fixed and life-like looking. He became a kind of guest, living in my apartment. I then broke my seclusion and started for a tour, with my servants and tents.

> The monk included himself in the party. Though I lived in the open, riding on horseback for miles each day, the illusion persisted. I saw the fat trapa [novice monk], now and then it was not necessary for me to think of him to make him appear. The phantom performed various actions of the kind that are natural to travellers and that I had not commanded. For instance, he walked, stopped, looked around him. The illusion was mostly visual, but sometimes I felt as if a robe was lightly rubbing against me and once a hand seemed to touch my shoulder.

> The features which I had imagined, when building my phantom, gradually underwent a change. The fat, chubby-cheeked fellow grew leaner, his face assumed a vaguely mocking, sly, malignant look. He became more troublesome and bold. In brief, he escaped my control.

> Once, a herdsman who brought me a present of butter saw the tulpa in my tent and took it for a live lama.

I ought to have let the phenomenon follow its course, but the presence of that unwanted companion began to prove trying to my nerves; it turned into a 'day-nightmare'...so I decided to dissolve the phantom. I succeeded, but only after six months of hard struggle. My mind-creature was tenacious of life.

There is nothing strange in the fact that I may have created my own hallucination. The interesting point is that in these cases of materialization, others see the thought-forms that have been created.

Like many of their other mystical powers and techniques, the ability of Tibetan Buddhists to create tulpas most likely derives from Bengali Tantrism, which embraces a technique for manifesting thought-forms known as kriya shakti ('creative power').

WHEREFORE THE WEREWOLF?

Not all thought-forms are created purposefully; sometimes they can be manifested unwittingly, but the result is no less dramatic - as discovered by occultist Dion Fortune.

In *Psychic Self-Defence* (1930), Fortune recalled a bizarre incident in which, via the power of her own thoughts, she had "...formulated a werewolf accidentally". She had been lying on her bed, her mind awash with intense feelings of anger, resentment, and vengeance after a painful disagreement with an acquaintance. Suddenly, she began thinking about one of Norse mythology's most vengeful monsters, the terrifying Fenris wolf, spawned by Loki, god of evil - whereupon she instantly felt a very strange sensation, as if something were drawing itself out of her solar plexus. Somewhat alarmed, she looked around - and there, lying next to her on the bed, was a huge, freshly-materialised but physically-substantial wolf!

Well-versed in the arcane lore regarding such entities, Fortune knew that it was imperative for her to gain and maintain control of this thought-engendered monster via the power of her own mental will until she could dispel it. And so, very courageously, she brusquely elbowed her uninvited visitor in its hairy ribs, ignoring its threatening growls, and pushed it off the bed.

The wolf abruptly vanished through the wall, but its presence remained in the house, as testified that evening by a member of her household staff, who saw its glowing disembodied eyes watching her from one of the corners of her room.

Realising that this lupine thought-form was not about to leave of its own accord, Fortune set about dispelling it herself. After visualising a long tenuous thread that linked the wolf's body to her own, she concentrated intensely upon drawing out the wolf's life-force along this thread. As she did so, the monster's shape became amorphous, its colour faded, and then suddenly, her psychic Fenris was no more.

THE KLUSKI CREATIONS

One of the 20th Century's most remarkable psychic mediums was a Polish engineer and writer, known to paranormalists worldwide by his preferred pseudonym, 'Franek Kluski' (his real name was Teofil Modrzejewski).

While in a trance state during his many public seances, Kluski materialised not only seemingly-authentic human apparitions, but also a number of extraordinary animal thought-forms. Among these was a bizarre humanoid (but not human) entity that became known as the Pithecanthropus, in dubious deference to modern man's direct ancestor, *Pithecanthropus* [now *Homo*] *erectus*.

ABOVE: Franek Kluski and his Pithecanthropus *(FPL)*

BELOW: Franek Kluski and his tulpoid nightjar *(FPL)*

One of its appearances occurred at a seance held by Kluski on 20 November 1920, which was attended and documented by French psychical researcher Dr Gustave Geley. According to Dr Geley, one of the sitters attending this seance felt the Pithecanthropus's large shaggy head, covered with coarse thick hair, press hard upon his right shoulder and against his cheek. When another sitter extended his hand, this eerie entity seized it and licked it slowly three times with its large soft tongue. Moreover, not only could it be seen and felt, it could even be smelled, filling the room with an odour similar to that of a deer or wet dog.

A second Pithecanthropus witness, Colonel Norbert Ocholowicz, likened this being to an ape in form, with low (though benevolent) intelligence, but incredible strength: "It could easily move a heavy bookcase filled with books through the room, carry a sofa over the heads of the sitters, or lift the heaviest persons in their chairs into the air to the height of a tall person". Sadly, alleged photos of the Pithecanthropus are less impressive, depicting a tall, lumpy figure resembling a stout man wrapped in a sack with a second sack over his head. Nevertheless, none of Kluski's apparitions has ever been exposed as a hoax.

Another of his photographed thought-forms was a winged, feathered creature. This entity materialised on his shoulders with its long wings outstretched during several seances held in 1919. According to the accounts of eyewitnesses present at these gatherings, it resembled a bird of prey, but a photo of this entity clearly depicts a nightjar - a nocturnal, insectivorous bird native to many parts of Europe, including southern England.

A third Kluski creation was a beast of somewhat wild, unpredictable temperament likened by its apprehensive observers at Kluski's seances to a lioness or maneless lion. According to an account in *Psychic Science* (April 1926) describing this mercurial thought-form, it was a rapacious beast as large as a very big dog and tawny in colour, with a slender neck, a mouth brimming with large teeth, and glowing feline eyes. Like the Pithecanthropus, it apparently enjoyed licking its human observers, with a moist and prickly tongue, and exuded a palpable odour, though in this case the smell was unmistakably (and potently!) feline. Indeed, even long after seances in which Kluski's 'lion' had materialised, the sitters remained "...impregnated with this acrid scent as if they had made a long stay in a menagerie among wild beasts".

Smallest of Kluski's thought-beasts was a weasel-like beast with a cold nose, which would run freely over the table, sniffing the hands and faces of everyone sitting around it.

FROM PUSS-IN-BOOTS TO POLAR BEAR

One of the most amusing, let alone amazing, thought-forms ever reported was documented in a *Fate* article (June 1960) by Nicholas Mamontoff, the son of one of its creators. Set in Russia, the incident took place many years earlier, in 1912, when a company of noblemen called the Brotherhood of the Rising Sun was encouraged by a visiting Chinese sage to induce the manifestation of a thought-form (or egrigor, in Russian) as a novel experiment.

As a means of confirming the accuracy (and safety) of the egrigor's appearance, the sage urged them to choose as unlikely and humorous but harmless an image as possible to be the focus of their mental concentration - and so it was that the image they selected was that of a red-haired cat wearing a pair of Russian boots!

And sure enough, after 30 minutes or so of intense concentration, a cloudy image duly appeared in front of them, which began to stabilise and coalesce into what eventually resembled a poorly-developed, stationary photograph of a red-haired cat standing motionless and wearing a pair of Russian boots on its hind feet. Once this incongruous egrigor was complete, the sage told the men to stop concentrating upon it, and within a short time their tulpoid 'Puss-In-Boots' had disappeared again.

No less bizarre was the curious case of the paranormal polar bear. While hospitalised a number of years ago, a psychical researcher was finding his enforced period of rest somewhat boring. And so, as a means of providing himself with an entertaining distraction, he decided to dream up a visiting polar bear that only he could see - rather like an ursine Harvey. Once conceived, however, this thought-engendered polar bear swiftly took on a life of its own, frolicking around the hospital ward in blatant disregard of its creator's increasingly-anxious attempts to keep it under control - and it was only with difficulty that he was finally able to make his white-furred phantom vanish for good.

Moreover, Richard Freeman of the CFZ, who may well be the world's only cryptozoological Goth, claims to have thought into being a giant spider tulpa in his lodgings during his three-year sojourn as an undergraduate zoology student at Leeds University. He fully describes this remarkable event in his book *Dragons: More Than a Myth?* (2005). Whether this daunting eight-legged thought-form still lingers there today, lurking in the shadows of some current unsuspecting student's lodgings, remains to be seen - literally!

ARE UFOS FROM OUTER SPACE OR INNER SPACE?

Not everyone believes that UFOs, if they are indeed something more than misidentified known objects, are extraterrestrial (or even interdimensional) spacecraft. Thomas Bearden, for instance, as mentioned earlier in this chapter, considered UFOs to be thought-forms.

According to the Swiss psychologist Carl Jung, the personal unconscious is ultimately composed of contents that have at one time been conscious but which have disappeared from consciousness through having been forgotten or repressed. The collective unconscious, conversely, can be looked upon as a kind of collective inherited memory, never having been in consciousness - and it is this, in Bearden's opinion, which is responsible for UFOs.

Bearden believed UFOs to be unconscious mental forms that materialise whenever the collective unconscious is subjected to stress. Pursuing this further, Bearden considered it no coincidence that Idaho businessman Kenneth Arnold's world-famous sighting of nine flying saucers near Washington's Mount Rainier on 24 June 1947, which precipitated the modern-day pandemic of UFO sightings, occurred during a period of appreciable world tension, caused by the Cold War between the U.S.A. and the U.S.S.R. in 1946-47.

A very different theory of UFOs as thought-forms is the 'earthlights' hypothesis proffered by Paul Devereux and colleagues. They consider that UFOs and certain mysterious BOLs (balls of light) are of terrestrial rather than extraterrestrial origin. Specifically, these phenomena are claimed to be the result of geomagnetic energies manifesting from geological faults, but may even interact with the electrical fields of the human brain, taking on whatever form is consciously or unconsciously bestowed upon them by the mind of the observer. In short, although UFOs and BOLs are not created directly by the observer's mind in the manner of 'true' thought-forms such as tulpas, their outward appearance is shaped by it.

MONSTERS OF THE MIND?

Yet even if we assume that tulpas, egrigors, psychic werewolves, Kluski's Pithecanthropus, earthlights, and other alleged thought-forms are indeed genuine mind-induced phenomena, from which portion of the human brain might they be originating?

A professional child psychologist, Stan Gooch is also a spirit medium and paranormal researcher, and in 1984 he devoted an entire book, *Creatures From Inner Space*, to this compelling conundrum, airing a very thought-provoking hypothesis. Gooch averred that the cerebellum, the portion of the brain controlling co-ordination and balance, also appears to be the origin of trance and paranormal activity and thus may well be the source of the energy involved in externalising these monsters of the mind - converting them from mental dreams, and nightmares, into physical (albeit often only temporary) realities. But this is not all.

In an earlier book, *The Neanderthal Question* (1977), Gooch noted that Neanderthal man *Homo sapiens neanderthalensis* possessed a larger cerebellum than that of modern man *Homo sapiens sapiens*, and is therefore believed by some researchers to have operated principally via paranormal or 'magical' powers. In contrast, modern man operates primarily via logic, dictated not by the cerebellum but by the cerebrum - a portion of the brain controlling sensory perception, learning, expression of the intellect and personality, language, and memory, and which is more highly developed in modern man than in Neanderthals.

Neanderthals are traditionally believed to have become extinct around 30,000 years ago (although there is some cryptozoological evidence in support of a more recent date). However, there is also a popular school of thought speculating that they did not truly die out, but interbred with modern man, thereby blending and submerging their characteristics among those of modern man. If this is true, we may have retained to varying degrees the Neanderthal capacity for paranormal activity, and those individuals with a greater than average share of Neanderthal ancestry might be expected to display paranormal skills beyond those of the norm.

Geographically, as Gooch noted, modern Asiatic humans have larger cerebellums than most modern westerners. Moreover, comparing the sexes, women have larger cerebellums than men. And sure enough, it is Asians and women who seem to exhibit the most profound paranormal skills - skills that include the mental generation of thought-forms.

The natural history of our planet may well be rich in diversity and fascination, but is effortlessly matched, it would seem, by the equally varied and compelling, shadowy fauna that constitutes the unnatural history of our mind.

Chapter 9
DENIZENS OF THE DREAMTIME
Australia's Mythical Menagerie

Mirriuula they used to call em. Great big dogs that grow. The more you look at em, the more they grow. They'll follow you and coax you away. They live in the river, in the water, and they've got real big eyes like saucers. The eyes is on the side of their face, like a fish. And they've got pointy ears. And they'll grow as big as a shetland [pony] while you're lookin at em.

I've heard a lot of stories about people being followed home at night by these big dogs. People just see em. Down the river fishin, a dog'll come out of the river, or a dog'll suddenly be behind em, followin em. Every time you look back, it's a bit bigger. Look back again and there's nothin there.

Frank Povah – *You Kids Count Your Shadows*

In the traditional beliefs of the native Australian (aboriginal) peoples, Alcheringa - the Dreamtime - was the Time of Creation. As described by Mudrooroo Nyoongah in *Aboriginal Mythology* (1994), it "symbolizes that all life to the Aboriginal peoples is part of one interconnected system, one vast network of relationships which came into existence with the stirring of the great eternal archetypes, the spirit ancestors who emerged during the Dreamtime".

In the Dreamtime, all of today's Australian animals existed in human form, as kangaroo-men, emu-men, koala-men, even starfish-men, and so forth, only later transforming into animals. However, there were also many much stranger beings - some monstrous, some humanoid or part-humanoid. These are discounted as fictitious by westerners and are largely unknown outside Australia. However, this vast continent's native people firmly believe that they still exist even today, and can occasionally be seen - if you know where, and how, to look for them.

As will be revealed here, some of these entities appear to be supernatural, others may be folk memories of real but now-extinct giant animals that were still alive when the native Australians' ancestors first reached this island continent over 30,000 years ago, and all offer fascinating insights into the ancient beliefs and traditions of Australia's indigenous people.

REPTILIAN RAINBOWS

Diamond python – one possible identity for the rainbow snake

Perhaps the most famous Dreamtime entity of all is the great rainbow snake. It is known to many different aboriginal tribes, and by many different names. These include: galeru or ungud (by the Ungarinjin of Western Australia's Kimberley Downs), karia (in southeastern Australia), kunmanggur (by Northern Territory's Murinbata tribe), langal (northwestern Australia), mindi (by Victoria's Kurnai tribe), muit (Northern Territory), pwiya (the western highlands of New Guinea), wollunqua (northern Central Australia's Warramunga tribe), yero (Queensland), and yulunggul or yurunggur (Arnhem Land).

Specific details concerning the rainbow snake's roles and activities vary from one region to another, but in general it is described as an ancestral Dreamtime creature of the sky that helped to create rivers and other water courses amid the dry land at the time of the world's creation. Since then, it has preferred to remain concealed in deep water holes during the dry season. Indeed, native Australians believe that each water hole is guarded by its very own rainbow snake. Whenever the rains begin, however, this enormous serpent ascends to the thunder clouds, its huge multicoloured body stretching across the sky to yield the rainbow. It is also a major fertility-linked spirit beast, and can be either male or female.

One of the most dreaded rainbow snakes was the mindi, which was said to be 10 miles long, with a huge head, a three-pronged tongue, and an immense mouth that spewed forth venom. Its merest gaze brought instant death, it suffused a foul stench, and left the terrible disease smallpox in its wake. Equally formidable was the yero, whose head was covered in red hair instead of the usual reptilian scales, and spouted forth entire waterfalls from its voluminous throat. Even more impressive was the wollunqua, claimed to be 150 miles long, and so tall that its head would vanish into the heavens if ever it should stand upright.

It is likely that these legends were inspired by some of Australia's numerous species of snake, notably the strikingly-marked diamond python *Morelia spilota*. They also suggest the possible late survival of certain extinct forms of giant Australian python-like snakes - such as *Wonambi naracoortensis*, a madtsoid, which may have attained 15-16 ft (larger than any modern-day Australian python) and could have been partly aquatic; and the memorably-named *Montypythonoides riversleighensis*, whose fossils have been found in northwestern Queensland.

SIX-LEGGED LIZARDS AND DREAMTIME DINOSAURS

Often confused with the yowie - a hairy bigfoot-like man-beast reported in modern times from many parts of Australia - is the whowie. This ancient Antipodean monster was variously claimed to be a giant lizard measuring up to 20 ft and resembling the real-life goanna (an Australian monitor lizard) but equipped with six legs instead of the normal four, or a gigantic six-legged insect with the head of a frog! Either way, it was a ferocious creature, fond of gobbling up any human that passed by its enormous cave on the banks of the Murray River in New South Wales's Riverina District. Fortunately, a combined effort by the water-rat human tribe and the other local animal-men tribes succeeded in smoking the whowie out of its cave, whereupon it was set upon and killed by its furious attackers. Worthy of note here is that a gigantic 20-ft monitor lizard, *Megalania prisca*, did indeed exist in Australia until at least as recently as 10,000 years ago. Perhaps encounters with this monster inspired legends of the whowie.

Another aquatic monster is the kurreah - a huge reptilian beast with bright scales and frills. In reality, however, it is nothing more than a crocodile, whose morphology has been exaggerated over countless generations of aboriginal story-telling in New South Wales.

Far more intriguing is the gauarge or gowargay, a bizarre emu-like creature but lacking feathers, which inhabits water-holes and will not hesitate to drag into a whirlpool anyone rash enough to bathe in its watery domicile, and drown him there. In his seminal cryptozoological tome *On the Track of Unknown Animals* (1958), Dr Bernard Heuvelmans noted that the morphological description of this peculiar beast is very reminiscent of certain bipedal dinosaurs, such as the ostrich-mimic dinosaur *Struthiomimus*. (Having said that, some dinosaur authorities nowadays believe that *Struthiomimus* may have been feathered, but in terms of posture this latter dinosaur would very likely have paralleled ratite birds such the emu.)

Equally tantalising is the similarity between *Tyrannosaurus rex* of North American prehistory and a bipedal reptilian monster known as the burrunjor, which is said by the local aboriginals to inhabit a remote expanse, also called Burrunjor, in Arnhem Land. There are even depictions of this terrifying 30-ft beast in native Arnhem Land cave art. Needless to say, however, it is highly unlikely that *T. rex* has a surviving cousin Down Under. Rather more plausible is that stories of the burrunjor stem from sightings of hefty (but nonetheless size-exaggerated) goanna lizards running on their hind legs - a common occurrence. (Having said that, there are a few genuine-sounding modern-day reports of huge bipedal reptiles on file from the coastal borderlands between Queensland and the Northern Territory.)

But how do we explain away the kulta? According to native lore in Central Australia, long before this region became a barren desert it was carpeted in lush vegetation, and was browsed by a huge but inoffensive creature known as the kulta, which possessed a very long slender neck and tail, a huge bulky body, and four sturdy legs.

This description immediately recalls the sauropods - those long-necked, elephantine dinosaurs exemplified by *Diplodocus* and *Apatosaurus*. Moreover, the fossilised remains of various sauropods have indeed been excavated in Australia, confirming the prehistoric existence of these long-necked dinosaurs here. They include *Rhoetosaurus brownei*, estimated to have measured more than 45 ft long, and *Austrosaurus mckillopi*, a brontosaur that may have exceeded 50 ft long.

Nor should we overlook the yarru or yarrba. In 1990, after learning from the elders of far-northern Queensland's Kuku Yalanji people about this traditional Dreamtime monster, which formerly inhabited rainforest water holes, missionary Dennis Fields asked one of the tribe's artists to paint a picture of it. This aboriginal artist had very little formal education, and no knowledge at all of palaeontology and what prehistoric animals looked like, only the descriptions of the yarru as preserved in his tribe's ancient lore – which makes it all the more remarkable that the creature painted by him bears an uncanny resemblance to a long-necked, flipper-limbed plesiosaur.

Is it conceivable that native Australians have sometimes found giant fossil bones belonging to such creatures as these, and that their myths of the kulta and yarru have arisen as a means of explaining where these bones came from? An alternative (albeit far more radical) possibility is that perhaps the sauropods and plesiosaurs in Australia did not die out alongside other giant reptiles by the end of the Cretaceous Period 65 million years, but persisted here into much more recent times, to be sighted and marvelled at by the first humans to reach this island continent from Asia.

BONES OF CONTENTION

If aboriginal belief in the kulta and/or yarru is indeed derived from the occasional discovery by native Australians of ancient fossil bones, it would not be unprecedented.

According to the lore of the Diyari (Dieri) people, the deserts of Central Australia were once covered in verdant pastures and giant eucalyptus trees that stretched up to Heaven itself. Their mighty trunks supported a thick canopy of vegetation, a veritable rooftop of foliage, inhabited by gigantic Dreamtime beasts called the kadimakara. Sometimes, however, these animals would descend the trees and explore the lands below, but once, while they were on the surface, the giant trees were destroyed, thus preventing the kadimakara from returning to their home in the canopy. From then on, they lived in permanent exile on the surface, spending much of their time wallowing in the waters of Lake Eyre until they died.

Scientists scoffed at such fanciful tales, until the Diyari showed them massive bones lying on the surface of salt pans along Cooper Creek and in the Lake Eyre Basin. These proved to be from immense herbivorous marsupials known as diprotodonts, resembling giant wombats. They had lived here many millennia ago, when this region had indeed been verdant, becoming extinct when the vegetation died. But this is

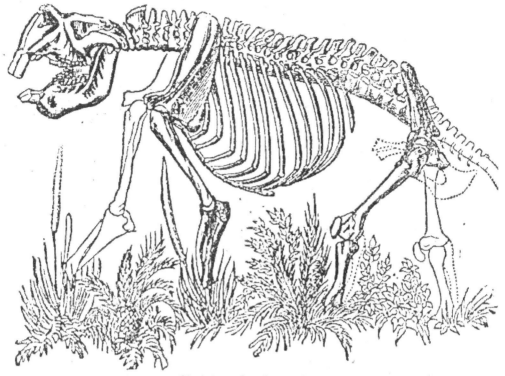

Skeleton of a diprotodont

not the end of the story.

Cave paintings dating back only ten millennia or so have been found that clearly depict diprotodonts, thus confirming that these giant marsupials and humans co-existed in Australia for a time. Hence it is clear that legends of the kadimakara do have some basis in fact.

So too, it would seem, does a similar mythical beast called the gyedarra. According to native Australians living near the Gowrie water holes, this creature was as large as a heavy draught horse, and their forefathers recalled seeing them in deep water-filled holes in this area's riverbanks, from which they emerged only to feed. This is a plausible diprotodont scenario. Moreover, when shown diprotodont bones, they stated that these were from the gyedarra. These extinct mega-marsupials may also explain the yamuti - giant wombat-like Dreamtime beasts in the lore of the Adnyamathanha people from South Australia's Flinders Range.

It is even possible that some reports of Australia's greatest cryptozoological celebrity, the bunyip, are based upon folk memories of swamp-dwelling diprotodonts.

IT'S SUPER EMU!

The Dreamtime lore of western Victoria's Tjapwurong people tells of gigantic flightless birds, said to be

Artist Philippa Foster's portrayal of a mirrii

Was the marrukurii based upon the now-vanished mainland thylacine?

"taller than the mountains", which they claimed were still alive only a few thousand years ago. Likening them to 'super emus', they referred to these veritable big birds as mihirung paringmal, and there are even depictions of them, portrayed with characteristically deep beaks, in native cave art from Queensland's Cape York Peninsula.

Like the kadimakara, the mihirungs seem to have been inspired by real but now-extinct species, as confirmed by findings in Australia of bones belonging to colossal flightless birds with deep beaks. Although superficially similar to emus, they were much taller and sturdier, and are now believed to be more closely related to waterfowl, and have thus been dubbed the demon ducks of doom. Known scientifically as dromornithids, they are referred to colloquially even by ornithologists as mihirungs - and one species, *Genyornis newtoni*, still existed when the first Asian voyagers reached Australia and co-existed with them for several millennia.

DOG DREAMING – MIRRII AND MARRUKURII

Two of the most interesting Dreamtime beasts are dog-like entities. The Wiradjuri still speak of magical canine creatures more than a little reminiscent of the paranormal Black Dogs featured in Western (particularly English and American) folklore, which they refer to as mirrii or mirriuula. A mirrii is black and very hairy, with bright red eyes and pointed ears, but is usually quite small when first seen.

The longer that someone stares at it, however, the larger it grows, swiftly becoming as big as a calf or pony, before suddenly vanishing in full view of its astonished observer. Although they are often merely inquisitive, mischievously following someone home before abruptly disappearing, they can be dangerous.

Sometimes a mirrii will emerge from a river or pool and attempt to befriend a passing human - but only with the sinister intention of luring him to his death in its watery depths.

Another dog of the Dreamtime is the marrukurii, which, according to aboriginal traditions prevalent in the vicinity of South Australia's Lake Callabonna, resembled a dog in outline, but was brindled with many stripes. They were believed to be dangerous, especially to human children, carrying away any that they could find to their own special camp at night, where they would savagely devour them.

When questioned, the native Australians denied that the marrukurii were either domestic dogs or dingoes. Is it possible, therefore, particularly in view of their brindled appearance, that these Dreamtime beasts were actually based upon memories of the striped Tasmanian wolf or thylacine *Thylacinus cynocephalus*? After all, this famous dog-like marsupial did not die out on the Australian mainland until about 2300 years ago.

Incidentally, this picture of a captive thylacine is a long-forgotten photograph that I discovered in an early animal encyclopedia and which, at least to my knowledge, was last published nearly 80 years ago (in sepia). As far as I am aware, it does not appear in any modern-day thylacine-related publication, and I have no details whatsoever concerning it.

THE SHADOW PEOPLE - FROM NYOL TO NINGAUI

According to the Wiradjuri people of New South Wales's central west, the elders always tell their children to count their shadows when playing, and be sure to tell them if they count an extra one - for that will surely mean a winambuu or a yuuri is playing with them.

Roughly equivalent to the Little People elsewhere in the world, the winambuu and yuuri (pronounced 'yawri') resemble small dwarf-like beings, only 3 ft tall and often hairy. They can be benevolent or antagonistic, depending upon their prevailing mood and the manner in which they are treated by humans, often acting as tricksters, but serving as guardians of certain localities too. Also spoken of in New South Wales, but this time by the Gumbangirr people, are the bitarr, who derive great pleasure from playing with Gumbangirr children.

Concealed to all but the sharpest of native eyes in the eastern Australian state of Victoria as they play amid the shadows of dusk are the nyols. These small humanoid entities have stony-grey skin, and spend their daytime underground like Antipodean gnomes, inhabiting subterranean caverns in deep rocks. According to Kurnai tradition, they can be good or evil, and will sometimes steal the memory of humans that they encounter.

The net-nets also hail from Victoria and inhabit rocky caverns, but mostly above-ground, and have brown skin with long claws instead of nails. In some ways the Australian counterpart of leprechauns, net-nets tend to make nuisances of themselves with humans – stealing things, and deceiving human hunters.

Ask any zoologist what a ningaui is, and if they are well-informed they will reply that it is a tiny shrew-like form of Australian marsupial mouse, the first known species of which were formally documented by science as recently as 1975. To the native Australian Tiwi people, conversely, ningauis are much more ancient, familiar entities - and it is from these that the ningaui marsupial mice derive their name. In traditional Tiwi aboriginal lore, the ningauis ('short ghosts') comprise a hairy race of 2-ft-tall Dreamtime beings with short feet and a passion for eating raw food, as they have no knowledge of how to make fire. They are active only at night, and inhabit dense mangrove swamps on Melville Island, off Australia's northern coast. The ningauis assisted in the earliest Kulama ceremonies, which are initiation rites into religious cults and feature the special preparation for eating of an otherwise poisonous yam known as the kulama.

STICK MEN AND NIGHT SPIRITS

Stranger in form than the aboriginal Little People are the various 'stick beings' of Dreamtime tradition. They include the desert-dwelling mimi of Arnhem Land in the Northern Territory. Said to have sported human form before the coming of the first aboriginals, these spirit people are nowadays tall but exceptionally thin, resembling animated sticks, and are thus able to live inside the narrowest rocky crevasses and amid the densest bush or scrub. Many ancient but finely-executed rock paintings exist in this region that depict mimi, portraying them dancing, running, hunting various creatures, and are usually painted only in red ochre. According to aboriginal lore, these are the work of the mimi themselves, i.e. self portraits. Furthermore, it is the mimi who supposedly taught the first aboriginal people in northern Australia how to paint, as well as how to hunt and cook kangaroos and other animals.

Nevertheless, mimi are not always benevolent, and today they are feared by native people here, because their diet not only includes yams, of which they are exceedingly fond, but also any unwary humans that they may choose to seize with their skeletal hands, especially if provoked. Consequently, when passing through mimi-frequented territory, it is best to choose a windy day. This is because these weird-looking entities are frightened to venture forth in such weather, in case their fragile thread-like necks should be broken by the blustery power of the wind.

An even more malevolent race of stick beings are the vampire-like gurumukas, frequenting Groote Eylandt ('Great Island') in the Gulf of Carpentaria. These spindly nocturnal spirits have long projecting teeth, and if one of them should encounter a native Australian walking alone at night, it will bite the back of his neck, causing him to die in great pain unless rescued and swiftly tended to by a medicine man.

Equally malign are the nadubi of Arnhem Land, which are equipped with barbed stingray-like spines projecting from their elbows and knees. These bizarre spirit beings also seek solitary humans, and if they should find one and succeed in stabbing a spine into his body, he will surely die unless the spine is removed immediately by a wise shaman.

The largest and most famous of all spirit stick men, however, are the quinkin, from Queensland's Cape York Peninsula. These giant entities represent the embodiment of human lust - on account of their excessively large (and often grotesquely-shaped) male sexual organs. As with the mimi, there are many prehistoric cave paintings depicting the quinkin, in Cape York's Laura rock galleries.

NEITHER MAN NOR BEAST - CLOSE ENCOUNTERS OF THE SPOOKY KIND

The most frightening monsters in any culture are those that appear partly, but not entirely, human, and this is certainly true of the Australian Dreamtime beings.

The yara-ma-yha-who is a truly grotesque spirit entity, which has red hair, red skin, huge eyes, lives in fig trees, and superficially resembles a small, toothless old man. However, it is also equipped with some decidedly non-human attributes. There are suckers on the ends of its long fingers and toes through which it sucks the blood of any unsuspecting human that it can leap upon. Its incredibly flexible jaws are not hinged at the back, hence they can open so wide that it can swallow its human victim whole. And its massive stomach is so obese that it can readily hold its victim until he is totally digested! Sometimes, however, it does not digest its victim, but regurgitates him and reswallows him several times. Each time, its victim becomes smaller, and redder - until at last he has transformed into a yara-ma-yha-who.

The aboriginals' ancestors travelled to Australia from southern Asia, which is home to a small, tree-dwelling, lemur-like primate known as the tarsier. Although harmless to humans, the tarsier does have enormous eyes and suckers on the ends of its fingers. Cryptozoologists have speculated that perhaps the yara-ma-yha-who is a distorted, much-exaggerated folk memory of the tarsier, passed down from the Australian aboriginals' Asian ancestors.

Similarly, just as there are many reports in southern Asia of giant bat-like entities - referred to in Java, for instance, as the ahool, and in Seram as the orang bati ('flying man') - native Australian lore also contains legends of veritable 'bat-men', known as the keen-keeng. Long ago, this tribe of half-humans inhabited a huge cave on the Western Australian border, and worshipped a fire god, to whom they sacrificed living humans. In their normal state, the keen-keeng were outwardly human, which greatly assisted them when luring victims to their cave, but they could be distinguished by their hands, which lacked the first two fingers of human hands. Their greatest difference, however, was their magical aerial ability - for whenever they chose, the keen-keeng could raise

The yara-ma-yha-who,
illustrated by Tim Morris

their arms above their heads and instantly transform them into a pair of large, powerful wings. This talent enabled these eerie entities to travel great distances when seeking potential sacrifice victims, but they were finally destroyed by two wise medicine men known as the Winjarning brothers.

Another semi-human monster vanquished by the Winjarnings was Cheeroonear, who lived with his wife and dogs in a dense forest near Nullarbor Plain, which overlaps the present-day Australian states of South Australia and Western Australia. According to William Ramsey Smith's *Myths and Legends of the Australian Aboriginals* (1930), Cheeroonear was:

> ...a being with ears and face like a dog, but without a chin. From the lower jaw there hung a flesh-like bag, shaped like the pouch of a pelican, and leading into the stomach. The ribs did not join in the centre to form a chest with one cavity, but were arranged so as to make two compartments. The compartment on the left side contained the lungs, and the one on the right side held the heart and its vessels, leaving the throat like a wide sack between the two, so that when it held water or food it looked like a tube...He stood eight feet high. His arms reached below his knees to his ankles. When he stretched or opened his fingers he could touch the ground. He could pick up objects from the ground without stooping.

Responsible for the disappearance of several people from human camps around the edge of the forest, Cheeroonear was finally ambushed by the Winjarnings, with the assistance of a dense fog sent by the God of the Dewdrops, and duly slain with their warrior boomerangs.

Happily, not all semi-human Dreamtime entities are dangerous or evil. The potkoorok of Victoria, for instance, is a shy, inoffensive man-frog, resembling a small human but with a wet pear-shaped body, long mobile fingers, and huge webbed feet. Highly reclusive, it actively hides away from human eyes in deep pools and rivers.

MARSUPIAL MISCHIEF –
DROP BEARS AND FEATHERED KANGAROOS

Finally, to end this chapter on a somewhat lighter note: Around the lumberjack camps of Wisconsin, Minnesota, and elsewhere amid the early days of North America's wild West, countless yarns and tall tales describing a vast variety of incredible but entirely fictitious creatures were spun and spread, as entertaining tomfoolery to while away the long dreary hours and also to fool gullible strangers. These stories ultimately yielded a truly marvellous menagerie unparalleled in its imaginative scope and drollness, whose very spurious species are commonly referred to collectively as 'fearsome critters', and have been well documented in many books down through the years.

Moreover, Australia can also boast a number of 'fearsome critters' - less familiar internationally, perhaps, but no less fascinating than their American counterparts. Within Australia itself, the most famous of these Antipodean curiosities is the drop bear. Closely related to the koala but larger and darker in fur colour, the drop bear shares its cuddly appearance, but not its inoffensive nature. On the contrary, the drop bear is greatly feared by anyone journeying through heavily-wooded outback territory, because it is known to lie in wait on overhead branches, and should anyone walk unsuspectingly beneath, this monstrous marsupial will drop unerringly down upon and dispatch its hapless victim with its lacerating claws and savage teeth. The only way to ensure safe passage through drop bear-inhabited terrain is to smear Vegemite behind your ears, which should be more than sufficient to deter even the most voracious drop bear.

Less daunting and much more exotic is the feathered kangaroo. The main claim to fame of this elusive creature is that its long white plumes are used to decorate the head-dress of certain Australian soldiers, namely the Australian Imperial Force (AIF) light horsemen. To the uneducated eye, these look very like emu plumes, but when asked, the AIF themselves are happy to confirm, with straight faces manfully employed, that they are indeed kangaroo feathers.

Mirrii dogs and marrukurii, nyols and ningauis, yero, yuurii, yarru, and yara-ma-yha-who, kadimakara, quinkin, whowie, mimi, potkoorok, and many more too - distant denizens of the Dreaming, but for whom there no longer seems to be any time in today's 'civilised', westernised world. Yet time is never still, and one day theirs too may come again.

Chapter 10
THE M FILES
Investigating Religious Marvels and Miracles

The world is full of wonders and miracles but man takes his little hand and covers his eyes and sees nothing.

Israel Baal Shem

The fitness of the Christian miracles, and their difference from these mythological miracles, lies in the fact that they show invasion by a Power which is not alien. They are what might be expected to happen when [nature] is invaded not simply by a god, but by the God of Nature: by a Power which is outside her jurisdiction not as a foreigner but as a sovereign.

C.S. Lewis - *Miracles*

In these modern scientific times, cynics and sceptics would have us believe that the age of religious marvels and miracles is long past - but nothing could be further from the truth. In multitudinous forms, they continue to be reported - daily, weekly, throughout the world. Some of them, such as visions, are very subjective, physically intangible, untestable, and intimately personal. Indeed, even in instances of shared visions, not everyone sees the same thing - the selectively-spied sun dances reported at Fatima are a case in point. At the other extreme are stigmata - spontaneously arising wounds that duplicate Jesus' crucifixion wounds - which are undeniably real, visible to all, but remain unexplained by current science.

And somewhere in between are all manner of fascinating, thought-provoking anomalies - statues that weep tears of human blood or drink copious quantities of milk, saintly relics credited with healing powers, records of astonishing talents possessed by religious figures, ranging from levitation and invisibility to fire immunity and physical incorruptibility following death. Ironically, the Church often tends to be as cautious and sceptical of deeming such phenomena miraculous as is science - the traditional foe of religious mysteries.

Yet who can blame them? After all, even if such phenomena are miracles, what do they mean? The devout believe that they are warnings of imminent doom if humanity does not turn away from its materialistic, secular pathway and pursue a more spiritual, godly course. The cynics dismiss them as profitable tourist attractions or as means of diverting the masses' attention from politically volatile situations. And the scientists all too often treat them as inconvenient exceptions to the stable, well-established laws of nature, and thus prefer for the most part to look away in the hope that these anachronistic oddities will quietly disappear.

As revealed in this chapter, however, miracles do not disappear that easily. They may change in form or locality, but they are as much a part of today's world as they have always been. So read on, and prepare, like our ancestors before us, to marvel at the marvellous!

MIRACULOUS ICONS

Since the first week of February 1997, crowds of pilgrims have visited the 11th-Century mountain-ensconced Kykko Monastery in Cyprus, to view what many Greek Orthodox Cypriots claim to be a modern-day miracle - and an omen of impending disaster. The focus of their concern, and prayers, is a 400-year-old icon of the Virgin Mary and the infant Jesus - for in disturbing contrast to the stoical, unemotional demeanour typically associated with statues and other graven images, this particular example has lately begun to weep tears!

Nor is this an isolated example of lachrymose icons. Numerous devotees flocked to the Church of the Nativity in Bethlehem during late November and December 1996, to witness for themselves a mystifyingly tearful centuries-old painting of Jesus. Built over what is claimed by local tradition to be the site of Jesus' birth, the church is shared by Roman Catholic and Greek Orthodox monks. Some eyewitnesses, including the Orthodox monastery's abbot, even stated that they saw the painted image of Jesus open and close its eyes.

Weeping Madonna statue at Brooklyn, New York
(FPL)

And on 1 September 1995, while visiting the Church of Mother Portaitissa and Saints Raphael, Nikolaos, and Irene, situated in a suburb of Toronto, Canada, a young woman called Pota Alexopoulos became the first of many pilgrims to allege that watery streaks of tears were freely running across the cheeks of the Virgin Mary, depicted in an ornate, 8th-Century colour reproduction of the Virgin and Child.

These are just a few modern-day examples of an extraordinary phenomenon substantiated by numerous others, some of which date back centuries, and recorded all around the globe.

One of the most celebrated weeping Madonnas is a small mass-produced plaster statue of the Virgin Mary that was initially owned by Antonietta and Angelo Iannuso, from Syracuse in Sicily. In August 1953, Antonietta was not only pregnant but also bedridden, due to a devastating combination of severe fainting attacks, seizures, and temporary periods of blindness. On 29 August, she experienced several fits, but later that day, when able to see again, she glanced at her Madonna statue - and saw to her amazement that it was crying.

This remarkable occurrence was duly confirmed by local visitors, and soon by sightseers from all over the world, journeying to Syracuse to witness with their own eyes Antonietta's miraculous Madonna - but the miracle was still not complete. Within a short time, Antonietta recovered completely from her diverse afflictions - and so too did a number of other sick people who visited her statue, including a mute girl who became able to speak, and a middle-aged man whose crippled arm duly regained its former power and capabilities.

When analyses of the Madonna's tears revealed that their chemical composition did indeed correspond with that of bona fide human tears, the Catholic Church officially proclaimed the statue's weeping as a miracle. It was subsequently placed both for protection purposes and for ease of visibility by future pilgrims in a glass display case in its own shrine, where it is today known as 'Our Lady of the Tears', and still attracts many devotees each year.

Other weeping icons include a Madonna statue at Maasmechelen in Belgium and an identical version in Brooklyn, New York; a coloured picture of a Madonna icon, initially owned by Pagona and Pagionitis Catsounis of Island Park, New York, but enshrined in the altar of St Paul's Greek Orthodox Church at Hempstead since late March 1960; and a Madonna statue at a Romanian Orthodox Church at Alexandria, near Bucharest, which restored the sight of a near-blind man praying before it in summer 1995, enabling him to read a wrist-watch.

Even more dramatic than weeping icons, however, are bleeding icons. A classic example, from 1980, featured another Sicilian Madonna statue, but which this time wept tears of blood. Eventually set into a rock wall near Niscima and sealed inside a glass case, it has been visited by considerable numbers of astonished worshippers and tourists alike. This remarkable artefact even converted the initially sceptical Bishop Alfredo Garsia into a fervent believer when, after originally banning visitors from venerating it, he finally decided to take a look for himself, and enthusiastically confirmed that it was indeed crying tears of blood.

In January 1971, a painting of the Virgin Mary owned by a lawyer from Maropati, Italy, began bleeding not only from its eyes but also from its hands, feet, and heart. Even when taken away by police and sealed overnight in a box, it still bled - and when the substance in question was analysed, it was found to be genuine human blood.

A statue of Jesus now housed in St Luke's Episcopal Church at Eddystone, Pennsylvania, has regularly exuded drops of blood from its palms since Easter 1975 - even when the hands were removed from the statue for closer examination! A plastic-coated postcard-sized picture of Jesus owned by Mrs Willie Mae Seymore of Roswell, New Mexico, unexpectedly began bleeding in May 1979, and its secretions were verified as genuine human blood. In 1985, several cases of bleeding icons were documented from Spain and Italy. On several occasions during November 2005, a white concrete statue of the Virgin Mary outside the Vietnamese Catholic Martyrs Church in suburban Sacramento, California, wept tears of blood which were photographed,; and in October 2007 a blood-weeping statue of Our Lady of Lourdes was reported from Jaffna, Sri Lanka.

As with weeping icons, the anomalous secretions of certain bleeding icons have been credited with profound healing properties. In the case of a bleeding Madonna statue at Baguio City, in the Philippines, the blood trickling from its heart was said have cured many seriously ill people, including its own sculptor, who had hitherto been suffering from cancer.

By definition, bleeding icons are controversial, but some are far more so than others. Named after the Italian coastal home town of Fabio Gregori, in whose garden it formerly stood, the Civitavecchia Madonna began weeping tears of blood in February 1995, and duly attracted vast numbers of visitors - until, on suspicion of fraud, it was seized by magistrates and its crimson secretions subjected to DNA tests. These verified that they were human blood - but of male origin. Gregori and other members of his family were then requested to donate blood samples for comparative analysis, but they all invoked their legal right to refuse to comply with this request. Having thus reached an impasse, in June 1995 the authorities returned the statue to Civitavecchia, where it was placed in the church of St Agostino. Since then, more than 20 miraculous cures have been attributed to this unassuming plaster figurine; sceptics, conversely, maintain that it is all a publicity stunt to attract tourism to the town.

Nevertheless, after 14 meetings that considered the testimony of a variety of different eyewitnesses, in February 1997 a commission of expert theologians concluded that there were sufficient grounds for classing the Civitavecchia Madonna's ability to shed tears of blood as a miracle, and their conclusions were forwarded to the Vatican's doctrinal department.

During 1911-15, religious paintings and even consecrated host at Mirebeau-en-Poitou, near Poitiers, France, would bleed copiously, but only when the Abbé Vachère was in the vicinity. The same occurred when he visited Aix-la-Chapelle. Investigations conducted by psychical researcher Everard Feilding indicated that it was possible, at least from a theoretical standpoint, that the Abbé was perpetrating hoaxes, but because he was an extremely devout person he would surely deem such activity blasphemous in the extreme and would therefore never even contemplate it.

In July 1995, the journal *Chemistry in Britain* published a paper by Pavia University chemistry researcher Dr Luigi Garlaschelli, in which he reveals how a convincing weeping or bleeding statue can be readily created. All that is needed is a hollow statue made of plaster, ceramic, or some other porous material. This should then be glazed or painted, rendering it impermeable to water, after which the statue is surreptitiously filled with water or some other liquid (through a tiny hole in the head, for example). This will be absorbed internally, through the porous material, but cannot exude externally - unless some of the glazing or impermeable paint is removed. If, therefore, some exceedingly faint, imperceptible scratches are made on or around the eyes, the internal liquid will slowly seep out through them, effectively creating the illusion that the statue is shedding tears. During his researches, Garlaschelli obtained an exact copy of the famous Syracuse weeping Madonna - and found that it was a plaster statue, glazed externally, with a hollow face. Just a coincidence?

Equally, the dark red-brown resin holding the eyes of Madonna statues in place will sometimes melt in very hot weather, thereby creating Madonnas that seemingly weep tears of blood. This was shown to be the solution to a blood-weeping Madonna of Fatima in Brunssum, southern Holland, during June 1995, and may also explain the bleeding Madonna at Grangecon in County Wicklow, Ireland, in May 1994 - but not those various cases in which the blood was analysed, and shown to be genuine human blood.

In short, there are some very straightforward physical mechanisms that can convincingly explain certain cases - and which may, moreover, be readily utilised to create fraudulent examples, presumably for financial gain or publicity (or both). However, as proposed by American parapsychologist D. Scott Rogo in *Miracles* (1982), it is conceivable that some instances are akin to poltergeist phenomena, i.e. they are induced by certain persons in close proximity to the icons. If true, this could explain those incidents implicating Abbé Vachère. Furthermore, argued Rogo, it is even possible that the flow of tears and blood by such icons is a physical consequence of the emotions or ailments of such persons being psychically transferred to the icons - as with Antonietta Iannuso?

SEVEN OF THE WORLD'S STRANGEST MIRACULOUS ICONS

1) Milk-drinking statues of Ganesha

Within a couple of days, beginning on 21 September 1995, statues of the elephant-headed Hindu god Ganesha amazed worshippers in India and elsewhere by drinking milk offered on spoons, but less than a week later this bizarre phenomenon had ceased as swiftly and mysteriously as it had begun. A second Indian outbreak took place during August 2006.

2) Moving statue at Ballinspittle

In 1985, many visitors to Ballinspittle in County Cork, Ireland, claimed that they saw a 3-cwt Madonna statue rock backwards and forwards in its grotto. Reports of similar activity by other Irish icons also emerged that year. Explanations include optical illusions, wish fulfilment, and mass hallucination.

3) Bleeding crucifix

In 1968, a 300-year-old wooden crucifix housed within a church at Pirto Alegre, in Brazil, abruptly began secreting a reddish substance, which, when analysed, was duly confirmed to be human blood.

4) Light-emitting communion wafer

During a weekend meeting of worshippers from the Cursillo movement (a lay spiritual group within the Catholic Church) at the Termonbacca Retreat Centre at the edge of the Creggan estate in Londonderry, Northern Ireland, in April 1998, many eyewitnesses there claim to have seen two bright rays beam forth from a communion wafer (the Host) that had been placed on a receptacle upon the altar. The rays, which began as Cursillo rector Don Moore dedicated the religious retreat to the Divine Mercy, reputedly continued to shine for hours, as worshippers brought others inside to observe this apparent miracle. A famous painting of the Divine Mercy shows two rays of light emitting from Jesus' body - one from his heart representing His blood, and the other representing the water that flowed from His side after He was pierced by a spear during the Crucifixion.

5) The bust and the bomb

On 6 August 1945, the day that the Japanese city of Hiroshima was devastated by the first atom bomb dropped during World War II, Pittsburgh businessman Allen Demetrius was amazed to see tears running down the cheeks of a 100-year-old bronze bust of a Japanese girl in his private collection of objets d'art.

6) The statue that closed its eyes

After a communion service on Good Friday 1989, the eyes of a 58-year-old plaster statue of Jesus inside Holy Trinity Church in Ambridge, Pennsylvania, were found to be closed. Why this was so amazing is that the statue had originally been made with its eyes open! Church officials were unable to explain this extraordinary occurrence.

7) Myrrh-weeping icon of murdered tsar

On 7 November 1998, a date that happened to mark the 80th anniversary of the Bolshevik coup in Russia that overthrew Tsar Nicholas II and led to the murder of the entire Russian royal family, an icon of Nicholas II in Moscow's Church of the Ascension inexplicably began exuding myrrh. Since then, the scent of myrrh emanating from it has been particularly strong on days when services have been held in memory of the tsar.

SAINTLY RELICS

The great Protestant Reformer John Calvin was extremely adept at publicising the undeniable fact that there were far too many saintly and other sacred Christian relics in existence for all of them to be genuine. And with, for instance, three separate corpses all claimed by their respective French churches to be the true corpse of St Mary Magdalene, four different Spears of Destiny (each said to be the lance used by a Roman centurion to pierce Jesus' side at the Crucifixion), numerous 'true' nails from Jesus' cross and 'true' thorns from His crown of thorns distributed throughout Europe's major religious centres, two separate heads of John the Baptist, and (in the words of Mark Twain) enough fragments of the 'true' cross to build a battleship, who could blame Calvin for his scepticism?

Even so, there are many relics that are genuine, and have been obtained from the corpses of saints by a holy process of dismemberment termed translation. Originally, relics (real and fraudulent) were obtained principally for purely financial reasons, to be resold for vast sums to pilgrims. Nowadays, however, fully-authenticated relics are sold at modest sums by the Church, as foci for public (but not personal) veneration. Moreover, stemming from their holy origins, a number of these relics have been credited with miraculous properties.

One well-publicised example is the preserved left hand of the 17th-Century martyr St John Kemble, a Roman Catholic priest who was executed at Hereford in 1679 and canonised in 1970. Retained in an oaken casket on the altar of the Church of St Francis Xavier in Hereford for over 200 years, the relic attracted worldwide interest in 1995. This was when, after the church's Father Christopher Jenkins suffered a severe stroke on 15 July and seemed likely to die with-

Painting by Fra Angelico featuring the Spear of Destiny or Holy Lance out even regaining consciousness, his assistant, Father Anthony Tumelty, placed the hand of St John on Father Jenkins's brow. Shortly afterwards, the 63-year-old priest awoke, and subsequently regained his previous good health. His doctors had earlier predicted that only the hand of God could save him - judging from its extraordinary effect, Father Tumelty had brought him the next best thing.

Of the four different versions on record, the example most widely accepted as the 'true' Spear of Destiny or Holy Lance (if, indeed, any one of them is) certainly has an eventful modern-day history. Supposedly owned by Charlemagne, several Saxon monarchs, Frederick Barbarossa, and the Habsburg dynasty, it eventually found its way into Vienna's Hofburg Museum, where it was exhibited until 1938 - when stolen by Hitler following Germany's annexing of Austria. The Spear of Destiny was very important to Hitler's plans for world domination, because he was well aware of the longstanding belief that whoever possessed this sacred relic remained invincible in battle, and that it was also allegedly infused with great healing properties.

In reality, these are merely legends. So presumably it is just coincidental that Charlemagne and Barbarossa both died shortly after dropping the spear, and that on the very same day when victorious American forces uncovered Hitler's hideaway for it in an underground vault at Nuremburg, several miles away in his Berlin bunker Hitler himself committed suicide?

Equally mystifying is the supposed dried blood of St Januarius (died c.305 AD), also called St Gennaro, preserved in two sealed phials at Naples Cathedral. On several different occasions each year, one of the phials is ceremonially displayed, during which time it mysteriously liquefies, forming bubbles and sparkling brightly. This acclaimed miracle has been witnessed by countless people, and even filmed on television, but never satisfactorily explained. Nor has the liquefaction of the blood of St Lawrence (=St Lorenzo) – martyred by the Romans in 258 AD - which is kept in a church near Naples, and is said to liquefy each year on 10 August (the saint's Feast Day).

However, a closely comparable artificial version to these, whose constituents include chalk and hydrated iron chloride, was created in 1991 by Dr Luigi Garlaschelli, also responsible for elucidating the mystery of weeping icons. This lends support to the long-held notion among scientists that the contents of the phials at Naples Cathedral are not blood; tellingly, they have never been formally analysed.

THE SHROUD OF TURIN AND THE VEIL OF VERONICA

Enshrined since 1578 at Turin's Royal Chapel of the Cathedral of St John the Baptist, and documented in detail within an earlier book of mine, *The Unexplained* (1996), the Shroud of Turin remains one of the world's most enigmatic artefacts, and is many things to many people. It is a miraculous icon to those who venerate it; a sacred relic to those who firmly believe it to be the linen sheet in which Jesus' body was wrapped after His crucifixion; and, even to sceptical scientists equally convinced that their radioisotopic datings place its origin firmly within medieval rather than biblical times, it lingers as a tantalisingly mysterious image whose identity and mode of creation continue to inspire all manner of fascinating conjectures.

In 1990, for instance, Shroud researchers Lynn Picknett and Clive Prince boldly claimed that this remarkable relic comprises an early attempt at photography achieved by none other than Leonardo da Vinci, and that its face is actually his own, superimposed upon someone else's body. Conversely, in June 1996 a team of scientists from the University of Texas Health Science Centre opined that the Shroud may date back to biblical times after all, alleging that previous radiocarbon dating techniques applied to the Shroud have been distorted by the presence of a microscopic layer (biofilm) of bacteria and fungi upon its surface.

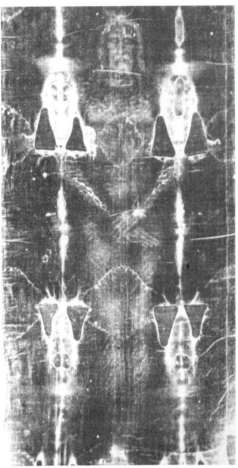

The Shroud of Turin *(FPL)*

Moreover, in 2005, research by retired research chemist Ray Rogers, published in the journal *Thermochimica Acta*, dismisses the medieval time-span of 1260-1390 AD obtained for the Shroud in 1988 using radiocarbon testing, by claiming instead that these tests had not been conducted upon any original cloth from the Shroud but rather upon a medieval repair patch on a section of the Shroud that had suffered damage. And in 2008, new research conducted by the Oxford Radiocarbon Accelerator Unit, which contributed to the earlier 1988 radiocarbon tests, led the Unit's director, Prof. Christopher Ramsey, to state in a BBC interview: "With the radiocarbon measurements and with all of the other evidence which we have about the Shroud, there does seem to be a conflict in the interpretation of the different evidence", and also to call upon the scientific community to probe further into the question of the Shroud's authenticity.

Clearly this most famous, and controversial, of sacred relics is a shroud by name and also by nature – destined to remain enshrouded in mystery. Its lasting potency, transcending devotion and disbelief alike, was well-expressed by journalist James Dalrymple (*Daily Mail*, 20 April 1998) after viewing it when placed on display in 1998:

Whatever the Shroud of Turin is, wherever it came from, and what it may mean, it is no longer a matter for priests, particle physicists and forensic sleuths. It now lies beyond cynicism and above the human hunger to reach out and somehow touch the divine. Against all the odds, it has triumphed.

Much less familiar than the Shroud of Turin, but surely no less intriguing, is the Veil of Veronica. Measuring just under 7 inches by 9.5 inches, this remarkable, near-transparent square of material bears an image of a bearded man that very closely resembles the face on the Shroud of Turin. Moreover, this same face appears on both sides of the Veil (which cannot be explained by any known preparation technique available in bygone times), yet disappears completely when viewed at certain angles.

It is said to be the cloth or veil that Jesus wiped his face upon as he carried the Cross to Calvary along the Via Dolorosa, before giving it back to a woman from Jerusalem who had handed it to him, and is claimed to possess miraculous healing properties. The woman supposedly travelled to Rome at a later date and presented the veil to the Emperor Tiberius.

By a curious coincidence, the name attributed to this woman is Veronica – which just so happens to translate as 'true image' – leading some researchers to dismiss the story as fiction. Nevertheless, the Veil itself is very real, but only rediscovered in 1999 after having been lost for almost four cen-

The cathedral of Jaén in Jaén, Southern Spain, has a copy of the Veronica, which probably dates from the 14th Century and originates in Siena. It is kept in a shrine by the high altar and is annually exhibited to the people on Good Friday and on the Feast of the Assumption.

turies. The Veil had last been reported in 1608, when Pope Paul V ordered the demolition of a chapel housing it, after which it was taken to the Vatican to be catalogued and stored, whereupon it vanished.

In 1999, however, Prof. Heinrich Pfeiffer, an expert in Christian art history at the Vatican's Gregorian University, announced that after spending 13 years searching through archives he had discovered this fascinating holy relic hidden away in the Capuchin monastery of the tiny Italian village of Manoppello, in the Apennine mountains.

According to the monastery's records, it had been donated there during the early 17th century by Dr Donato Antonio De Fabritiis, a Manoppello nobleman, who in turn had purchased it from Marzia Leonelli, a soldier's wife needing money to secure her husband's release from jail. Not all scholars, however, believe this to be the original Veil of Veronica, and there is an example stored in the chapel of St Peter's Basilica in Rome that has a rival claim to this status.

As with the Shroud, moreover, there is much controversy as to the true age of the Veil, but because it is so small there is concern that any tests conducted upon it could well destroy or severely damage it. Consequently, it too remains a chronologically contentious relic.

THREE LESSER-KNOWN SACRED RELICS

1) Pinning the tail on the donkey?

During Calvin's time, what was purported to be the tail of the donkey that carried Jesus into Jerusalem was exhibited at Genoa. Calvin, however, was suspicious of its authenticity, for good reason - bearing in mind that what was avowed to be the entire skeleton of this selfsame holy donkey, complete with tail, was preserved at Vicenza!

2) Gabriel's feathers?

The Escorial Palace in Spain may still possess a feather claimed in Renaissance times to be from one of the Archangel Gabriel's wings. Some ornithologists have speculated that it could conceivably be one of the exquisite 2.5-ft-long tail feathers of a male quetzal bird from Guatemala, which may have been brought back to Europe as a curiosity by the Spanish conquistadors. According to one author who saw the feather in 1787, however, it was rose-coloured, whereas the quetzal's are vivid emerald. Another possibility, as explored by me in my books *Mysteries of Planet Earth* (1999) and *Extraordinary Animals Revisited* (2007), is that it is a feather from a red-plumed New Guinea bird of paradise.

However, there may be a rather more mundane explanation for it. As noted by Lorraine Boettner in *Roman Catholicism* (1962), elsewhere in Spain a certain cathedral once exhibited what was claimed to be part of one of Gabriel's wings, which had somehow been left behind when he visited Mary at the Annunciation. When examined, conversely, it was found merely to be an ostrich feather, albeit a very magnificent one. Could a rose-dyed ostrich plume explain El Escorial's Gabriel feather?

In July 2008, I learned that another Gabriel feather is said to be preserved at the Basilica di Santa Croce in Gerusalemme, Florence, Italy; and that at least two more allegedly exist – or existed – in a monastery at Mount Athos, Greece (as documented in 1882 by D.M. Bennett in his book *A Truth Seeker Around the World*). I am currently pursuing both of these leads for further information.

Interestingly, a fictional account of a fraudulent Gabriel feather (a parrot's plume) appears in 'The Tenth Tale [told on the sixth day] - Friar Cipolla and a Feather of the Angel Gabriel' from *The Decameron* (c.1350 AD) by Giovanni Boccaccio.

3) The Holy Tunic

Not to be confused with the Shroud of Turin, or the Holy Robe (said to be Jesus' coat, and retained at the Convent of Argenteuil), the Holy Tunic is claimed to be the shirt that Jesus wore on His way to Calvary, and is credited with miraculous healing powers. On 19 April 1996, it was publicly exhibited for only the third time during the 20th Century when it was displayed for four weeks within a sealed glass case by Germany's Trier Cathedral. As the tunic has been preserved in a rubber solution, which prevents any attempt to carbon-date it, its age and possible origin cannot be investigated.

INCORRUPTIBILITY

The most astonishing relics of all, however, are not relics in the strictest sense of the word, for they comprise the entire corpses of certain saints. Unlike normal corpses, however, which soon decompose, these particular examples have remained amazingly fresh, or incorrupt. Such saints, therefore, are the incorruptibles, whose earthly remains appear inexplicably immune to the passage of time and the ravages of nature. The most famous incorruptible is undoubtedly St Bernadette of Lourdes. She died in 1879, but when her corpse was exhumed in 1909 and again in 1919, there was no sign of decomposition. Today, her corpse is on display, and although it has never been embalmed or treated via any other means of preservation, it is still so fresh that she appears merely to be sleeping. Even more astonishing, however, was the well-preserved state of the corpse of St Cecilia, who was martyred in 177 AD, yet was still incorrupt when inspected in 1599. Even the savage wound on her neck inflicted during her partial beheading was still readily visible.

Perhaps the most amazing case of an incorrupt holy corpse, however, comes not from Christianity but from Buddhism. In October 1992, Chinese and Tibetan officials confirmed that the corpse of the Panchen Lama, one of Tibet's most revered leaders, who had died in 1989, was actually sprouting hair around his ears! True, unlike the examples previously discussed here, his corpse had been preserved. Even so, we can only assume that whichever embalming fluid was employed in this process (which has never been made public), in faithful homage to a certain lager it refreshes the parts other embalming fluids cannot reach!

The mechanism responsible for incorruptibility of corpses is unknown. Saponification (the conversion of internal body tissue to a soap-like substance beneath a hardened outer skin) may be involved, and the precise composition of the encompassing soil and humidity may be important in exerting retarding ef-

The incorruptible corpse of
St Bernadette of Lourdes
(FPL)

fects. Nevertheless, this cannot explain why some corpses remain incorrupt whereas others, often lying next to them, decay in the normal manner. In anomalies researcher Dr Lyall Watson's book *The Nature of Things* (1990), his collaborator Ion Will offered a very thought-provoking insight into the nature of saintly relics:

> When things are venerated, they become venerable. They change. It can happen to anything. St Catherine of Siena wore a little bit of leather as her wedding ring and insisted that it was Christ's foreskin. It does not matter whether it was or not, belief is all that counts. If you add up all the fingers of St Mark in Venice, you get the impression that he must have had eight arms, but they all work miracles of healing.

Certainly, there is little doubt that the power of devotional faith can indeed achieve some astonishing, scientifically inexplicable results. Perhaps, therefore, that is the true miracle of saintly relics - that in reality, they are not the source but rather the catalyst for miracles.

STIGMATA - BLESSED WITH THE WOUNDS OF CHRIST?

For almost 800 years, there have been cases of stigmatics - people who have mysteriously gained bleeding wounds known as stigmata, which replicate the wounds inflicted upon Jesus at the Crucifixion. Stigmata reputedly appear of their own accord, frequently coinciding with some Easter-related date or festival in the Church calendar. In most cases, they heal shortly afterwards, sometimes at a remarkably rapid rate, and without experiencing any of the infections normally associated with open wounds - but reappear unbidden at the same time each successive year.

The precise number of stigmata that appear varies from person to person. Some only develop stigmata in their hands or feet or both, duplicating the wounds produced by the nails that held Jesus to the Cross.

Certain stigmatics, however, also display a stigma in one flank, mirroring the injury allegedly suffered by Jesus when He was speared by a Roman centurion. There are even cases whereby the stigmatic develops a ring of stigmata around the brow, thereby replicating the wounds engendered by the crown of thorns, or whip marks on the back, mirroring those inflicted upon Jesus by His Roman persecutors. The first known stigmatic was none other than St Francis of Assisi, but according to noted stigmata authority Ted Williams - author of *The Marks of the Cross* (1981) – over 400 cases have been recorded. Most have been female European Roman Catholics, and all have memorable case histories, as the following selection reveals.

ST FRANCIS OF ASSISI

It was 14 September 1224, and St Francis of Assisi had now completed a month-long period of fasting on Italy's Monte La Verna. Although still weary from this, the latest in many fatiguing trials of religious devotion to which he had willingly committed himself during the past 16 years, St Francis entered into deep contemplation concerning the Passion of Christ.

And as he did so, he experienced a glorious vision, in which he saw a radiant angelic being descend from Heaven. It was a seraph, borne upon six fiery wings, but assuming the form of a crucified man. The seraph began conversing with St Francis, but before it ascended into Heaven again, powerful beams of light streamed from its hands and feet, piercing those of the saint. And from that moment on, St Francis bore the crucifixion wounds of Jesus.

THERESE NEUMANN

It is important to note that a considerable number of stigmatics have histories of significant emotional and/or psychological trauma, and Therese Neumann was certainly no exception.

She was born at Easter 1898 in Konnersreuth, Bavaria, and in her youth she had allegedly been the victim of attempted rape on two separate occasions. Moreover, by her early 20s she had suffered a terrifying onslaught of debilitation, including severe head and spinal injuries, paralysis, blindness, and convulsions. However, she experienced miraculous cures on the anniversaries of important days in the life of St Thérèse of Lisieux, and on 5 May 1926, following various holy visions, she gained the first of her stigmata, in her side. After further visions during the early hours of Good Friday, she gained four more stigmata, in her hands and feet. From then on, they would appear almost every Friday for the rest of her life (she died in 1962), but would be totally healed by Sunday. Blood would also pour liberally from her eyes, judging from several dramatic photographs.

Even so, not everyone is convinced that Neumann was genuine, noting that there do not seem to be witnesses who actually saw blood flowing from her stigmata, and the only person who took photos of her when blood was flowing from her eyes was her own brother.

PADRE PIO

One of the most celebrated modern-day stigmatics was Padre Pio da Pietrelcina (1887-1968), an Italian friar who spent most of his life at the ascetic Capuchin monastery of San Giovanni Rotondo, Foggia, following a traumatic childhood plagued by ill health and his own proclivity for excessive religious self-denial.

In September 1915, during the anniversary week of St Francis of Assisi's stigmatisation, Padro Pio had

Padre Pio *(FPL)*

been visiting his parents, when on 20 September he developed stinging pains in his hands after an intense prayer session.

Exactly three years later, on 20 September 1918, he had been contemplating the Passion while gazing at a rather graphic crucifixion statue of Jesus in the monastery, when his hands, feet, and left side spontaneously gained stigmata, which were so painful that he screamed in agony. Their presence, and their pain, would stay with him for the rest of his life.

For although the stigmata in his hands and feet bled each Friday and also during the evenings immediately preceding and following each Friday but scabbed over during the remainder of each week, they never disappeared, and the stigma in his chest bled continually.

On 23 September 1968, however, his stigmata vanished - later that day, Padre Pio died.

Many stigmatics have been associated with miracles other than the appearance of their stigmata, as exemplified by Padre Pio. Extraordinary healing powers have been attributed to him, and he was also apparently gifted with the ability of bilocation, sometimes visibly appearing at some far-distant locality while simultaneously remaining in his monastery, though more often revealing his presence elsewhere via a mysterious, highly-fragrant scent - the odour of sanctity?

Worth noting here is that at his own requiem Mass, the church where his body had lain in state became suffused by this inexplicable perfume.

ETHEL CHAPMAN AND HEATHER WOODS

Born into a religious Anglican family in 1921, Ethel Chapman developed her stigmata relatively late in life - shortly after reading about the Crucifixion in an illustrated Bible during Easter 1974. Later interviewed by Ted Williams for a BBC documentary on stigmata, those in her hands actually began to develop during the course of the interview, forming the first eruptive marks normally preceding the onset of the bleeding wounds themselves.

Even more dramatic, however, were the stigmata of 43-year-old Heather Woods from Lincoln, who was interviewed on film by Ted Williams for an ITV religious programme in April 1993. The stigmata on Heather's hands and feet developed fully during the day, from round marks to bleeding wounds, and a cruciform (cross-shaped) stigma arose on her brow like a welt - first white, then red.

Equally amazing was Ethel Chapman's revelation that not only had she experienced accurate premonitions of the deaths of certain fellow residents at the Leonard Cheshire home in Springwood where she spent the last years of her life, but during the night before the death of each person, her stigmata had actually begun to bleed. In one instance, the person's death had been wholly unexpected, taking the home's staff and fellow residents by surprise - but not Ethel Chapman's stigmata, which had duly bled during the previous evening.

CLORETTA ROBINSON

American schoolgirl Cloretta Robinson (alternatively named Robertson in some accounts) gained world-wide fame at the age of 10 when, in 1972, she became the world's first black, Baptist stigmatic. Hailing from West Oakland in California, Cloretta's history began quite inauspiciously, when she read a religious book written by John Webster, entitled *Crossroads*, which dealt with the Crucifixion of Jesus.

Just a week later, however, on 17 March, a stigma spontaneously appeared in her left palm while she was in her classroom. During the next 19 days (leading up to and including Good Friday), several others also arose - on her right palm (Day 4), the upper surface of her left foot (Day 6) and her right foot (Day 7), the right-hand side of her chest (Day 7), and the centre of her brow (Day 14). These stigmata bled inter-mittently during that period, gradually decreasing in frequency, and were observed not only by Cloretta and her family but also by teachers at her school and various members of staff at West Oakland Health Center, including Dr Laretta Early, who studied her case. Unlike most other cases, however, Cloretta proved merely to be a temporary stigmatic; her wounds have not reappeared since.

GIORGIO BONGIOVANNI

Hauling the stigmata phenomenon well and truly into the modern age, Italian stigmatic Giorgio Bongio-vanni is a longstanding UFO enthusiast. In September 1989, he visited the famous Marian shrine of Fatima in Portugal - after claiming that, 5 months earlier, a 'shining being' had instructed him to make this visit, and had promised that when he arrived there he would receive an important sign from the Vir-gin Mary.

Sure enough, not long after reaching Fatima, Bongiovanni's hands each gained a distinctive, cruciform stigma, followed later by the development of stigmata in his feet and his side. More recently, he has gained a cruciform stigma in his brow too. Bleeding regularly, Bongiovanni's stigmata do not vanish, but they heal extremely rapidly. Bongiovanni offers an intriguing, singular link between religious and ufological scenarios, inasmuch as he claims to have experienced visions of Jesus and the Virgin Mary in which they have made their appearance in UFOs.

THE BLEEDING MIND?

The Bleeding Mind is the title of an authoritative work from 1988 by Ian Wilson, in which he provides persuasive evidence for believing that the spontaneous appearance of stigmata could actually be psycho-somatic - physically induced by the power of the stigmatic's own mind.

For instance: when serving in India several years earlier, a certain army officer was regularly tied up at night with ropes by his fellow soldiers, because of his tendency to sleep-walk. As a result, he became very disturbed emotionally, and in 1946 his case was publicised by British psychiatrist Dr Robert Moody, who had discovered that despite his former hardships, he still sleep-walked - but that was not all. To his astonishment, Moody had observed that during a period of supervised sleep-walking, the officer would spontaneously develop bleeding weals and indentations unmistakably resembling the long-vanished wounds that would have been inflicted by his fellow soldiers' ropes all those years earlier.

Another of Moody's patients was a woman whose body would spontaneously develop marks whose shape precisely mirrored those that her father would regularly inflict upon her with an unusually-carved stick when she was a young girl.

Perhaps the greatest clue concerning the likelihood that stigmata are generated by the stigmatic's own mind, however, can be found in the precise nature of the stigmata themselves. Invariably, they appear in the sufferer's hands, thereby corresponding with popular paintings and other depictions of Jesus' crucifixion. In reality, however, as revealed by biblical historians and other scholars, when people were crucified, they were nailed through their wrists, not their hands - which could not have supported their weight.

Similarly, some stigmatics develop a ring of small bead-like stigmata encircling their brow, thereby mirroring depictions of Jesus' crown of thorns. Again, however, such depictions are probably inaccurate, historically speaking. The placing of the thorns on Jesus' head was actually intended as a painful, additional punishment. Yet it is unlikely that a crown of thorns would have been very painful. Hence it is far more reasonable to suppose that His captors would simply have crushed a rough, shapeless mass of thorns onto His head instead.

The undisputed fact, therefore, that the nature of the stigmatics' wounds duplicate the inaccurate (but widely available) depictions of Jesus' wounds by artists, rather than the accurate (but less well-known) versions revealed by historical research, offers compelling evidence for believing that the spontaneous formation of stigmata is engendered by the stigmatic's own, inherent mental powers, and not by any mystical external influence.

As with all unexplained phenomena, there are a number of stigmata cases that have been conclusively exposed as fraudulent, or are very likely to have been. The vast majority of cases on file, however, appear sufficiently convincing to be genuine. Judging from these latter examples, therefore, it seems plausible, albeit astonishing, that certain individuals - especially those who have suffered severe trauma or are given to religious hysteria (particularly relevant in relation to various devout historical stigmatics) - can modify the shape of their own flesh via some still-undetermined mental process.

To quote Ian Wilson:

> If all this thinking is valid, then the implications are truly extraordinary. What is important is not so much the finding that stigmata are not miraculous marks of divine favour...Rather, the truly significant feature is that the flesh really does change, in an extraordinarily dramatic way, in response to mental activity, and that the power of mind over matter is phenomenally more powerful than previously thought possible - though it appears that individuals have to be in a highly stressed state for this process to happen.

As for the mechanism responsible for this astounding capability, parapsychologist D. Scott Rogo, writing in *Miracles* (1982), speculated that stigmatics are likely to possess great psychic powers, and opined: "The sufferer literally directs psychokinesis onto his own body. This forces lesions to open in the flesh".

FOUR OF THE WORLD'S MOST UNUSUAL STIGMATA CASES

1) A stigmatic photographs Jesus.

Born in Kenya during the early 1960s, Sister Anna Ali is a Catholic stigmatic nun whose hands bear stigmata that bleed every Wednesday at midnight. She also states that every Thursday she is visited by Jesus, and she has a good-quality photograph of Him that she claims to have snapped in August 1987. However, it portrays a bearded face with long dark hair that looks much more similar to popular idealised European depictions of Jesus than to the face of a Palestinian Jew.

2) Transparent stigmata?

Lourdes, photographed by the author during his visit in 1982 *(Dr Karl Shuker)*

In 1375, St Catherine of Siena developed five painful stigmata. She prayed for the stigmata themselves to vanish, but for their pain to remain - and this is indeed what happened. Following her death in 1380, however, her corpse was examined, and the stigmata were detectable as areas of transparency within her tissues.

3) Nails in the flesh?

According to medical specialist Dr Dei Cloche, a black spot resembling the head of a large nail arose at the centre of each of Tyrolean stigmatic Domenica Lazzari's stigmatised hands. Similarly, Thomas de Celano's account of St Francis of Assisi's life, written shortly after his death and on the authority of Pope Gregory IX, claimed that the saint's hands and feet contained nail-like inclusions formed from his flesh but matching the appearance of the real nails that would have inflicted Jesus' wounds at the Crucifixion. These flesh-engendered 'nails' may have been composed of keratin, the horny substance that comprises hair as well as finger- and toe-nails

4) Stigmatic statue

In June 2002, a statue of an angel cast in bronze, which was suspended between two blocks of flats in a Glasgow housing development, was seen to exude a reddish liquid from the wrist of its upraised right hand, in a manner startlingly similar to that of a human stigmatic. There is no word in the report (*News of the World*, 29 September 2002), however, as to whether the liquid was ever analysed.

VISIONS AND IMAGES

Throughout the Christian world, visions of the Blessed Virgin Mary (or BVM, for short) have often been reported, and they still are today.

Perhaps the most famous history on file is that of Bernadette Soubirous, the 14-year-old girl who claimed to have witnessed an illuminated, speaking BVM vision in a grotto near the French village of Lourdes on 18 separate occasions, beginning on 11 February 1858. Resembling a girl dressed in shining white apparel, the vision identified itself as the Immaculate Conception, and pointed to a patch of earth which, when dug into by Bernadette, revealed an underground spring, whose waters are widely believed by devotees to possess healing properties. Lourdes has since become an internationally-renowned focus for Christian pilgrimages, particularly by chronically-sick people hoping to be cured of their afflictions.

Early depiction of Hildegarde von Bingen experiencing visions

During 1917, three children saw a series of speaking BVM visions at Fatima, Portugal, which culminated in a gathering of around 70,000 people on 13 October, to attend what the children stated would be the Virgin Mary's final appearance here. Although only the children reputedly saw the Virgin, some (but not all) of the other observers in the amassed crowd alleged that they had seen the sun begin to dance, rotating dramatically, then abruptly plunging down from the sky towards them before veering back up into the heavens.

These astonishing if selectively-spied 'sun dances' have featured in several other Marian incidents too, including the numerous BVM visions reported at Medjugorje in Bosnia-Herzegovina during the early 1980s; and, more recently, at Agoo in the Philippines. This was when an enormous visitation of sightseers, estimated at between 300,000 and a million, converged on the small town after a boy seer called Judiel Nieva announced that the Virgin Mary would appear there at 1.15 pm on 6 March 1993. Many observers afterwards averred that they had seen the silhouetted figure of a lady with a dark waistband.

Other BVM visions in modern times were experienced by two teenagers in 1990 and 1993 at Litmanova in Slovakia; by 18-year-old Fiona Bowen and others in Inchigeela, Ireland, beginning in August 1985 and still occurring in August 1992; and also by retired Canadian mechanic John Greensides from Marmora in Ontario and friends since visiting the BVM shrine at Medjugorje a few years ago. In each case, news of their experiences has annually attracted gatherings of several thousand pilgrims to the site of their visions.

There are many recorded accounts of visions of Jesus, angels, and other holy figures too, but the most astonishing visions of all must surely be those of the German mystic Hildegarde von Bingen (1098-

1179). During her remarkable life, she claimed to have experienced breathtaking visions of Heaven itself, whose blinding light flowed through her entire brain, instantly elucidating for her the writings of philosophers and seers, and enabling her to write extensively upon the most esoteric scientific and moral matters, compose music of extraordinary beauty, and become a valued advisor to popes and rulers. And whereas most visionaries enter a trance to perceive their visions, Hildegarde saw hers while fully awake.

Also on file is a fascinating photograph of what may be the Virgin Mary, depicting a glowing haloed female figure with a second haloed figure standing in front of her. It was snapped on 3 September 1989 by Legeti Karoly, while visiting a Catholic church in the village of Karacsond, northern Hungary. As yet, investigators have been unable to offer any satisfactory orthodox explanation for the image captured in Karoly's perplexing photograph.

And speaking of images: over the centuries, countless people have reported seeing miraculous images resembling the face or figure of Jesus, the Virgin Mary, and other holy persons. Such images are known as acheropites, and can sometimes appear in the most unlikely places. One such case, which effectively links BVM visions with acheropites by featuring both, involved a Carmelite Postulant nun called Sister Teresita, and was documented in detail by John M. Huffert in *Russia Will Be Converted* (1950).

During September 1948, while walking in the garden of the convent at Lipa City in the Philippines, Sister Teresita saw the first of several BVM visions. Moreover, the vision spoke to her, telling her to gather the rose petals. And when Teresita looked down, she saw to her amazement that the ground was indeed strewn with dozens of rose petals that had not been there before. During the next two months, inexplicable showers of rose petals also fell inside and outside the convent, and when examined, images of Jesus and the BVM appeared upon them. Moreover, whereas normal rose petals soon wither, these were still fresh and fragrant a year later. In recognition of this miracle, a basilica was erected over the spot where Teresita first saw her vision, and the petals have been seen by countless people.

No less unusual is the acheropite termed 'Christ in the Outback'. In November 1995, while walking through his wheat farm near Beverley in Western Australia with his 20-year-old son Adrian, Julian Webb heard a strange voice that seemed to originate from a pile of rocks and which spoke the following words to him: "Let I, the Lord, show you the way". Two days later, accompanied by his son again, Julian went back to these rocks, and this time they saw an image on a slab of granite. The image clearly depicted the face of a man with long dark hair, beard, and long slim moustache - a face instantly recalling portraits of Jesus. Moreover, when professional photographer Tony McDonough later took a picture of it, he was unnerved to find that as he scrutinised it through his camera's viewfinder, it seemed to acquire a flesh-like appearance, even though the slab of granite bearing it was greyish. As might be expected, the image of Christ in the Outback attracted considerable media attention - a little too much, in fact, as it turned out. On 24 January 1996, Julian Webb was attacked by two men who tied him up in a shed, then stole the slab bearing this enigmatic acheropite. As yet, the local police have failed to trace either the men or the slab.

Another celebrated acheropite of Jesus is the 'Christ in the clouds' photo. First published in Ashland, Kentucky, by the *Daily Independent* newspaper, where it caused a sensation, this picture depicts an amazingly Christ-like face and shoulders image. It was snapped by a bomber pilot from Chicago while photographing some other bombers' aeroplanes flying high in the clouds during the Korean War. The pilot claimed that he had not perceived the image while taking the photos, only noticing it in one of them after they had been developed. A more recent acheropite to have attracted notable media coverage is the giant image of the Virgin Mary that appeared on the tinted windows of the Seminole Finance office block in Clearwater, Florida, on 17 December 1996. Rainbow-hued, standing 50 ft high, and spanning two storeys of the building, it recalls the robed head, shoulders, and upper body of the Virgin.

Since its appearance, the building has become a virtual shrine, besieged by worshippers and sick pilgrims seeking a cure, who leave candles and handwritten prayers. According to one believer, Sister Christian of the Order of Saint Anne, it is a divine symbol, a Christmas miracle: "God is telling us it is time to change our ways". Sceptics maintain that it is simply caused by the sun reflecting off water left behind by sprinklers.

Father Ren, Laurentin, author of *The Apparitions of the Blessed Virgin Mary Today* (1991), has spent over 20 years studying the Lourdes phenomenon and has devoted almost as long to two other cases featuring religious visions. Consequently, his opinion concerning visions can hardly be dismissed lightly, so we should take heed of his words: "Without the receptivity of a subject there would be no apparitions, but, as a corollary, subjectivity can lead to illusory apparitions".

Examined from a medical standpoint, the perception of visions is popularly categorised nowadays as a manifestation of brain disorders, notably neuroelectrical aberrations occurring in certain cerebral regions termed the temporal lobes - associated with speech, thought, information, emotions, and memory. Dr Tim Betts, from Birmingham's Queen Elizabeth Psychiatric Hospital, has many case histories on file of patients suffering from temporal lobe seizures who concomitantly experienced all manner of extraordinary visions. Equally, medical specialists tend to explain Hildegarde von Bingen's heavenly visions, invariably accompanied by extreme physical debilitation, as the rather more prosaic product of severe migraine.

STRANGE TALENTS

Certain holy figures in history have been set apart from their fellow humans by virtue of their inexplicable talents or abilities.

ODOUR OF SANCTITY

Her companion nuns in the convent at Rovereto, Italy, were never at a loss to know the whereabouts of Sister Giovanna Maria della Croce - for wherever she went, she exuded the most exquisite, indescribably sweet scent. Known as the odour of sanctity, this unearthly fragrance suffused for a considerable time afterwards any clothes that she had been wearing, any cell that she visited, the straw mattress that she lay upon at night, and anything or anyone that she touched. It even preceded her when she left her cell to attend Communion or other services, but was always at its most powerful following Communion.

The scent had first appeared in around 1625, when she made her final vows at a mystical wedding ceremony during which she became 'betrothed' to Christ. Due to the embarrassment that it caused her, however, she made fervent attempts during the remainder of her life to disguise this divine scent - she even placed vile-smelling objects in her cell, but the odour of sanctity effortlessly transcended their wholly ineffective stench.

Nevertheless, there would also appear to be a less fragrant, if no less mystifying, aspect to the odour of sanctity. According to the Reverend Dr James Bentley's *Restless Bones: The Story of Relics* (1985), devotees smelling the stink of rotting flesh lingering upon the decayed relics of a deceased saint (incorruptibles excepted) often claim that it is exquisite and even call it the odour of sanctity, whereas non-devotees are repulsed by its foul stench.

How can we explain this odiferous dichotomy? Again, it would seem that the power of religious faith can profoundly influence or even transform our normal sensory perceptions.

LEVITATION

St Joseph of Copertino, a 17th-Century Italian Franciscan monk, became quite infamous for his apparently involuntary spasms of levitation.

On one such occasion, he burnt himself when an attack of levitation abruptly ended by depositing him upon some lighted candles standing on his monastery's altar during Mass; and on another occasion he greatly perturbed Pope Urban VIII by rising several feet off the ground, directly in front of his monastery's esteemed visitor. Other saints with renowned levitational powers included St Teresa of Avila and St Adolphus Liguori. And as recently as 1911, Father Suarez performed an act of levitation at Santa Cruz in Argentina.

FIRE IMMUNITY

The talent exhibited by St Francis of Paola, St Polycarp of Smyrna, St Catherine of Siena, and Anne of Jesus, among others, may have been rather more earthbound, but was certainly no less marvellous - for all of these devout figures were inexplicably immune to the scorching, death-dealing power of fire. Born in the 15th Century and canonised in 1519, St Francis of Paola became famous for his incredible ability to handle fiery or red-hot objects with complete impunity.

On one occasion, he prevented the spread of flames by keeping them at bay with his bare feet; on another, he held a large chunk of red-hot iron in his hands in order, so he claimed, to warm himself. And when tested by two church dignitaries, he provided a vivid demonstration of his extraordinary power by plunging his hands into the heart of a blazing fire and pulling out some burning sticks and fiery coals.

St Polycarp of Smyrna revealed his ability in an even more dramatic way. Tied to a stake and set alight as a punishment for denying the divinity of the Roman emperor ruling Smyrna in Turkey during AD 155, he remained entirely unscathed by the roaring flames surrounding him on every side. Tragically, however, even his fiery immunity could not protect him from the emperor's wrath, who instructed the executioner to stab the saint to death instead.

INCENDIUM AMORIS – THE FIRE OF LOVE

St Stanislaus Kostka and St Philip Neri were two 16th-Century saints who displayed an equally anomalous heat-related condition termed *incendium amoris*, or 'the fire of love'.

Filled with such emotional ardour or love for their Saviour that their body temperature soared as if fever-ridden, they were often obliged to apply cloths soaked in icy water or even to walk bare-chested on snowy winter days in order to quench their internal furnace.

LEFT: St Stanislaus Kostka
RIGHT: St Philip Neri

HOLY INVISIBILITY!

Not only was Francis of Paola gifted with the ability to withstand fire, he could also become temporarily invisible!

The most famous demonstration of this astounding talent occurred when the king of Naples sent a considerable number of soldiers into the church where the saint was known by all to be taking refuge. Yet although they searched it thoroughly, they could not see him. During their search, however, some of the soldiers actually touched him or brushed by him - yet they were still unable to detect him!

Sometimes, a saint's powers of invisibility can be very selective. For instance, on one occasion during the 15th Century, Queen Yolande, consort to King Juan of Aragon, insisted on entering the cell of the Spanish monk St Vincent Ferrer, much to St Vincent's displeasure. Consequently, he made himself invisible, but only to her - and so, when she entered his cell, all of his fellow monks could clearly see him, but the queen could not. Greatly abashed, she had no option but to apologise and make her exit.

St Francis of Paola

Intriguingly, invisibility may not be restricted to medieval holy personages. According to California-based researcher Donna Higbee, there are many people alive today who suddenly, but unwittingly, become temporarily invisible - discovering to their alarm that no-one nearby can either see or hear them.

Indeed, so common is this alleged phenomenon that it has even been assigned its own name - HSII, or Human Spontaneous Involuntary Invisibility.

DIVINE LIGHT

According to her biographer Thomas À Kempis, St Lidwina of Schiedam (died 1433 AD) was often engulfed by divine light:

> ...she was discovered by her companions to be surrounded by so great a divine brightness that, seeing the splendour and struck with exceeding fear, they dared not approach nigh to her. And although she always lay in darkness, and material light was unbearable to her eyes, nevertheless, the divine light was very agreeable to her, whereby her cell was often so wondrously flooded...that to the beholders the cell itself appeared full of material lamps or fires.

BILOCATION

Most marvellous of all, however, must surely be the talent of bilocation - to appear simultaneously in two widely separate locations (see also Chapter 12). In 1742, no less august an authority than the Roman Church itself announced that St Martin de Porres (1579-1639) had made 'impossible' appearances in countries as far-distant from Europe as China and Japan, judging from many reports that a mysterious figure, a dark friar matching St Martin's appearance, had often been seen working there. Similarly, during Easter 1226 in Limoges, France, St Antony of Padua performed the incredible feat of kneeling in prayer in full view of his parishioners in the Church of St Pierre du Queyroix while simultaneously reading a lesson in the presence of the monks at a monastery also situated in Limoges, but some distance further away. More recently, Padre Pio, the stigmatic monk and healer documented earlier here, was also credited with this amazing ability.

XENOLALIA AND GLOSSOLALIA

Speaking in tongues is another talent exhibited by holy figures down through the ages. It dates back at least as far as the feast of Pentecost, when Jesus' disciples were transformed by the Holy Spirit, and spontaneously began speaking in many different languages hitherto unknown to them but familiar to strangers. This phenomenon is known as xenolalia, and has been witnessed on occasions during modern-day sessions of hypnotic regression.

Distinct, but related, is glossolalia - in which the speaker gives voice to a wholly unrecognisable or seemingly non-existent language. Glossolalia occurs today among certain Christian sects, such as the Pentecostals, Quakers, Shakers, and members of the Church of Latter-Day Saints, and was once deemed to be the language of angels. Phonetic experts, however, do not consider glossolalic outpourings to constitute languages of any form, but merely to comprise sounds belonging to language types, lacking syntax and vocabulary.

It is perhaps ironic but fitting that the simplest yet most succinct definition of miracles dates not from today's frenetic age of limitless information online, but rather from a much earlier, less sophisticated, yet possibly more enlightened time. Penned during the early 5th Century AD, it derives from *De Civitate Dei* ('The City of God'), by St Augustine of Hippo: *"A miracle does not happen in contradiction to nature, but in contradiction to that which is known to us of nature"*.

Chapter 11
ALIEN ARTEFACTS
A Martian Manuscript and Some VERY Odd Orbs

My name is Ulo and I write this message to you my Friends on the Planet of the Sun you call Earth. Where I live I will not say. You are a fierce race and prepare travel. No one from any other planet ever has landed on earth, and your reports to the contrary are faulty. Men cannot travel far in space vehicles owing to sudden changes in speed direction and many other reasons. They are machines, part at our 'control', part 'auto-control' to avoid objects in way. It is impossible to receive radio over far distances owing to natural waves in space unless key of several frequencies is used, but we can receive single frequencies from near transmitter recorder in space vehicles.

Opening section of the message found inside the Silpho Moor disc

Typically, physical evidence for the occurrence of a UFO and/or visiting alien entities takes the form of photographs, radar scans, or the detection of various energy forces, such as electricity, magnetism, microwaves, or radiation. Occasionally, however, evidence of a more exotic nature is obtained, as revealed by the following selection of mysterious artefacts - all of which have been claimed at one time or another to be of alien, extraterrestrial origin.

A MANUSCRIPT FROM MARS?

On 2 March 1965, John Reeves was walking in the woodlands near his home at Weeki Wachee Springs, Florida, when he saw a landed UFO.

He swiftly concealed himself in some nearby foliage, but was seen by a 5-ft-tall humanoid entity wearing

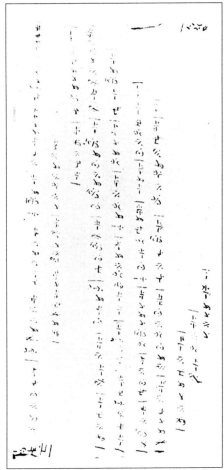

The Martian manuscript *(FPL)*

a grey-silver suit and a transparent glass dome over its head. The dome revealed that the alien had very dark skin, a slightly pointed chin, and a human-like nose and mouth, but its eyes were set further apart than a human's. It walked to within 15 ft of Reeves, held a device at chin level, and pointed it directly at him. The device flashed twice, and Reeves assumed that it had taken some photos of him. Then the alien walked back inside its craft, taking off into the sky.

Once it had gone, Reeves inspected the landing site, and spotted two pieces of folded tissue-like paper lying there. When he examined them, he discovered that they bore several vertical columns of a strange script that he likened to Oriental writing or shorthand.

When these were subsequently deciphered, they were found to comprise an incongruously mundane message: "Planet Mars - Are you coming home soon - We miss you very much - Why did you stay away too long?". An interplanetary telegram?

Perhaps the most mystifying facet of this case, however, was revealed in 1988 by investigating psychiatrist Dr Berthold Schwarz. Back in 1957, eight years before Reeves's encounter, a nine-year-old girl had suddenly picked up a piece of paper and spontaneously jotted down some indecipherable characters - an inexplicable action which had so intrigued her father that he kept the piece of paper afterwards as a curiosity.

Ten years later, events came full circle, when her father saw a magazine article that reproduced the Martian manuscript - and discovered to his amazement that the first 13 characters in it were identical to those that his daughter had scribbled on the piece of paper a decade earlier.

WORDS FROM THE PLANET VENUS?

Possibly the most (in)famous example of supposed interplanetary writing was the message mysteriously reproduced upon one of George Adamski's photographic plates on 20 November 1952. According to Adamski, the plate had been taken from him by a visiting Venusian whom he had met near Desert Center, California, and was returned on 13 December, now containing the message. Although the faint, spidery characters initially seemed unfamiliar, ufological researchers later realised that they were very similar to various primitive characters discovered by Professor Marcel Homet while conducting archaeological expeditions in the Amazon jungle during 1949-50, and which were widely reproduced shortly afterwards in media accounts of the expedition. Just a coincidence - or the true source of Adamski's space script? Today, many ufologists doubt that this 'Venusian' writing is authentic.

THE SAUCER OF SILPHO MOOR

Equally suspect are the hieroglyphic-like notations of Ulo and Tarngee. They were elegantly inscribed upon a series of 17 thin sheets of copper foil contained within a copper tube that was found inside an enigmatic discoid object popularly dubbed the Silpho Moor saucer.

As documented by Janet and Colin Bord in *Life Beyond Planet Earth?* (1991), this strange metallic disc measured 18 in across, bore hieroglyphics on its external surface too, and supposedly landed upon moorland near Scarborough in North Yorkshire some time during November 1957. The extensive writing that it bore on its surface and upon the copper sheets was painstakingly analysed and translated by Philip Longbottom, and was found to comprise a lengthy message composed by two different alien authors - Ulo and Tarngee.

Ulo would not identify their home planet, but made it readily apparent that it was technologically superior to Earth. Much of the message was devoted to the dangers of atomic weapons and warfare, but was hardly the most sophisticated prose in print. By contrast, the message's linguistic structure was, in Longbottom's opinion, highly complex, and: "...would seem to be out of all proportion to a hoax, however elaborate...I firmly believe that this is not a 'made-up' language, but one in constant use". Notwithstanding his opinion, however, this case is often dismissed as a hoax, but even if it was, who was responsible, and what purpose did it serve? An early 'ban the bomb' protest, perhaps?

Three years earlier, a still-undeciphered parchment-borne script had been found inside a small metal box that had in turn been discovered at the site where a biker called Wardon had spied the brief landing and departure of a large shed-shaped UFO at Dudzele in Belgium.

FUNNY FOIL

Several cases of strange metallic foil-like objects associated with UFO visitations are on record. For example, following a UFO sighting on 7 September 1956 in Chosi City, Chiba Prefecture, Japan, pieces of strange 'metal foil' were collected on the ground. Similar items, resembling thin strips of chaff but of a metallic composition, have also been discovered in parts of New Jersey following UFO sightings, as documented in *The Flying Saucer Menace*.

Such cases as these, however, may simply be due to the disintegration of earthly satellites, rather than to UFOs.

A STRANGE SPHERE FROM FORT GEORGE ISLAND, FLORIDA

APRO is the Aerial Phenomena Research Organization, founded in Tucson, Arizona, in 1952 by noted American ufologists Coral and Jim Lorenzen. During 1974, its bimonthly *APRO Bulletin* published accounts of two very strange spherical objects that may well have been of unearthly origin - but were they of extraterrestrial origin?

The first account appeared in the bulletin's March-April issue, and concerned the strange metallic ball, measuring just under 8 in across, that was spotted on 26 March 1974 by 21-year-old Terry Betz, while walking in the grounds of his parents' home on Fort George Island, near Jacksonville, Florida. Although he took it back home as a curiosity, Betz paid little attention afterwards to his unexpected find - until, some time later, he happened to be playing his guitar near to the window seat in his bedroom on which he had placed the sphere. As he played, he was startled to discover that certain notes induced the sphere to vibrate, just like a tuning fork, humming at low frequency "...like a motor was running inside".

When examined at Jacksonville Naval Air Station by a metallurgist, the sphere's outer surface was shown to be composed of stainless steel, and its outer shell was estimated to be about 0.5 in thick - but what was inside it? According to civil engineer Dr James A. Harder, one of APRO's scientific consultants, a normal sphere when x-rayed should resemble a doughnut.

Conversely, when Betz's anomalous orb was x-rayed at Jacksonville, the resulting x-ray plates revealed two discrete internal spheres. This suggested that the outer stainless steel shell was far less dense than the internal material, which was therefore responsible for most of the sphere's total weight. The sphere also possessed four separate magnetic poles.

Later in 1974, this baffling object was the star exhibit at the New Orleans conference of the *National Enquirer*'s UFO panel, and elicited much discussion. Sadly, however, its identity, remained as opaque as its interior. Among the ideas put forward were a 'time and tide' marker (but it seemed too heavy for this), and a sea-bottom marker (suggested by its phono-induced vibrations; but, if so, how can its terrestrial provenance be explained?).

OHIO'S ODD ORB

Also in 1974, but this time within its September-October issue, *APRO Bulletin* published an account by Professor Theodore Spickler concerning a second strange sphere. This one had been found near Lowell, in Ohio, on 7 September 1974, following some dramatic events.

Walking up a wooded hillside in search of squirrels to hunt on the farm of acquaintance Ollie Wagner, Keith Hammerman suddenly heard a loud sonic boom, followed by a sharp crack, then a swishing sound. As he peered around, greatly alarmed and anxious to discover what could have caused these noises, he perceived a cloud of smoke through a patch of low scrub trees 50 ft ahead him, and he could also smell a pungent, sulphurous odour.

Thoroughly frightened by now, Hammerman ran back at once to his car, and drove away to Wagner's farm, where he recalled the eerie happenings that he had experienced, and gave Wagner directions to visit the smoking site if he so chose. Later that day, Wagner decided to investigate the site for himself, and searched through the scrub meticulously for 3 hours, following the direction of the strange smell pervading the area. He claimed that this smell was similar to, but nonetheless distinguishable from, that of real sulphur.

Guided by the smell, Wagner finally spied its source - a shiny bluish-grey sphere almost buried in the ground's moss-covered white clay. After digging it out, he ascertained that it was 11.5 in across, about 20 lb in weight, and sported a thick weld-like rim or seam around its equator.

The sphere also bore two latch-like protrusions on opposite sides, and two highly-melted cylindrical protrusions along the ends at 90° to the rim. One of these latter protrusions was sealed, but the other contained a 0.25-in hole leading into the interior, which was apparently hollow, as the sphere emitted a clear ringing sound when struck.

The sphere's near-buried state, and the presence of pieces of its metallic surface firmly embedded in the hole from which it was removed, clearly demonstrated that it had hit the ground with great force and at considerable speed. Indeed, the bottom of the hole was 15 in beneath the level of the surrounding, undisturbed ground level.

Other people living in the area also claimed to have heard the sonic boom reported by Hammerman, and this was presumably the sound of the sphere's descent from above - a descent that seemingly involved appreciable heat, judging not only from the melted state of the two protrusions, but also from its ablated outer surface. For this appeared to have liquefied, flowed over the layer directly underneath (leaving bubbles in its wake), and then solidified again. Moreover, its bluish-grey colour, as noted by Professor Spickler, is consistent with that of steel which has undergone a heat treatment akin to atmospheric entry. A thin layer of clay was baked onto the sphere's bottom, and an appreciable amount of steam, resulting from the clay's high water content, was rising up from the impact site.

As for the sphere's identity, Spickler opined that the possibilities ranged from a ballistics ordnance to a primitive foreign satellite, but conceded that these were wholly speculative.

MENDOZA'S MYSTERY CAPSULE

Measuring approximately 12 ft by 4 ft, an anomalous fusiform (cigar-shaped) object was discovered at San Miguel, an arid region near Mendoza, Argentina, in January 1965 by some local herdsmen, and was subsequently photographed from above by an aeroplane sent from the Provincial Aeronautical Bureau. Local people vehemently believed the capsule, which emitted an intense brilliant-white luminosity, to be a crashed UFO, and claimed to have seen archetypal 'little green men' walking around it in glowing outfits resembling the wetsuits of divers. Eventually, it was taken away by police officials, and was later passed on to the Centre for Space Investigations at Cordoba. The fate of the capsule's viridescent occupants (if the locals are to be believed), conversely, is not recorded.

PANCAKES FROM OUTER SPACE?

One of the most bizarre cases of alleged extraterrestrial artefacts features the 'space pancakes' donated by three alien visitors to Wisconsin farmer Joe Simonton of Eagle River.

According to Simonton's testimony, a silver elliptical UFO descended into his farm's yard at 11.00 am on 18 April 1961, and he could see three humanoid entities sitting inside. He later described them as resembling Italians, with black hair, swarthy complexion, standing about 5 ft tall, and wearing dark blue clothing. After noticing that they were cooking food near a griddle-like apparatus, he made them understand via gesticulations that he would very much like to taste some. So they gave him four small flat brown objects, each measuring 3 in wide and resembling a perforated, biscuit-textured pancake. A few moments later, the UFO ascended into the sky, and soared away at considerable speed.

Anxious to learn more about the gastronomic quality of intergalactic fast food, Simonton promptly ate one of the pancakes, but was very disappointed by its insipid quality, commenting afterwards that it tasted like cardboard. Duly dissuaded from consuming the others, he decided instead to keep one of them as a souvenir, and submit the remaining two for formal analysis to the United States Department of Health Education and Welfare.

This proved equally disappointing, however, inasmuch as the results claimed that the pancakes were composed of mundanely terrestrial constituents - flour, sugar, and grease, but there was one intriguing twist to the tale. The pancakes did not contain any salt - which, according to traditional mythology, is a common attribute of Faerie food too. Even so, as suggested by ufologist William H. Spaulding among others, it is possible that Simonton was the victim of a hoax engineered by secret governmental agencies spreading disinformation.

LOLLADOFF PLATE

Dr Shuker holding a copy of *Sungods in Exile*

Purchased in northern India during 1945 by the Oxford scientist after whom it is named, the Lolladoff plate measures just under 9 in across, is composed of an exceedingly hard but unidentified substance, and can apparently gain or lose weight at will. One of its surfaces is plain, but the other one is covered with obscure symbols. This mystifying disc is documented in an enigmatic book entitled *Sungods in Exile* (1979), edited by David Agamon and supposedly authored by the late Dr Karyl Robin-Evans - described as an Oxford scholar who visited Tibet in 1947, met a physically-degenerate tribe called the Dzopa, and died in 1974.

The owners of discus-shaped artefacts similar if not identical to the Lolladoff plate, the Dzopa claimed to be descended from a race of alien entities from the Sirius star system who crash-landed in this bleak land during 1017 AD and ultimately interbred with the native people there! Not surprisingly, some researchers became increasingly suspicious about this book's authenticity, and it is now well-established that it was indeed a hoax (Dr Robin-Evans never existed), albeit a successful one for quite a time (in 1994, Agamon, aka Gamon, obliquely confessed to it in a letter published by *Fortean Times*).

An eclectic assortment indeed is the selection of alien bric-a-brac perused above - always assuming, of course, that any of these items can truly boast an extraterrestrial provenance.

Yet even if they really are from other worlds, what can they tell us of these worlds? It is, after all, rather like attempting to anticipate the conceptions a visiting alien might gain about us if it were to peruse the heterogeneous terrestrial flotsam and jetsam that we breezily refer to as litter - and what a frightening thought that is!

Chapter 12

SEEING DOUBLE

Face to Face with Real-Life Doppelgängers

"Listen, my friend," said the Shadow to the learned man. "Now I am as lucky and powerful as any one can become, I'll do something particular for you. You shall live with me in my palace, drive with me in the royal carriage, and have a hundred thousand dollars a year; but you must let yourself be called a shadow by every one, and may never say that you were once a man; and once a year, when I sit on the balcony and show myself, you must lie at my feet as it becomes my shadow to do. For I will tell you I'm going to marry the Princess, and this evening the wedding will be held."

"Now, that's too strong!" said the learned man. "I won't do it; I won't have it. That would be cheating the whole country and the Princess too. I'll tell everything - that I'm the man and you are the Shadow, and that you only wear men's clothes!"

"No one would believe that," said the Shadow. "Be reasonable, or I'll call the guards."

"I'll go straight to the Princess," said the learned man.

"But I'll go first," said the Shadow; "and you shall go to prison."

And that was so; for the sentinels obeyed him of whom they knew that he was to marry the Princess...

In the evening the whole town was illuminated, and cannon were fired – bang! – and the soldiers presented arms. That was truly a wedding! The Princess and the Shadow stepped out on the balcony to show themselves and receive another cheer.

But the learned man heard nothing of all this festivity, for he had already been executed.

Hans Christian Andersen – 'The Shadow'

According to the mythology of cultures throughout the world, every person has an exact double, most commonly referred to as a doppelgänger (from Jura folklore). To see one's own double can often be an augury of imminent death - as evocatively depicted by the English Pre-Raphaelite artist Dante Gabriel Rossetti in his magnificent painting 'How They Met Themselves' (housed in Cambridge's Fitzwilliam Museum). Sometimes, however, a doppelgänger will warn its counterpart of impending doom, enabling him to avoid this.

A sinister variation on the doppelgänger theme is when someone's double takes the place of the original person in question – a scenario chillingly utilised by Hans Christian Andersen in his fairy tale 'The Shadow', in which a learned man's shadow becomes an independent fleshed-out double of its former master and ultimately replaces him.

Doppelgängers have traditionally been dismissed by science as fantasy. Yet there are a startling number of cases on file documenting the apparent reality of such entities, and including some extremely famous persons from history.

HE LOOKS FAMILIAR!

The concept of doppelgängers may well seem positively medieval in today's ultra-scientific age. Yet as recently as November 1994, the *British Journal of Psychiatry* published a report reviewing no fewer than 56 cases of people who had apparently encountered doppelgängers.

One episode featured a man in a cafe who suddenly spied his exact double standing in the street, watching him. The man raced outside, but the doppelgänger had gone. Another case concerned a retired medical doctor who saw his own double walk across a room; and a third featured a former pilot who saw himself several yards away for about 10 minutes.

Surely one of the most amazing of all modern-day cases, however, must be that of Adrian Brown from Winton in Dorset. In 1990, Brown worked as a driver for a local security firm, and every night for the past five months his job had entailed driving in his van to a number of different sites owned by the firm in order to check that all was well with them. Working to a strict timetable, he followed exactly the same route each night and thus visited the same site at the same time each night - until, that is, he was unavoidably delayed by 20 minutes one evening when driving towards one of these sites, a gravel pit.

Finally nearing the gravel pit, Brown had reached a roundabout in the road when, to his great surprise, he saw a white van coming towards him on the other side that seemed identical to his own, and was even decorated with precisely the same lettering. Moreover, it had reached the exact spot that his own van would have reached by then if he had been on schedule as normal, instead of being delayed - but the similarities did not end there.

As the mysterious duplicate van drove by him, Brown looked into it - and was profoundly shocked to see that the man driving it was himself! Not just someone who looked similar to him, or was wearing the same clothes, but an exact, perfect double, who had even mirrored his own action by turning to look at him as he drove by.

Although Brown completed his journey to the gravel pit, checked it, then returned home, he felt extremely frightened, because he knew that the only other van owned by his firm was off the road, awaiting repairs, and no-one else at the firm looked like him anyway. Voicing this episode's perplexing paradox, Brown speculated: "What if I'm not really here and the guy driving the other van is the real Adrian?".

An even closer meeting with oneself occurred in 1885 and featured the famous French short stories writer Guy de Maupassant. While seated in his study one day, writing a horror story entitled 'The Horla', de Maupassant was somewhat taken aback when an unheralded, uninvited man suddenly walked into the room and sat down directly opposite him. But before de Maupassant had time to enquire the reason for this intrusion, his unbidden guest began to speak - and to the writer's amazement, the words that the

Johann Wolfgang von Goethe

stranger duly dictated were the very ones that he had intended to pen in his story! Gazing up into the man's face, however, gave de Maupassant an even bigger shock - for the face that he saw was his own! In response, his doppelgänger stood up, turned away, and left the room as silently and mysteriously as he had arrived.

Some cases of doppelgängers are even more remarkable than those mentioned here so far, because the person concerned does not espy a precise double, but sees instead an accurate image of himself as he will be in years to come. Perhaps the most famous example on record featured the renowned German dramatist-poet Johann Wolfgang von Goethe, in which his uncanny meeting with himself was almost dream-like in nature, but nonetheless provided an amazingly authentic glimpse of his future form. As recalled in his autobiography, he had been riding on horseback to Drusenheim one day in 1771:

...when one of the strangest experiences befell me. Not with the eyes of the body, but with those of the spirit, I saw myself on horseback coming toward me on the same path dressed in a suit such as I had never worn, pale-grey with some gold. As soon as I had shaken myself out of this reverie the form vanished. It is strange, however, that I found myself returning on the same path eight years afterward...and that I then wore the suit I had dreamt of, and not by design but chance.

In his *Encyclopedia of Ghosts* (1984), paranormal researcher Daniel Cohen revealed that a Californian artist called Catherine Reinhardt regularly witnessed a doppelgänger that always looked about five years older than she did. Perhaps the most telling incident occurred when Reinhardt, then aged 28, saw her double at a cocktail party, and noticed that it was walking with a slight limp. Four years later, Reinhardt severely injured her leg in a car accident, and for ever afterwards she walked with a slight limp.

WHEN DOUBLE MEANS TROUBLE - DOPPELGÄNGERS AND DEATH

In stark contrast to the eerie but apparently harmless examples documented above, some real-life doppelgängers have readily reprised their sinister role in folklore and superstition, inasmuch as their appearance has been a portend of death for their hapless counterparts.

In 1796 Catherine the Great's ladies-in-waiting were alarmed to see her walking into the throne room - because they had only just left her, asleep in bed. When they went back to her and told her what they had seen, the formidable Russian empress lost no time in going to the throne room personally - only to see her doppelgänger arrogantly sitting upon her throne! When it saw Catherine, it stepped down and walked towards her, but the empress ordered her guards to fire their guns at this intangible interloper. However, their bombardment had no effect upon it, and it remained in the room for a time before finally disappearing of its own accord. Shortly afterwards, Catherine died.

In his *Miscellanies*, the 17th-Century historian John Aubrey reported another fatal date with a doppel-

Percy Bysshe Shelley

gänger, featuring the Earl of Holland's daughter, Lady Diana Rich:

...as she was walking in her father's garden at Kensington, to take the fresh air before dinner, about eleven o'clock, being then very well, met with her own apparition, habit and every thing, as in a looking-glass. About a month after, she died of the smallpox. And it is said that her sister, the Lady Isabella Thynne, saw the like of herself also, before she died.

In 1822, visiting Italy, the poet Percy Bysshe Shelley decided to sail across the Bay of Spezia, and when the yacht arrived he eagerly stepped on board, looking forward to his voyage - until his eyes fell upon his fellow passengers.

Seated among them was his own doppelgänger. Shelley never completed his cruise - a violent storm unexpectedly arose, in which the yacht foundered, and Shelley was drowned.

Occasionally, however, a doppelgänger has actually prevented, rather than presaged, the death of its mortal counterpart. This happened on two separate occasions in the life of an American soldier called Alex B. Griffith. The first time was in summer 1944, when as an infantry sergeant he was leading a foot patrol through enemy territory in France.

Suddenly, he was amazed to see a precise duplicate of himself standing just up ahead, waving its arms at him and silently calling. Interestingly, none of his men could see anything.

Confused, but heeding the wraith's apparent warning not to progress any further, Griffith and his party paused, and as they did so an American supply vehicle trundled by and on to where Griffith's doppelgänger had been standing - whereupon a carefully-concealed German machine-gun opened fire, killing the vehicle's driver.

Twenty years later, Griffith's doppelgänger reappeared, this time as Griffith, now a civilian, was hiking with his family through the tail-end of an exceedingly windy storm.

Just like before, the apparition waved its arms, but what made its appearance especially interesting this time was that it still resembled Griffith as he had been when a sergeant, instead of mirroring him as he was now, i.e. two decades older. Remembering how the doppelgänger had saved his life the first time, Griffith took note of its warning and turned back with his family. A few moments later, in the very clearing where they would have then been walking had they not turned back, an enormous tree, uprooted by the power of the recent storm, abruptly crashed to the ground.

CRISIS APPARITIONS

There are many cases on file whereby a person experiencing extreme anguish has apparently generated a doppelgänger, but has not seen it himself. Instead, its observers have been friends or members of his family, separated from him by a considerable distance - thereby greatly heightening their astonishment when visited in this way by his double. This type of doppelgänger is commonly termed a crisis apparition.

In 1960, Mrs John Church, residing at that time in India, woke up one evening after feeling sure that she had heard someone call her name. Gazing around, she was amazed to see the familiar figure of her brother standing looking at her - because she was well aware that he was working at that time in Goshen, New York, as a charter-service pilot. Yet here he was, standing beside her bed, and was even dressed in his distinctive pilot's uniform. Before she had chance to speak to him, however, he seemed to grow pale, and slowly vanished directly before her unbelieving eyes - but that was not the end of this extraordinary episode. She later learnt that during the same night that she had seen his double in her room, her brother was actually piloting an aeroplane, and at the very moment that she had heard him call her name, he had almost crashed.

While in Paris during 1895, playwright August Strindberg became seriously ill, and desperately desired to see his family in Sweden. Not long afterwards, he received a letter from his mother-in-law, anxiously enquiring whether he was ill - because she had recently spied his double standing behind the piano at their Swedish family home!

THE KA

According to ancient Egyptian tradition, a person's mortal, physical body possessed two immortal components. One was the soul, which was carried to the Next World or Underworld when the person died. The other was the ka, or double - an undying exact replica of the person, which protected him throughout his life. Following his death, it remained alongside or within his mummified corpse (or even inside an attendant statue of the person) when it was interred in the tomb. Indeed, some large tombs contained a specially-designated area, the house of the ka, in which nourishment was regularly provided for the ka's sustenance, and sacred offerings brought by priests, during its eternal vigilance.

ERIKSON GORIQUE AND THE VARDØGER

A vardøger (from the Norwegian for 'before-goer') is a special kind of doppelgänger, which makes the acquaintance of people before its 'original' appears on the scene. One of the best-known examples featured a U.S. importer called Erikson Gorique, who visited Oslo, Norway, for the very first time in July 1955 - only to discover that wherever he travelled, the local people with whom he came into contact claimed to know him already, and even to have spoken with him.

The receptionist at the hotel in which he decided to book a room addressed him by name and recalled that he had reserved a room there earlier in the year - when in reality he had never communicated with the hotel before. A wholesaler visited by him claimed to have already conducted some business with him earlier in the year, and even knew the addresses of his New York office and warehouse.

Wholly bemused, Gorique was forced to draw the decidedly disturbing conclusion that his visit to Oslo had been preceded by that of a vardøger - identical to him in every way, and fully conversant with his entire sum total of knowledge and experience.

EMILIE SAGÉE

One of the classic cases of doubles is that of a French teacher called Emilie Sagée, who lost countless jobs during her career - due to her disconcerting but seemingly unconscious ability to be in two places at the same time, often within sight of one another. During 1845-46, she worked in Livonia, Latvia, employed at a highly exclusive school called the Pensionat von Neuwelcke - scene of her most famous doubling. One day, while picking flowers in the school's garden, she also materialised in a nearby classroom in front of its 42 pupils - all of whom could clearly see her outside too, still gathering flowers! On another occasion, her double appeared alongside her while she was teaching at a blackboard, and precisely mirrored her every movement in front of a class of 13 amazed girls.

Of particular note is that whenever Sagée duplicated herself like this, both 'twins' were pale and insubstantial, almost as if her bodily substance were not sufficient to create two fully-fledged versions. Similarly, both were somewhat lethargic, as if her physical strength had also been divided between the two, instead of being doubled.

BILOCATION

According to the Catholic tradition, devout persons such as saints and other holy people possess the consciously-controlled ability to be not just visually but also physically present in two different places simultaneously - a truly baffling phenomenon termed bilocation.

Perhaps the most famous exponent of bilocation was a 17th-Century Spanish Franciscan nun called Sister Maria Coronel de Agreda. For although she spent her whole life within her Spanish convent, she was also reported more than 500 times in North and Central America, where the Spanish Catholic Church was expanding its missionary work. Described by eyewitnesses there as a 'lady in blue', she appeared during the day and taught Christianity to the Jumano Indians of New Mexico, but always vanished again at night.

St. Anthony of Padua

Greatly intrigued by such stories, which he had gathered while working at the Isolita Mission in New Mexico, Father Alonzo de Benavides sought out Sister Maria when he returned to Spain in 1630 - and was startled to discover not only that she freely admitted to visiting this region of the Americas via bilocation but also that she was inexplicably knowledgeable concerning the area. She was even well-versed in the local native names of people and places there, all of which were wholly unknown to people outside the New World at that time. The most amazing fact of all exposed by Father Benavides, however, was that a chalice used at Mass by Jumano Indians claiming to have been converted by the mysterious 'lady in blue' had indeed originated from Sister Maria's convent at Agreda!

Another saint gifted with the power of bilocation was St Anthony of Padua. At Easter in the year 1227 AD, he reputedly knelt in prayer in the Church of St Pierre du Queriox in Limoges, France, in full view of his parishioners, while simultaneously chanting a lesson of Tenebrae in the presence of his fellow monks at his own monastery in Padua!

In modern times, the Italian Capuchin monk Padre Pio of Pietrelcina (1887-1968) gained worldwide renown as a healer and a stigmatic (bearing mysterious bleeding wounds in his hands and feet comparable to Jesus' crucifixion wounds – see Chapter 10). However, he was able to bilocate too - because he reputedly made a number of unexpected appearances around the world while simultaneously remaining at the monastery of San Giovanni Rotondo at Foggia in southeastern Italy, where he spent much of his life. Once, during a conversation with Dr Sanguinetti, the medical director of the hospital at the monastery, who was one of his longstanding friends, Padre Pio was asked his opinion as to the likely nature of the mechanism facilitating bilocation. In reply, he stated that a person sent by God to another place via bilocation is consciously aware of the transference and is genuinely in two places at the same time, and that the transference occurs via "...a prolongation of his personality".

Judging from the remarkable diversity of doppelgänger activity and appearances surveyed here, it seems unlikely that any single solution can explain all of these and the many other cases on file. Some, such as the meeting between Adrian Brown and his van-driving double, the vardøger of Erikson Gorique, and the older doppelgängers of Goethe and Reinhardt, may involve time-slips or some other form of temporal distortion, whereby the future and present mysteriously overlap.

In cases featuring crisis apparitions, conversely, the double seems to comprise an unconscious stress-induced projection of the person himself - thereby analogous (or even homologous?) to astral projection. Indeed, some researchers contend that a crisis apparition is nothing less than the person's soul, rather than a mere image.

As for bilocation, in which both versions of the person appear to be corporeally tangible, this implies some form of physical duplication rather than a wholly visual phenomenon. Yet how could such a biologically profound process possibly be achieved?

And what are we to make of cases featuring doppelgängers like those of Catherine the Great and Sergeant Griffith, which seem to exhibit a life of their own, moving and acting independently of their 'original' and even interacting with it?

All that we can say is that the occurrence and production of doubles involve mechanisms whose secrets currently lie far outside the accepted boundaries of scientific knowledge and comprehension. That they do exist seems undeniable, but how they exist is another matter.

Nevertheless, one thing is certain - the next time that anyone angrily exclaims: "How do you expect me to be in two places at the same time?", all that you need to do to prevent any further argument is to murmur the word 'bilocation', and then show them this chapter!

Chapter 13
WEST AFRICAN WEIRDNESS
The Secret Animals of Senegambia

Ati, bwana! There is a story you will not believe, because you are a white man. White men laugh at the stories told by the black man. They say this is not so, and that is not so. We have not seen this or that, so how can it be? They say, Ho, Ho! Black men are like little children, telling tales to each other in the dark. But remember, bwana, white men have been in this country for a time that is less than the life of one man, so how can you know all the things that have been known to black men for a hundred lifetimes and more?

Roger Courtney – *A Greenhorn in Africa*, quoting an elderly African hunter, Ali

Whereas many mystery animals have been well documented from North, East, Central, and southern Africa, far fewer have been publicised from West Africa - especially from its westernmost corner, comprising Gambia and its encompassing neighbour, Senegal. In reality, however, these two small countries (sometimes referred to collectively as Senegambia) apparently harbour a sizeable array of bizarre, unidentified beasts rarely if ever brought to widespread cryptozoological attention...until now.

I owe a great debt of thanks to a longstanding colleague, naturalist Owen Burnham, who spent his childhood and teenage years in Senegal, for very kindly supplying me during our longstanding correspondence with information regarding the creatures documented here.

While living in Senegal, Owen became formally accepted as an honorary member of the native Mandingo (Mandinka) tribe, and thus learnt much about this land's mystery animals and also those of Gambia that has remained unknown to other Westerners.

Owen Burnham in Africa

One such creature, the Gambian sea serpent, launched my own career in cryptozoology when I investigated its case in detail during the mid-1980s, and has now become very well known and well-documented in the literature.

However, Owen also learnt of several other mystery beasts that have received far less publicity, and so it is with these hitherto little-documented yet no less interesting cryptids that the following chapter is concerned.

MYSTERY STONE PARTRIDGE

This enigmatic Senegalese bird was originally documented by me as follows in a *World Pheasant Association News* article (May 1991) on gallinaceous mystery birds.

The stone partridge is represented in Senegal by its nominate subspecies *Ptilopachus petrosus petrosus* – a familiar sight to Owen Burnham. However, he remains perplexed in relation to the covey of stone partridges that he spotted at Fanda, Senegal, in 1985. Unlike this country's normal brown-headed, buff-breasted specimens, these were very finely but noticeably mottled with white upon their head and neck, and their breast was whitish. They were also rather smaller in size, but most unexpected of all was their habitat. Eschewing the rocky terrain or scrubland normally frequented by *Ptilopachus*, this covey was dwelling within a small but dense area of undergrowth in a rice field, many miles from the nearest expanse of stony ground. Owen saw a second covey of this strange form of stone partridge at Kouniara, and this time they were living in thick woodland, comprising a mixture of real forest and palm trees.

Yet despite their radically different habitat, their behaviour was similar to that of typical stone partridges, scurrying rapidly across the ground – though in this case over fallen trees and through the forest, rather than over rocks and through scrub.

Local hunters had informed Owen that such birds existed, but he had not believed this until he had encountered them himself. In view of their morphological differences and markedly distinct habitat, could these stone partridges constitute a separate subspecies, isolated topographically from the nominate race? Bearing in mind the tragic, continuing destruction of Senegal's wildlife habitats, especially forests, it is to be hoped that this mystifying bird form can be thoroughly investigated in the near future, to enable it (if still surviving) to be saved not only from continued scientific obscurity but also from ensuing extinction.

Stone partridge

GIANT BUSHBABY

Bushbaby – Does Senegal harbour an undiscovered giant species?

Related to the Madagascan lemurs and the Asian lorises, as well as to Africa's own pottos and angwantibos, the bushbabies or galagos constitute 19 currently-recognised species of primitive primate. Nocturnal and arboreal, they are characterised by their large ears, long tail, and fairly small size. Currently, the largest species are the three aptly-named greater bushbabies, with an average total length of 3 ft, of which over half comprises the tail.

However, Senegal may be harbouring a rather more sizeable surprise. In June 1985, while exploring the heart of the Casamance Forest, Owen spied a mysterious creature resembling a giant form of bushbaby. It was the size of a half-grown domestic cat, with pale grey fur, and was accompanied by two or three young ones. Several years later, a similar animal was also reported from another West African country, the Ivory Coast. And in 1994, an assistant of bushbaby taxonomist Dr Simon K. Bearder, from Oxford Brookes University in England, encountered and even photographed a strange cat-sized creature in Cameroon that once again was superficially reminiscent of a giant bushbaby. Further details concerning these perplexing extra-large prosimians can be found in my book *The New Zoo: New and Rediscovered Animals of the Twentieth Century* (2002).

HAIRY MAN-BEASTS OF FOREST AND STREAM

Another mystifying entity reported from Senegambia, and also from Guinea, but unrecognised by science is the fating'ho. Although still believed in by the more elderly members of native Senegalese society, younger people here tend to discount them as mere superstition or folklore, but occasionally something happens to make them think again.

For instance: one day in or around November 1992, one of Owen's longtime Senegalese friends, a youthful native entomological researcher called Malang Mane, was conducting research in a densely forested area of northern Guinea at an altitude of about 3600 ft when he saw something that drove all thoughts of insects far from his mind. Without warning, and completely silently, a man-sized entity stepped out of the undergrowth only a short distance ahead of him. It was covered in long, shaggy black hair, had a noticeably large head, and emitted a guttural grunting sound. Most significant of all, however, was the fact that this veritable man-beast was walking on its hind legs, and was not holding onto any branches or foliage for support, i.e. it was fully bipedal, just like humans. Too shocked and frightened to move, Malang watched it approach to within a few feet of him before it ran away again.

Malang is very familiar with the West African chimpanzee, and he was certain that the creature was not a chimp, bearing in mind that he had observed it in detail at very close range. Nor was it a gorilla, which is not native to this region of West Africa anyway. Only then did he realise that he must have seen one of the elusive, legendary fating'ho.

Similar man-beasts have been reported elsewhere in Africa too, and some cryptozoologists have suggested that they may be surviving australopithecines - primitive hominids that officially became extinct at least a million years ago. Like many West African 'monsters', however, the fating'ho seems to inhabit a twilit world midway between mythology and mystery, for it combines various ostensibly physical features with certain purportedly preternatural ones, thus frustrating traditional attempts at cryptozoological classification.

Some eyewitnesses, for example, claim that these entities will sometimes disappear into thin air in full view of their human observers. It is also believed that they can fire arrows at humans that are not tangible, but are 'spirit arrows' instead. These reputedly cause disfiguring ulcers to break out on their victims' skin, which never heal again.

The fating'ho is not the only mysterious man-beast reported from Senegal. Also on file is the wokolo, which is chiefly differentiated from the fating'ho morphologically by its yellow eyes (those of the fating'ho are red) and long pointed beard. However, whereas the fating'ho prefers dense forests, the wokolo is more commonly encountered near streams.

GUIAFAIRO AND KIKIYAON - ENCOUNTERS OF THE EERIE KIND

Two of the weirdest and most grotesque monsters reported from Senegambia - or anywhere else, for that matter - must surely be the guiafairo and the kikiyaon.

Said to remain hidden by day within the hollow trees and cave-ridden rocky outcrops rising above the hot savannahs, it is during the evening that the guiafairo takes to the wing, earning itself a fearful but memorable title - 'the fear that flies by night'. Few people who have been unfortunate enough to receive a visitation from this dire entity can agree upon its precise appearance. Some claim that it is grey in colour and winged, with a human face and clawed feet - a form of giant bat? Yet others aver that it is phantasmal, with no permanent, corporeal form, and can even materialise through locked doors. All confirm, however, that its arrival is accompanied by a vile, nauseating smell that engenders a suffocating, mind-numbing fear never forgotten by those who experience it - always assuming that they do survive. Some of the guiafairo's victims have died soon afterwards from a creeping, paralysing malaise, almost as if their fear has itself acquired a lethal, physical reality.

No less deadly, or dreadful, than the guiafairo is the kikiyaon, which is said by the Bambara tribe to inhabit only the darkest expanses of forest, and rarely emerges from this stygian gloom. On those occasions when it is seen, however, it is likened to a monstrous owl, with a pair of immense wings, huge talons on its feet, and, most notable of all, a razor-sharp spur projecting from the tip of each of its two shoulder joints. Yet whereas its wings are feathered like those of normal owls, the body of this awesome apparition is clothed in short, greenish-grey fur, and it is even said to possess a short tufted tail.

Most native people believe the kikiyaon to be a truly supernatural, rather than merely an elusive natural, creature. They claim that evil sorcerers utilise this entity to kill people, either physically or spiritually, and can even directly transform themselves into a kikiyaon. Yet it can give voice to some very substantial cries. These include a deep far-reaching grunting call that has been likened to (albeit not conclusively identified with) that of Pel's fishing owl *Scotopelia peli*, a sizeable owl that is native to Senegambia. However, there is another cry that does not seem to resemble that of any known species of owl here, and has been compared to the hideous shrieks of someone being slowly strangled! Intriguingly, this is precisely the description applied to the voice of another still-unidentified mystery beast - the devil bird of Sri Lanka (see my book *From Flying Toads To Snakes With Wings*, 1997).

Who knows? Perhaps a real, reclusive creature, possibly even an undescribed species of owl, originally inspired belief in the kikiyaon, but was gradually 'transformed' by superstition and folklore into the bizarre monster claimed to exist here today. It certainly wouldn't be the first time that a seemingly impossible creature has ultimately been shown to have a somewhat less dramatic and hitherto unrecognised but unequivocally genuine animal at its source.

WERE-HYAENAS AND SABRE-TOOTHS

Is the booaa a mysterious giant hyaena?

Another Senegalese mystery beast that may be more substantial than surrealistic is the booa. Although only rarely seen, when it is observed the booa is usually likened to a giant, abnormally-coloured form of hyaena. In contrast, it is very frequently heard, especially at night. Indeed, its name is onomatopoeic, being derived from the hideous screaming cry that reverberates loudly through the still evening air when one of these creatures is in the vicinity.

As with the kikiyaon, some Senegalese people are convinced that the booa is actually a transformed sorcerer, i.e. a were-hyaena. They claim that if a booa is shot and its trail of blood followed, it will surely lead to a human house, inside which a man or woman will be found, bleeding profusely from gunshot wounds. (This scenario closely echoes many medieval Western accounts of werewolves.) There is a similar Senegalese belief regarding the mo solo - said to be a type of were-leopard (not to be confused with the leopard-man cults).

However, reports of the booa also readily call to mind numerous accounts from East Africa, especially Kenya, of a seemingly allied but corporeal mystery beast variously termed the chemosit, kerit, or Nandi bear. Many descriptions of this infamously ferocious, forest-dwelling creature have likened it to a huge form of hyaena, of aberrant colouration and with a relatively short face. Perhaps the Nandi bear has an occidental counterpart in Senegal?

Due to poaching and political unrest, in quite recent times some of Senegal's forests have been destroyed, and its more exotic, rarer animals have become extinct. In addition, it is possible that some particularly secretive species have actually died out here even before their very existence was recognised by science.

During discussions with native hunters in Senegal's depleted Casamance Forest, Owen has learnt that they can still readily recall a huge but very mysterious form of cat, which they refer to as the wanjilanko.

According to their descriptions, it was striped, possessed very large teeth, and was so ferocious that it could even kill lions. Tragically, however, it appears to have died out, as too have the lions that it once attacked.

Reports of huge striped cats with very large teeth and savage temperament have also been recorded elsewhere in West Africa. In Chad, for example, such a creature is known as the mountain tiger or hadjel, whereas further east, moving into the Central African Republic, local tribes speak variously of the gassingram or vassoko. Their descriptions invariably recall *Machairodus*, the officially extinct African sabre-toothed tiger. In addition, when illustrations of this prehistoric stalwart's likely appearance in life have been shown to native hunters, they have readily identified them as pictures of their lands' striped, toothy mystery cats (see my book *Mystery Cats of the World*, 1989, for additional details).

The prospect of sabre-tooth persistence into modern times must rate as very slim indeed. Nevertheless, there are few places on earth more capable of sustaining such survival beyond the reach of scientific detection than the remote, little-explored jungle-lands of West Africa.

The wanjilanko, illustrated by Tim Morris

Also needing an explanation are Senegalese stories of a strange long-necked red lion known as the chakpuar, and peculiar 'cat-wolves' referred to as the guomna and sing sing. To quote one of Owen's communications to me concerning the sing sing:

> The "cat-wolf" is a strange concept that I have invented really to explain the oddities of the Sing Sing which seems to have the speed and stealth of a cat but the tenacity and stamina of a dog. It appears to have a head like a wolf and non retractable claws. The pelage is said to be somewhat brindled, like that of a laughing hyena [f the spotted hyaena *Crocuta crocuta*] without the spots. Its tail is short and ringed. Again, this creature inspires fear in hardy hunters and is rarely talked about in case discussing it causes it to appear suddenly from the depths of the forest.

Except for the short tail, this description recalls the striped hyaena *Hyaena hyaena*, which is indeed native to Senegal. As this species is normally nocturnal, and therefore not readily seen, it may have engendered a heightened, exaggerated sense of fear among the local people, thus explaining their dread of it and its elevation in their minds to the status of a veritable monster - the sing sing.

THE TANTALISING TANKONGH

While visiting Guinea, another West African country that may still contain some intriguing zoological surprises, Owen learnt of yet another unidentified beast, the diminutive tankongh. This extremely shy beast is said by local hunters to resemble a small zebra, yet lives only in the high mountain forests and is

Water chevrotain

rarely seen. However, Owen was once shown a pair of tiny dull grey hooves and some pieces of black and cream mottled skin – the remains of a tankongh that had been killed and eaten.

Owen mentions that according to local reports, this mysterious animal has a pair of small canine tusks, which makes me think of the water chevrotain *Hyemoschus aquaticus*. This is a small, hornless, but tusked ungulate adorned with stripes and spots, which is native to Guinea's lowland forests and swamp margins. Could this known but exceedingly elusive mammal be the identity of the tankongh, or could the latter even be an unknown related species adapted for a montane existence?

And what of the un-named, uncaptured toad, also hailing from Guinea, that reputedly gives birth to live young? Is this a new form?

It was Pliny the Elder who said: "Ex Africa semper aliquod novi" - 'There is always something new out of Africa'. Judging from the cryptic creatures documented here, all currently lurking within that dusky borderland between reverie and reality, the intrepid cryptozoologist would do well to heed his words, and pay a keen-eyed visit to this mysterious continent's all-too-long-overlooked Western quarter. Who knows what extraordinary revelations may still await disclosure here?

Chapter 14

GOODNESS GRACIOUS!
GREAT BOLS OF FIRE!

Spooklights, Min-Mins and More

Of lights that goe before men, and follow them abroad in the fields, by the night season.

    There is also a kind of light, yt is seene in the night season, and seemeth to goe before men, or to follow them, leading them out of their way onto waters, and other dangerous places. It is also very often seene in the night, of them that saile in the Sea and sometime will cleave to ye mast of the shippe, or other high partes, sometime glide round about the shippe, and either rest in one part till it goe out, or else bee quenched in the water. This impression seene on the land, is called in Latine, Ignis fatuus, foolish fire, that hurteth not, but onely feareth fooles. That which is seene on the Sea, if it bee but one, is named Helena, if it bee two, it is called Castor and Pollux.

    The foolish fire, is an Exhalation kindled by meanes of violent moving, when by cold of the night, in the lowest region of the ayre, it is beaten downe, and then commonly, if it be light, seeketh to ascend upward, and is sent downe agayne; so it banceth up and downe: Els if it move not up and downe, it is a great lumpe of glewish or oyly matter, that by moving of the heat in it selfe, is enflamed of it selfe, as moyst hay will be kindled of it selfe. In hote and fennie Countries, these lightes are often seene, and where as is aboundance of such unctuous and fat matter, as about Churchyards, where through the corruption of the bodies there buried, the earth is full of such substance: wherefore in Churchyards, or places of common buriall, oftentimes are such lights seene, which ignorant and superstitious fooles have thought to bee soules tormented by the fire of Purgatorie. Indeed the devill hath used these lightes (although they be naturally caused) as strong delusions, to captive the mindes of men, with feare of the Popes Purgatorie, whereby hee did open injury to the bloud of Christ, which onely purgeth us from all our sins, and delivereth us from all torments, both temporal and eternall according to the saying of the wiseman, The soules of the righteous are in the bands of God, and nor torment toucheth them. But to returne to the lights, in which, there are yet two things to bee considered. First, why they lead men out of their way. And secondly, why they seeme to follow men and goe before them. The cause why they lead men out of the way, is, that men, while they take heed to such lights, and are also sore afraid, they forget their way, and then being once but a little out of their way, they wander they wot not wither, to waters, pittes, and other very danger-

ous places. Which, when at length they hap the way home, will tell a great tale, how they have beene led about by a spirit in the likenesse of fire. Now the cause why they seeme to goe before men, or to follow them, some men have said to bee the moving of the aire, by the going of the man, which aire moved, should drive them forward, if they were before, and draw them after, if they were behind. But this is no reason at all, that the fire, which is oftentimes three or foure miles distant from the man that walketh, should bee mooved to and fro by that aire which is moved through his walking, but rather the moving of the aire and the mans eyes, causeth the fire to seeme as though it moved: as the Moone to children seemeth, if they are before it, to run after them: if shee bee before them, to run before them, that they cannot overtake her, though shee seeme to be verie neere them. Wherefore these lights rather seeme to move, than that they be moved indeed.

William Fulke – *A Goodly Gallerye: William Fulke's Book of Meteors* (1563)
[this is the earliest published account of will o' the wisps in
the English language]

Balls of light, or BOLs to aficionados, must surely comprise one of the most widespread, and wide-ranging, of all mysterious phenomena. As will be revealed in this survey of spherical inexplicables, the variety of examples on record and explanations proposed is quite exceptional.

BALL LIGHTNING

It is virtually impossible to provide an adequate definition of any phenomenon as notoriously variable as ball lightning. Nevertheless, its most familiar, commonly reported guise is a luminous, yellow-red spherical mass, rarely more than 18 in across, which sometimes emits a hissing or buzzing noise, often exists for just a few seconds, leaves behind a smell variously likened to ozone or sulphur, and is frequently associated with electrical storms. It sometimes exhibits the anomalous ability to pass through solid objects without leaving any trace, and can actively avoid objects in its path, displaying a distinct 'sense' of direction.

Documented worldwide for centuries, ball lightning has sometimes occurred at the most unexpected times, and in the most unlikely places. What may be the earliest recorded instance of ball lightning was documented by St Gregory of Tours and occurred in the 6th Century, during a ceremony marking the dedication of a chapel. Without warning, an incandescent fireball materialised just above the procession of attending dignitaries, who were so terrified by its impromptu appearance that they threw themselves en masse onto the ground.

In April 1915, a man from Columbia in Missouri was standing inside his house, looking through one of his windows at the thunderstorm in progress outside, when he suddenly heard a single, sharp blast that sounded like a rifle shot. Almost immediately, the telephone in the room where he was standing made a strange clicking sound, and as he looked at it he was astonished to see a small bubble-like ball of light appear out of the mouthpiece. As he watched, this unheralded BOL floated towards him, but after briefly rolling about on the windowsill, his mysterious visitor disappeared. Nor is this an isolated example. Two months later, telephone operators at Drifton, Pennsylvania, were amazed (and not a little alarmed) to see several tiny "balls of fire" emerging from the plugs of their switchboards.

One sultry, thundery evening during the Caen offensive of June 1944, British zoologist Professor John L. Cloudsley-Thompson was commanding the leading squadron of the Desert Rats' leading tank regiment, advancing through a cornfield. Constantly shelled and mortared, surely the last sight that he expected in the midst of such wartime chaos was a ball lightning visitation, but during their advance he noticed to his left front "...a fireball lifting slowly from the ground and ascending into the sky". In fact, that was Cloudsley-Thompson's second BOL encounter. His first took place one evening when he was aged 8 or 9, on holiday with his family by the sea. Suddenly: "a ball of fire, about half the size of a football, appeared to zoom into the ground, with a whizz, from a distance of about twice my height".

One clear frosty November night in 1950, Miriam Hopper from Cotgrave, Nottinghamshire, and her future husband were sitting in a room lit only by a gas fire when a blue grapefruit-sized specimen of ball lightning appeared at one side of the bay window, and rolled silently along the entire length of a decorative plate rack. It followed every corner of the rack, yet left no trace of its passage on any of the plates, before vanishing at the other side of the window.

Engraving of a ball lightning occurrence inside a barn
at Salagnac, in Correze, France, during 1845

On 19 March 1963, Kent University electronics specialist Professor R.C. Jennison was flying through a severe electrical storm aboard an Eastern Airlines flight to Washington from New York, when he and one of the air hostesses spied a glowing BOL floating down the aircraft's aisle! Just over 8 in in diameter, it had emerged from the pilot's cabin, and remained about 2.5 ft above the ground as it sedately made its way towards the toilet at the end of the aisle. This was a particularly noteworthy sighting, because by featuring an electronics expert as a principal eyewitness, it greatly assisted in raising the credibility status of ball lightning within what had hitherto been a highly sceptical scientific community.

One afternoon in February 1989, Doris M. Humphrey from Rockville, Maryland, was watching television in her bedroom when she observed several small BOLs dancing and hovering along a long electrical cord connecting the television to the wall outlet. They even darted down into the wastepaper bin, inside which one loop of the cord had fallen, and then soared back out again, still tracing the cord's length towards the wall. White in colour, these animated globes ranged in size from a marble to a ping-pong ball, but before reaching the wall their 'attention' was apparently diverted by the nearby presence of an artificial fireplace containing electrical logs - dancing around them, even though these logs were not switched on. Finally, the globes coalesced into a single, larger BOL that vanished behind the logs.

Some very unusual examples of ball lightning are also on file. In 1907, while standing on a street corner in Burlington, Vermont, William Alexander and two other men heard a huge explosion, and then spied a 6-ft-long torpedo-shaped specimen of ball lightning, suspended about 50 ft above the street buildings, and sending forth tongues of fire from its dark surface. A "round, bronze glistening ball with gleaming rays shooting from the top and sides...reminding one of an ornament at the top of a Christmas tree" materialised inside Mary Hunneman's house in Fitzwilliam, New Hampshire, on 10 August 1937, at the same moment that a crash of thunder resounded outside. And on 15 August 1975, a Miss M.E. MacArthur observed a truly bizarre example of ball lightning outside her home in Argyllshire. The main portion was 9 in long, opaque, and dome-shaped, but attached to this and trailing behind it was a crinkly ribbon-like 'tail'. It was white in colour, at least 3 ft long but only an inch or so in width, and was suspended slightly higher in the air than the dome. A short time later, this unique specimen abruptly vanished: "...snuffed out like a candle, leaving no trace".

It is ball lightning's diversity and its unique combination of features that has caused so many problems for scientists attempting to assess and explain this remarkable manifestation of Nature. At one end of the spectrum of proposed solutions is the possibility that ball lightning does not even exist, that it is simply an optical illusion - specifically, an after-image on the eye's retina caused by a flash of lightning. However, this cannot explain multiple-eyewitness cases, in which everyone has reported seeing precisely the same thing, behaving in precisely the same way. Also, ball lightning can yield physical evidence of its reality, sometimes charring items that it touches, or leaving distinctive odours in its wake.

Other notions on offer include violent chemical reactions in the atmosphere compressing forked lightning into spherical form; burning orbs of gas; nuclear reactions; composite nanobatteries; or even tiny meteorites of antimatter or minuscule black holes as plausible sources or identities. The most popular proposal, however, is that specimens of ball lightning are glowing spheres of plasma, created by natural electromagnetic forces. Although this still does not provide a comprehensively satisfying solution, in recent years small globes of plasma remarkably similar in appearance and behaviour to ball lightning have been artificially created in scientific laboratories - thereby suggesting that this is the correct avenue of investigation to pursue in the continuing search for this mystifying phenomenon's origin and identity. However, in 2007 Brazilian scientists Drs Antonio Pavão and Gerson Paiva produced small glowing orbs by oxidising silicon vapour, thus indicating an alternative, or even an additional, origin.

BEAD LIGHTNING

Far less familiar, but no less mysterious, than ball lightning is bead lightning. One of the best examples on file was reported in *Popular Astronomy* by J.F. Steward, from Colorado. One day in 1894, while driving in the country through a sizeable thunderstorm, Steward spied a pair of parallel flashes of lightning. One vanished instantly, but the other "...broke into a multiplicity of what seemed to be spheres. This string of beads, as it seemed, appeared to be formed of material furnished by the original flash".

WILL O' THE WISPS

According to traditional Western folklore, those famous BOLs known as will o' the wisps or the ignis fatuus ('foolish fire') are mischievous Faerie entities that glow brightly at night, and enjoy luring unsus-

Engraving of will o' the wisp activity *(FPL)*

pecting humans astray, sometimes even to their doom, when travelling through treacherous marshy terrain (similar BOLs spied flitting among the graves in cemeteries are termed corpse candles or spooklights). Science, meanwhile, has other ideas on the matter.

Although scientists have traditionally discounted will o' the wisps and comparable BOLs as nothing more novel than burning methane (marsh gas), the mechanism by which methane could ignite spontaneously has remained unknown. During 1993, however, German investigators Dr Dieter Glindemann and his aptly-named colleague Dr Günter Gassmann from the Helgoland Biological Institute in Hamburg proposed that the answer to will o' the wisps was actually the spontaneous combustion of another gas, called diphosphane.

During their research, they discovered that diphosphane can be produced via the chemical reduction of phosphates by micro-organisms that inhabit not only oxygen-depleted marshes but also the digestive tract of dead mammals, including humans. Hence this gas could seep forth from marshes and from recently-buried corpses in cemeteries, and as diphosphane spontaneously burns when it meets air, the result will be small BOLs, or will o' the wisps.

MIN-MINS AND QUINNS LIGHT

Named after the Min-Min Hotel in western Queensland, where they were often seen by Australia's early European settlers who used this hotel as a staging post, min-mins are ghostly BOLs reminiscent of will o' the wisps and other spooklights. Although generally white, they do sometimes change colour, and are looked upon with dread by the native aboriginals, who believe that they can steal a person's mind and drain away his life-force.

Some scientists have attempted to explain away the min-mins as the headlights of distant vehicles that appear closer than they really are via a mirage-like inversion effect. Yet this ingenious explanation cannot reconcile their longstanding familiarity to the aboriginals, which considerably predates the development of motor vehicles.

Another 'solution' is that these glowing BOLs are barn owls, whose white feathers make them appear luminous in moonlit conditions. However, there are a number of startling cases on file in which min-mins have actively followed, and even teased, their human observers - hardly the most typical behaviour of the notoriously shy barn owl. One such case was included in my book *The Unexplained* (1996), and featured a New South Wales sheep drover who was pestered by a blue min-min that appeared over his shoulder while he was working, and continued to follow him as he rounded up his sheep. Finally, the drover lost his temper, and chased the infuriating BOL on horseback, but after failing to catch up with it he returned to his flocks - only for his min-min tormentor to reappear over his shoulder!

Often reported in Australian bush areas, the Quinns light is a phosphorescent BOL about the size of a large bird, which dances round in giddy circles until it finally vanishes without warning. Some frustrated pursuers have even fired at it, but without effect - thereby confirming that it is neither a single animal nor a swarm of insects.

DANCING BLUE SPOOKLIGHTS IN SILVER CLIFF'S GRAVEYARD

During the 1960s, the graveyard at Silver Cliff, an old mining centre in Colorado, regularly featured in American newspaper headlines - due to the pulsating blue BOLs that eerily danced around its graves. Fungi-induced phosphorescence from rotting wooden tombstones, glowing mineral ores in the ground, and even reflected lighting from nearby houses were all deemed responsible at one time or another, but none of these suggestions was ever proven - and today, the BOLs dance no more, so their identity is likely to remain a mystery forever.

EARTHQUAKE LIGHTS AND MARFA LIGHTS

Photo of the Marfa lights *(FPL)*

Geophysicists have long known that earthquakes are sometimes accompanied by anomalous forms of illumination, lighting the sky above the epicentre but of undetermined origin and identity. Some earthquake lights resemble radiating beams, others are BOLs. One of their keenest investigators is Paul Devereux, author of *Earth Lights* (1982) and *Earth Lights Revelation* (1989). He believes that they are the product of geomagnetic energies manifesting from geological faults in the earth, and also that they may interact with the human brain's own electrical fields, acquiring whatever form the observer's mind bestows upon them.

Among the more notable examples of what Devereux considers to be earthlights are the famous Marfa lights, whose highly animated aerial displays just east of Marfa in Texas, near the Mexican border, attract great numbers of sightseers. Larger than many other examples of BOLs, they are about half as big as a basketball, and resemble flickering yellow lanterns or headlights - but as with the Australian min-mins, they were first reported long before motor vehicles were added to mankind's list of modern-day inventions.

LUMINOUS BUBBLES OF LIGHT

One of the least-known yet most fascinating types of BOL is the luminous bubble of light - generally smaller and less substantial than its more familiar (if equally inexplicable!) counterparts. During the very hot, late afternoon of 17 August 1876, with sheet lightning flickering in the distance, two ladies were walking on the cliff at Ringstead Bay, Norfolk, when they became surrounded by a veritable phalanx of dancing bubbles of iridescent light.

Like a swarm of luminous but wholly silent bees, they swirled all around their bemused observers, from ground level up to a few feet above their heads, each moving independently of the others, and all deftly eluding the ladies' attempts to reach out and touch them. Intriguingly, their numbers continually fluctuated - from a few dozen to several thousand, then back to a few dozen again, in just a few minutes - until finally they all disappeared.

FOO FIGHTERS

During World War II, aircraft pilots often encountered small BOLs, usually white or red, that danced around the outside of their craft in an uncannily playful, inquisitive manner - sometimes even pacing them as they flew through the sky on their missions - as if curious to learn more about these giant man-made constructions. Becoming known as foo fighters, they attracted a great deal of attention, with special units set up to investigate them, because they were initially believed to be secret weapons developed by the enemy - until it was realised that British, American, German, and Japanese pilots were all seeing them!

Britain's foo fighter investigators comprised the Massey Project, named after its head of command, Lieutenant General Massey.

The U.S.A. employed its American 8th Army for the same purpose. And Germany set up its Sonder Buro (Special Bureau) #13 (codenamed 'Operation Uranus'), with Georg Kamper as director, to assess Luftwaffe reports. Yet none of the teams obtained a conclusive explanation for these enigmatic BOLs.

Foo fighters around a World War II aeroplane *(FPL)*

ARE BOLS INTELLIGENT?

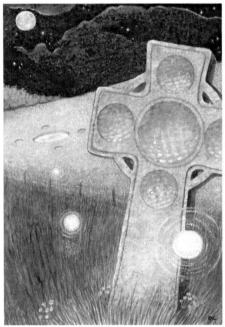
Painting of Celtic cross and BOLs, by artist Philippa Foster

Whereas some BOLs seem wholly meteorological and/or geomagnetic in origin, there are others, such as the min-mins and foo fighters, whose interactions with humans appear uncannily intelligent, even premeditated. Veteran mysteries investigator Vincent Gaddis presented many such cases in his book *Mysterious Fires and Lights* (1967), and was one of several researchers bold enough to contemplate seriously the incredible possibility that these particular BOLs are in fact sentient life-forms - composed largely of pure energy but possessing at least a rudimentary level of intelligence.

Gaddis and company have even suggested that some UFOs may actually be living organisms too, highly specialised for an exclusive existence amid the rarefied world of our planet's all-encompassing atmosphere (see Chapter 22 for further details).

It is supremely ironic, or apt, depending upon your point of view (and the amount of time that you may have spent investigating these glowing enigmas!), that in many BOL cases, the solution remains as intangible and elusive as one of the most infamous examples of their own kind. Namely, the fickle, evanescent will o' the wisp - luring us ever onwards along false, deceptive trails, and always remaining just out of reach of our trusting, outstretched arms.

Chapter 15
IN THE WAKE OF THE USOS
Unidentified Submarine Objects

But Ned Land was not mistaken, and we all perceived the object he pointed to. At two cables' lengths from the Abraham Lincoln, on the starboard quarter, the sea seemed to be illuminated all over. It was not a mere phosphoric phenomenon. The monster emerged some fathoms from the water, and then threw out that very intense but inexplicable light mentioned in the report of several captains. This magnificent irradiation must have been produced by an agent of great shining power. The luminous part traced on the sea an immense oval, much elongated, the centre of which condensed a burning heat, whose overpowering brilliancy died out by successive gradations.

"It is only an agglomeration of phosphoric particles," cried one of the officers.

"No, sir, certainly not," I replied. "Never did pholades or salpae produce such a powerful light. That brightness is of an essentially electrical nature. Besides, see, see! it proves; it is moving forwards, backwards, it is darting towards us!"...

There was no doubt about it! this monster, this natural phenomenon that had puzzled the learned world, and overthrown and misled the imagination of seamen of both hemispheres, was, it must be owned, a still more astonishing phenomenon, inasmuch as it was a simply human construction.

We had no time to lose, however. We were lying upon the back of a sort of submarine boat, which appeared (as far as I could judge) like a huge fish of steel.

Jules Verne – *Twenty Thousand Leagues Under the Sea*

Just after 8.00 pm, on a previously uneventful evening in late summer 1954, 80 miles out at sea from New York, Captain Jan P. Boshoff stood upon the bridge of a Netherlands government ship called the 'Groote Beer', and gazed through binoculars in amazement at an incredible craft that by all the known laws of science could not exist - but it did!

Only a few moments earlier, Captain Boshoff had been called to the bridge by a member of his crew to observe an extraordinary object that had just begun to surface from the depths of the sea. And here it was. Resembling a flat, moon-like disc, it appeared greyish at first, but as he stared at it, this curious craft's lower portion became brighter, and he could discern a series of bright spots around its edges that looked like lights. While he was still peering at it, however, the strange object and its fluid medium abruptly parted company - because this unidentified structure suddenly rose completely out of the sea. Speeding up and away from the ship, it soared swiftly through the sky until it was soon lost to sight.

Boshoff's third officer, Cornelius Kooey, had been monitoring their mysterious visitor's movements with a sextant, and he reported that the object's speed had been approximately 32 minutes of arc in 1.5 minutes of time.

This is not an isolated case. On the contrary, it is just one of many on file whose mystifying subjects are aptly referred to as USOs - Unidentified Submarine Objects (also Unidentified Submerged Objects).

Surprisingly, however, USOs have previously attracted little attention from investigators of mysteries - with the notable exceptions of American zoologist Ivan T. Sanderson, who painstakingly examined this subject in his book *Invisible Residents* (1970); and British Fortean writers Janet and Colin Bord, whose chapter on USOs in their book *Modern Mysteries of the World* (1989) is an invaluable source of reliable, thought-provoking data.

Yet although far less familiar than UFOs, USOs are certainly no less fascinating, or mysterious, and have more than a little in common with them too, as will be revealed here.

UNEXPLAINED, AND UNDER THE SEA

Just like UFOs, USOs come in many forms. Sometimes reminiscent of the fictional 'Nautilus' piloted by Captain Nemo in Jules Verne's *Twenty Thousand Leagues Under the Sea*, the most straightforward (or least bizarre?) version can be conveniently dubbed the pseudo-submarine.

A classic example was the 90-ft craft that surfaced on 28 December 1976 just off Coomlieyna Beach, near Ceduna in South Australia - much to the consternation of a group of 20 or so aboriginals who were fishing there at the time. According to their statements, the object resembled a submarine in general appearance, and bore a white conning-tower, but it had a black centre-line and a red stripe at the water-line. When the fishermen reported their sighting, it caused great surprise, because there were no submarine craft known to be operating in this area at that time. Moreover, as later confirmed by a Royal Australian Navy spokesman, what makes it even more intriguing is that their description did not correspond with any known type of submarine belonging to any of the world's navies!

On 12 January 1965, the captain of a DC-3 flying at an altitude of 500 ft between Whenuapai and Kaitaia, New Zealand, spied a symmetrical, streamlined pseudo-submarine that measured about 100 ft long but not more than 15 ft across even at its widest point. As it was stationary in 30 ft of water in Kaipara Harbour, he had initially mistaken it for a stranded greyish-white whale. Upon closer observation, however, it seemed to be metallic, with the suggestion of a hatch on its upper surface, but no other external structures. The captain's report attracted appreciable interest, because this area was wholly inaccessible for normal submarines, on account of its encompassing mudflats and mangrove swamps. So what was it - and how had it arrived there? We still do not know - and nor do we know what happened to it afterwards. Rather like maritime leprechauns, USOs have a disconcerting habit of disappearing as soon as their observers take their eyes off them!

They can be just as difficult to approach, too. During four consecutive days in 1963, 13 U.S. Navy vessels situated off Puerto Rico tracked by sonar a USO endowed with incredible capabilities. Whenever a U.S. submarine attempted to draw near, the USO would effortlessly elude it by racing away at speeds exceeding 170 mph, and diving to depths of 27,000 ft - superlative feats far beyond the capacity of any known type of submersible.

Equally alien was the gleaming fusiform pseudo-submarine spied by Captain Julian Lucas Ardanza and fellow officers during the evening of 20 July 1967, while aboard the Argentinian steamer 'Naviero', 120 miles off Cape Santa Marta Grande, Brazil. About 50 ft away from the 'Naviero', and

Some USOs recall the fictional Captain Nemo's 'Nautilus' submarine

105-110 ft long, the cigar-shaped USO emitted a strong blue and white glow but was utterly silent, and conspicuously lacking in standard submarine accoutrements, such as a conning-tower, periscope, railing, or other superstructures. Travelling at a speed of 25 knots, it was unlike anything previously seen by the sailors, but Chief Officer Carlos Lasca opined that it may be a submersible UFO - a comment well worth remembering, as will be seen.

WHAT GOES UP...

Perhaps the most astounding pseudo-submarine of all, however, was encountered late at night on 1 August 1962 at the French fishing port of Le Brusc, by three fishermen. A submarine-like vessel was cruising slowly along the surface, and several humanoid figures resembling frogmen clambered out of the water and onto the vessel's surface before disappearing inside - after which this supposed submarine ascended completely out of the water and hovered above the surface! And as it hovered, it emitted a series of bright red and green lights, as well as a strong white beam that scanned like a scrutinising searchlight in the direction of the bewildered fishermen aboard their boats.

Suddenly, however, the searchlight vanished. So too did the gleaming red and green lights - only to be replaced by a uniform orange glow, as the pseudo-submarine began to rotate, spinning faster and faster. Without warning, it lurched upwards, soaring silently through the sky until the men were unable to distinguish its light amid the glittering panoply of stars.

This is when reports of USOs and UFOs begin to blur, because in cases where USOs rising from the sea have become airborne, they often greatly resemble 'conventional' UFOs.

For instance: a cigar-shaped craft glowing blue and sporting a row of portholes with yellow lights emerged from the sea off Puerto La Cruz, Portugal, and flew off into the sky - much to the amazement of the commander and crew aboard a Norwegian tanker passing by.

On 4 August 1967, in the Gulf north of Arrecife, Venezuela, the sea began bubbling violently within a circle measuring about 18 ft across, near to where Dr Hugo Sierra Yepez was fishing. Gazing at this zone of disturbance, Dr Yepez was very startled when a large flat disc rose up through it, and hovered just above the surface, dripping with water. Greyish-blue in colour, the disc had a revolving section with a number of triangular window-like structures. As Yepez watched, this mysterious vessel rose upwards in a curved path and streaked out of sight.

Later that month, Ruben Norato was on the beach at Catia La Mar, Venezuela, when the water started churning dramatically. Suddenly, three huge plate-like discs ascended from the epicentre of this activity, and flew rapidly off into the distance. And in 1959, a triangular USO, whose sides were about 12 ft long, rose out of the Baltic Sea and circled over the barracks of some frightened soldiers near Kolobrzeg, coastal Poland.

...MUST COME DOWN!

Just as there are an abundance of reports recounting the unheralded emergence of USOs out from their liquid medium and up into the sky, so too are there ample accounts on file of supposed UFOs descending from the sky and plunging down into the sea or inland freshwater bodies.

Probably the most famous example made its debut at 11.30 pm on 4 October 1967 above Shag Harbour, Nova Scotia. It took the form of a hovering 60-ft-long UFO equipped with a row of reddish-orange lights, which switched off and on sequentially, and were aligned at an angle of 45°, dipping to the right. Suddenly, however, in full view of several eyewitnesses, it flew down towards the harbour's water, extinguishing its row of reddish lights and replacing them with a single white light, before bobbing up and down on the surface of the waves.

Numerous people, including a constable of the Royal Canadian Mounted Police, set out in boats in pursuit of the harbour's remarkable visitor, but by the time that they had reached the spot where it had landed, . the UFO had become a USO, having vanished beneath the surface. All that remained was a strange, 80-ft-wide patch of bubbling yellowish froth.

Even more remarkable was the 'ball of fire, trailing white smoke' that inexplicably defied being extinguished as it plunged in and out of the sea many times before finally disappearing from sight.

It was reported by crew aboard the Honduran freighter 'Aliki P', and what makes this case so noteworthy is that it occurred at around the same time and in much the same area as the 'Groote Beer''s previously-mentioned encounter with the emerging discoidal USO.

Perhaps the most dramatic conversion from UFO to USO, however, occurred during late February 1963, when the radar aboard a frigate sited 30-50 miles off Norway's north coast

Terry Rose's photo of a possible USO in Seattle's harbour *(FPL)*

suddenly detected a UFO. The frigate was one of several vessels from the British Royal Navy North Atlantic Fleet engaged in exercises here, and its radar revealed the UFO to be solid, 100-120 ft across, and flying at 35,000 ft. Its spontaneous arrival eliminated conventional aircraft as possible identities, and after it was recorded by a second ship in the fleet but did not respond to radio contact, jet planes were sent to intercept this enigmatic craft. As they approached it, however, the UFO abruptly plummeted down to the sea at appreciable speed.

A few seconds later, it vanished from the radar screens, but then the ships' sonar equipment began recording a mysterious trace - a very large unidentified object travelling in the same direction as that of the UFO just moments before, but beneath the water surface.

Clearly, the UFO had dived into the sea, and was now speeding onwards and downwards, zigzagging into deeper water, where it soon evaded further detection by the ships' sonar.

In contrast to UFOs, very few USO photographs exist. One possible exception is a picture snapped by Terry Rose from a ship in Seattle's harbour during 1966.

As can be seen, it shows to the right of the ship a strange glowing orb seemingly in the sea nearby. Is it merely a photographic or optical artefact, or could it be a genuine USO? Over 40 years later, the jury is still out.

181

FRESHWATER USOS

USOs are not confined to marine localities. Quite a few records hail from rivers and lakes.

During the sunny afternoon of 16 May 1981 at the Thompson River, Kamloops, in Canada's British Co-lumbia, a fisherman was puzzled to hear a strange noise that he likened to water being poured into a fry-ing pan. Looking down, he noticed a patch of water bubbling, about 100 yards from where he was stand-ing - and as he continued to watch, a mysterious craft that he termed a "typical flying saucer" emerged out of the river and flew away, showering him with tiny pellets of an undetermined nature as it did so.

One of the most remarkable freshwater USOs must surely be the Scandinavian ice-breaker sighted on 30 April 1976. Other than being dark (possibly grey) in colour, little detail of its appearance was recorded, but there was no doubt as to its reality, or its power - for with consummate ease this extraordinary craft tore a long channel through the frozen surface of Sweden's Lake Siljan as it made its way across the large inland body of water.

This USO's activity is of particular note, because over the years many unexplained holes and channels have been reported in frozen lakes and rivers throughout Scandinavia. One of the most spectacular was a veritable chasm measuring 90 ft by 60 ft across, which pierced a 3-ft-thick sheet of ice encompassing Lake Uppramen, also in Sweden, in 1968.

Moreover, on 8 December 1983, a sizeable hole measuring 10 ft by 8 ft was found in the ice surfacing Finland's Säkkiä Lake by local villagers. Is it just a coincidence that only a few hours earlier, the villag-ers had spied some mysterious whirling lights overhead in the sky?

THE MECHANICAL 'SEA MONSTER' OF BLACK FISH BAY

In Jules Verne's *Twenty Thousand Leagues Under the Sea*, the fictitious Captain Nemo had deliberately designed the 'Nautilus' to resemble a sea monster, so that it would be ignored by the world's naval au-thorities. Fact, however is often stranger than even the most imaginative fiction, as suggested by an in-credible incident reported on 3 July 1893 by the *Tacoma Daily Ledger* newspaper.

A fishing party consisting of H.L. Beal, J.K. Bell, Henry Blackwood, William Fitzhenry, W.L. McDon-ald, and two other men had departed from Tacoma, Washington, on 1 July 1893, and had camped that evening at Black Fish Bay, Puget Sound. There, during the early pre-dawn hours of the next day, they and a nearby camp of surveyors received an awesome visitation, as described in the following excerpts from one of the fishermen's statement:

> ...in an instant a most horrible noise rang out in the clear morning air, and instantly the whole air was filled with a strong current of electricity that caused every nerve of the body to sting with pain, and a light as bright as that created by the concentration of many arc lights kept constantly flashing...As both light and sound came from off the bay, I turned my head in that direction, and...right before my eyes was a most horrible-looking monster.
>
> By this time every man in our camp, as well as the men from the camp of the surveyors, was gathered on the bank of the stream; and...the monster slowly drew in toward the shore, and as it approached, from its head poured out a stream of water that looked like blue fire. All the while the air seemed to be filled with electricity...One of the men from the surveyors' camp incautiously took a few steps in the direction of the water, which reached the man. He instantly fell to the ground and lay as though dead.

So too did W.L. McDonald, when he was splashed by water thrown over him by the monster's movements. Sensibly, all of the men from both camps fled into the nearby woods for safety, but they could still hear its thunderous roars and see the flashes of brilliant light emitted from its body before it finally sank beneath the bay's waters:

> This monster fish, or whatever you may call it, was fully 150 feet long, and at its thickest part I should judge about 30 feet in circumference. Its shape was somewhat out of the ordinary insofar that the body was neither round nor flat but oval, and from what we could see the upper part of the body was covered with a very coarse hair. The head was shaped very much like the head of a walrus, though, of course, very much larger. Its eyes, of which it apparently had six, were as large around as a dinner plate and were exceedingly dull, and it was about the only spot on the monster that at one time or another was not illuminated. At intervals of about every eight feet from its head to its tail a substance that had the appearance of a copper band encircled its body, and it was from these many bands that the powerful electric current appeared to come. The bands nearest the head seemed to have the strongest electric force, and it was from the first six bands that the most brilliant lights were emitted. Near the center of its head were two large horn-like substances, though they could not have been horns for it was through them that the electrically charged water was thrown.
>
> Its tail from what I could see of it was shaped like a propeller and seemed to revolve, and it may be possible that the strange monster pushes himself through the water by means of this propeller-like tail. At will this strange monstrosity seemed to be able to emit strong waves of electric current, giving off electro-motive forces, which causes any person coming within the radius of this force to receive an electric shock.

Happily, the two men who had received a shock did recover, but even if we are willing to assume that it was not simply the product of a newspaper reporter's somewhat fevered imagination, this 'monster' was clearly not a living creature. It was obviously a mechanical craft, presumably designed purposefully to look like a horrifying monster to scare away the curious-minded. Its origin, and the identity of those who constructed it, conversely, remain totally unknown, and it seems destined to remain the most bizarre USO ever reported.

ARE USOS FROM OUTER SPACE, OR INNER SPACE?

As with conventional UFOs, there are many explanations on offer for USOs, ranging from the mundane or prosaic to others that are, quite literally, out of this world.

Just as some UFOs are certainly highly advanced aircraft specifically designed by various countries' military scientists for spying purposes and thereby shrouded in governmental secrecy and disinformation, it is very likely that a fair proportion of the pseudo-submarine USOs are sophisticated, classified spy submersibles - explaining why they appear unfamiliar (allegedly?) to official navy spokesmen. According to American UFO expert Jerome Clark in *Unexplained!* (1999), for instance, the mysterious submarines that plagued Scandinavian territorial waters during the 1960s and 1970s were actually Soviet craft on spy missions.

A far more controversial yet fascinating possibility, mooted in Dr Bernard Heuvelmans's book *In the Wake of the Sea-Serpents* (1968), is that some mystery submarines, such as various inexplicably fast 'phantom u-boats' with periscopes reported during World War I, were really long-necked sea monsters, similar in form to Nessie!

As for reports of UFOs plunging down into the sea or inland bodies of freshwater, there is little doubt that some of these could simply be small meteorites, examples of ball lightning, or even normal aircraft that had somehow malfunctioned and were spied crashing downwards, out of control. Certain cases, conversely, such as those documented here, genuinely suggest mechanical, controlled crafts, homologous in all respects to other UFOs - except that they can travel effectively underwater as well as through the sky. But who can say that other UFOs cannot do so too? After all, if a given UFO is only seen flying through the sky, there can be no way of knowing whether it also possesses subaquatic abilities.

As for those USOs that have been observed emerging out of the sea or freshwater and soaring off into the sky, there does not appear to be any conventional explanation on offer - other than the seemingly self-evident conclusion that they are mechanical craft of a truly alien nature, beyond the current scope of publicly-recognised scientific knowledge, and which may (or may not) be one and the same as airborne UFOs.

If USOs really are UFOs that can move through aquatic and aerial media alike, then perhaps they are piloted by fluid-friendly extraterrestrials journeying here from other star systems. Alternatively, to coin a veritable contradiction in terms, could they be piloted by native aliens, i.e. a race of highly-advanced entities indigenous to planet Earth, but having evolved for a subaquatic rather than a terrestrial existence? To quote Ivan T. Sanderson:

> If a superior technological type of intelligent civilization(s) developed on this planet under water, they would very likely have gotten much farther ahead than we have, having had several millions, and possibly up to a billion years' headstart on us, life as we know it having started in the sea.

Taking this thought-provoking notion even further, in *Atlantis Rising* (1973) Brad Steiger presented a more precise line of speculation for consideration:

> Perhaps the UFO enigma might be traced to the inner space of our seas rather than the outer space of our solar system, and the mystery of Atlantis might be solved by seeking Cosmic Masters under the Atlantic rather than turning over rubble and artifacts on Thera [=Santorini, a Greek island believed by some to be a remnant of this fabled lost continent].

Yet whatever the answer, one thing seems likely. USOs are technologically far superior to anything presently developed by humanity. So if they are indeed the craft of a mysterious race of beings concealed beneath the seas of our planet, these beings apparently mean us no harm - otherwise we would have been conquered by them long ago. Ironically, it is quite conceivable that the reverse is true. In the prophetic words of Janet and Colin Bord:

> If underwater civilizations do exist, all they need fear from mankind is that we will pollute them out of existence.

Little wonder, perhaps, that USOs - and UFOs - seem so interested in our activities?

Chapter 16
IN SEARCH OF ANCIENT WONDERS
Myths and Marvels from Bygone Worlds

I met a traveller from an antique land
Who said: "Two vast and trunkless legs of stone
Stand in the desert....Near them, on the sand,
Half sunk, a shattered visage lies, whose frown,
And wrinkled lip, and sneer of cold command,
Tell that its sculptor well those passions read
Which yet survive, stamped on these lifeless things,
The hand that mocked them, and the heart that fed:
And on the pedestal these words appear:
"My name is Ozymandias, king of kings:
Look on my works, ye Mighty, and despair!"
Nothing beside remains. Round the decay
Of that colossal wreck, boundless and bare
The lone and level sands stretch far away".

Percy Bysshe Shelley – *'Ozymandias'*

H ad I not become a zoologist, I may well have sought a career in archaeology, as I have al-ways been fascinated with ancient civilisations and the many extraordinary monuments, edi-fices, and other spectacular creations that once existed in those bygone realms.

This chapter brings together a number of my favourite examples drawn from long-departed worlds, in-cluding several that have previously attracted little attention and documentation.

THE COLOSSUS OF RHODES – A BEHEMOTH IN BRONZE

During ancient times, the patron deity of Rhodes was Helios, the Greek sun god. And for over 50 years during the 3rd Century BC, travellers approaching this Mediterranean island by sea may well have been forgiven for thinking that its divine protector had descended from the heavens to honour Rhodes with a personal visit. For there, standing by the harbour, was the gigantic, burnished figure of Helios himself, more than 100 ft tall, gleaming brilliantly in the bright sunlight, and holding up a massive torch to the sky.

In reality, it was of course a statue - but no ordinary one. Nothing like this resplendent gargantuan effigy had ever been seen before - nor would anything like it ever be seen again - in antiquity. Not for nothing did Antipater of Sidon in c.140 BC include it within his famous poem listing the Seven Wonders of the Ancient World. Certainly this sculptural masterpiece must indeed have been a marvel to behold - for this was the Colossus of Rhodes.

Formerly, the term 'colossus' had been applied to all statues, regardless of size. Following the creation of the Colossus of Rhodes, however, its usage became restricted to statues of gigantic size. And today, anything of immense proportions is said to be colossal.

The Colossus was created by a renowned sculptor known as Chares of Lindos, a talented pupil of the even more eminent sculptor, Lysippus, and took twelve years to prepare, reaching completion in c.282 BC. Erected upon a base of white marble and shaped in stages, beginning with the feet (which were themselves larger than most entire statues previously sculpted) and with each successive portion cast directly upon the portion immediately beneath it, the complete statue stood 110 ft tall. Externally, it was composed of countless bronze sheets, but in order to fortify it internally, Chares incorporated an inner framework of iron bars set into squared blocks of stone.

Inevitably, the creation of such a stupendous statue was an exceedingly expensive undertaking. According to a description of the Colossus contained in Pliny's major work *Naturalis Historia* (Vol. 34), it cost 300 talents, equivalent today to approximately £1.5 million. The source of this vast sum came from the selling of war machinery left behind on the island following the celebrated Siege of Rhodes - an unsuccessful attempt in 305 BC by Demetrius Poliorcetes to take over the island following its people's refusal to support him in his ongoing war against Ptolemy of Egypt.

It is supremely ironic that despite its awe-inspiring stature, the precise appearance of the Colossus is no longer known. None of the ancient historians who documented it gave any details, except for its size and composition. Nevertheless, based upon modern knowledge of sculptural colossi (such as the even taller Statue of Liberty), some reasonable deductions can be offered.

A statue as tall as the Colossus of Rhodes would certainly have possessed a simple columnar shape, because this is by far the most stable configuration. If, as some scholars believe, it did indeed hold an immense torch in its carved hand, its arm would have pointed directly upwards. This is because any attempt to sculpt an arm stretched out horizontally or held out at a sizeable angle from the body would have rendered it exceedingly unstable. And in keeping with other works by Chares and Lysippus, the figure itself may well have taken the form of a young athletic man, probably naked.

As for the popular belief that the Colossus actually stood astride the entrance to Rhodes's Mandraki Harbour, with its left foot planted on one side of the harbour and its right foot planted on the other, thereby allowing ships to sail in between its legs, this is a fallacy - dating back no further than the 14th Century AD. In reality, the size of Mandraki Harbour was such that if this had indeed been true, the statue's legs

The Colossus of Rhodes

would have been more than 1300 ft apart - and not even the mighty Colossus of Rhodes was that big!

Moreover, there is even some dispute as to whether the Colossus actually stood by the harbour entrance at all. A plausible alternative location for it, suggested by renowned British archaeological expert Dr Reynold Higgins, was further inland, near the Temple of Helios.

For 56 years, the Colossus of Rhodes astonished all who gazed upon its burnished glory - but in 226 BC a powerful earthquake finally brought this lofty titan to its knees, literally. During the fateful earthly upheaval, the Colossus toppled over, snapping at its knees and crashing to the ground. Here, its broken but still-incredible remains would lie for almost 900 years (an oracle had forbidden its restoration), but would continue to incite wonder from all who beheld them. Countless spectators came to stare in amazement inside the huge caverns exposed at the junctions where its limbs had broken off from its torso, revealing its internal packing of stone and iron.

In 654 AD, however, Rhodes was conquered by invading Arabs, who carried away the fallen Colossus to the Middle East, ultimately selling its broken remains to a Jewish merchant from Emesa as scrap metal. An ignoble end indeed for a sun god simulacrum.

Today, not a single fragment remains of this ancient wonder. Hope flickered briefly in 1987, when a curious object resembling a giant fist, carved from stone and measuring 6 ft across, was found 700 yards out to sea, in Rhodes's harbour. Could this be internal packing from the torch-bearing fist of the Colossus? Sceptics denied any link with the famous statue, as the 'fist' was stone, not bronze. Conversely, Dr Susan Walker from the British Museum revealed that the Colossus's bronze was merely sheet metal, over a framework of iron bars set into stone blocks. Even so, the 'fist' was later identified as part of a large amount of rock dumped there by a mechanical crane in 1985.

Yet even though the physical reality of the Colossus is no more, its memory - and its name - lives on, perpetuating for all time a veritable wonder of wonders, whose like we shall certainly never see again, even in this modern world of marvels and miracles.

TALOS – THE FIRST ROBOT?

Interestingly, long before the Colossus of Rhodes had been created, its subject, the Greek sun god Helios, had already become associated with a bronze giant, but this one was of a much more animate nature. In the dialect of Crete, Helios became Talos - and according to Cretan legends incorporated into classical Greek mythology, Talos was a gigantic living man cast entirely from bronze (or brass in some versions) by the fire god Hephaestus. Talos contained a single internal vein running from his neck to his feet, and was sealed at one ankle by a huge bronze nail. This vein was filled with ichor, a magical substance present only within the very blood of the gods themselves, thereby rendering Talos immortal.

After Zeus had seduced the maiden Europa while assuming the form of a great bull, he carried her off on his back across the sea to the island of Crete. When they arrived there, he placed Talos on guard, to ensure that no-one abducted her, and Talos thereafter ran around the island three times every day to keep a constant watch for anyone who may try to rescue her, hurling huge boulders out to sea at any approaching ship. Eventually, Europa became Queen of Crete, but Talos remained, ever vigilant. According to a different version of the Talos legend, he was given by Hephaestus to Minos, King of Crete, as a gift, but once again he guarded the island by running around its perimeter three times a day.

One of the creepiest scenes in any fantasy film appears in 'Jason and the Argonauts' – a thoroughly enjoyable, albeit decidedly Hollywoodised, treatment of the epic Greek legend, filled with classic stop-motion special effects created by the master of screen monsters, Ray Harryhausen. The scene in question is when Jason and his fellow Argonauts aboard the Argo reach the island of Talos, who seems to be nothing more than a giant bronze statue in crouching position, on top of a massive chamber packed from floor to ceiling with untold treasures.

Ignoring Jason's strict instructions not to take anything from the island, two of the Argonauts, Hercules and Hylas, plunder the chamber, but as they re-emerge and look up at the enormous 'statue' on top of it, to their horror it suddenly turns its head and looks down at them! And as they watch, terror-stricken, Talos swiftly comes totally alive and clambers down from the chamber, poised to step on them like tiny ants as they flee before him, racing back to their ship to alert the other Argonauts of the monster that their greed has unleashed upon them all. Happily, Talos is rendered immobile once more, when, guided by the voice of the goddess Hera, Jason successfully prises out of his heel the cork that retains his body's vital ichor, which gushes out from his vein, bringing Talos's immortality to an abrupt end.

A descendant of Talos
with the author

In the original Jason myth, conversely, Talos himself removes the ichor-retaining nail from his ankle after being bewitched by the sorceress-priestess Medea (who was accompanying Jason and the Argonauts back home after they had seized the Golden Fleece at Colchis), and thereby brings about his own death.

Unlike the Colossus of Rhodes, which unquestionably once existed, there is no evidence whatsoever to suggest that Talos was ever anything more substantial than a figure of legend. Nevertheless, as an animate metallic humanoid entity in the annals of world mythology, he may well lay claim to being the world's first robot – were it not of course for the fact that in his murderous pursuit of any hapless visitors to his island domain, he clearly paid scant regard to Isaac Asimov's celebrated Three Laws of Robotics!

DOCTOR MIRABILIS AND THE SPEAKING HEAD OF BRASS

One of the greatest yet most paradoxical English intellectuals of the Middle Ages was Roger Bacon (1214-1292), a Franciscan monk whose eclectic expertise combined philosophy and science with alchemy and magic. Little wonder that he became known as Doctor Mirabilis, to whom all manner of astounding feats were attributed - so many, and so diverse, in fact, that it is often difficult when perusing his life's history to distinguish reality from legend. One such claim relating to this extraordinary 'miracle worker' concerns the Brazen Head.

It is said that Bacon harnessed his vast knowledge of magic to create a human head cast from brass and endowed with the gift of speech, which would advise him on how to construct a mighty wall of brass to encircle the entire island of Great Britain, thereby repelling any future invaders from Europe. The task proved so arduous, however, that when the Head was complete, Bacon fell into a deep sleep, leaving a servant to pay heed in case his creation spoke. While Bacon slept, the Head did speak a number of words, but tragically for Bacon (and for Great Britain!), his servant did not consider them very important and hence did not rouse his master - until, in rage, the Head exploded in a flash of fire. Bacon awoke at once, and when he saw what had happened he furiously chastised his servant, but it was too late - which may explain why Britain is not encircled by a wall of brass today!

Several other legendary man-made speaking heads are also on file, some of which may even have been real. The following excellent summary of these long-forgotten marvels is from a truly fascinating yet nowadays largely-unknown antiquarian repository of obscure but enthralling facts entitled *The World of Wonders: A Record of Things Wonderful in Nature, Science, and Art* (1882), edited by A. Taffs:

> The speaking heads of the ancients were constructed for the purpose of representing the gods, or of uttering oracular responses. The speaking head of Orpheus, at Lesbos, is one of the most famous, and had the credit of predicting, in the equivocal language of the heathen oracles, the bloody death which terminated the expedition of Cyrus the Great into Scythia. Odin, who imported into Scandinavia the magical arts of the East, possessed a speaking head, said to be that of the sage Minos, which uttered responses. The celebrated mechanic, Gerbert, who filled the papal chair as Sylvester II (A.D. 1000) constructed a speaking head of brass. Albertus Magnus is said to have executed a head in the thirteenth century, which not only moved, but spoke. It was made of earthenware; and Thomas Aquinas is said to have been so terrified when he saw it, that he broke it in pieces, upon which the mechanist exclaimed, "There goes the labour of thirty years!" In these cases it is probable that the sound was conveyed by pipes from a person in another apartment to the mouth of the figure. Lucian, indeed, expressly informs us that the impostor Alexander made his statue of Esculapius [=Aesculapius, the Greek god of medicine] speak by the transmission of a voice from behind, through the gullet of a crane, to the mouth of the figure; and this method was probably general, for we read that in the twelfth century, when Bishop Theophilus broke to pieces the statues at Alexandria, he found some which were hollow, and which were so placed against a wall that the priest could conceal himself behind them, and address the ignorant spectators through their mouths.

In short, ventriloquism on a grand scale!

ST PETER'S STONE COCKEREL – THE STATUE THAT CAME TO LIFE

Many religious marvels and mysteries have been reported down through the centuries, but few can be stranger than St Peter's stone cockerel.

Carved from stone, this remarkable bird formed part of a statue representing St Peter repentant, in the Caiaphas chapel at Sacro Monte, in Piedmont, Italy, and, as one might expect of a statue, it maintained a rigidly immobile stance - until one fateful day in 1653. Confronted by an inebriated itinerant who had staggered aimlessly into the chapel, the cockerel abruptly 'came alive' - flapping its petrologically-plumed wings, and turning about on its pedestal, before just as suddenly regaining its characteristic, petrified pose.

If the drunk had been the only person to witness the cockerel's inexplicable bout of animation, it is unlikely that this Pygmalionesque incident would have left any trace within the historical records, for obvious reasons. However, several other persons present inside the chapel also observed the statue's un-anticipated outburst of activity.

Those who accept the authenticity of this event consider it to be a religious marvel, but might it not instead have been a marvel of the mechanical kind? As noted by Dr Joe Nickell in *Looking For a Miracle* (1993), in which he rigorously seeks plausible scientific solutions to many religious mysteries, animated icons that in reality comprised skilfully constructed automata were not unknown even as far back as biblical times. These were epitomised by the statue of an Egyptian goddess dating from the 1st Century AD that would pour out wine from a vase when a fire was placed upon the altar on which the statue stood.

Moreover, as again commented by Nickell, such devices as fake idols and automata "...may also have been used in churches even as late as the Renaissance", their purpose being to maintain the populace's belief in miracles and demonstrate the power of religion. This might explain reports dating from that period regarding talking and moving statues - possibly even wing-flapping, turnabout cockerel statues too?

Indeed, as confirmed in John Cohen's fascinating book *Human Robots in Myth and Science* (1966), there was no shortage of ingenious and often intricately complex mechanical figures in existence long before the prodigiously automated age in which we now live. One pertinent example is a mechanical peacock constructed by General de Gennes in 1688 that would walk and even eat. Another is the astonishing automated duck invented less than a century later by the master toy-maker Jacques de Vaucanson. This robotic marvel not only could eat, but accurately mimicked every gesture made by a real duck when eating, and also digested its food internally before defecating a pellet from its mechanical anus. It could drink, quack, and paddle too, just like its living counterparts. Compared to this, a cockerel statue that merely flaps its wings and turns around is hardly a radical concept.

In January 2006, I visited Egypt to see for myself not only its many world-famous but also some of its less well-known ancient wonders that I had read so much about – from the Pyramids and Great Sphinx of Giza, the Temples at Karnak and Luxor, the resplendent long-entombed relics of Tutankhamen and other treasures on display in Cairo's Egyptian Museum, and the Necropolis of Thebes on the West Bank, to the cryptozoological Narmer Palette, Memnon's singing Colossus, and the head of Ozymandias.

THE NARMER PALETTE PALAVER

Surrounded by spectacular sarcophagi, mummies, and other necrological relics of every conceivable size, age, and nature, the last thing that I expected to encounter during my visit to the Egyptian Museum in Cairo was an artefact of cryptozoological controversy. However, while walking around Gallery 43 on the museum's ground floor, this is precisely what happened. Suddenly, I found myself in front of a large glass case containing a greyish-green, shield-shaped exhibit, and as I looked in surprise at the pair of bizarre beasts carved upon one side of it I realised that I was looking at the extraordinary Narmer Palette - one of Egypt's oldest, and most enigmatic, historical objects.

Composed of dark schist, measuring 25 in high and 16.5 in wide, and richly adorned on both sides with elaborate, finely-wrought carvings, this remarkable artefact was discovered during 1898 by archaeologist James E. Quibell in the Upper Egyptian city of Nekhen (nowadays Hierakonpolis) while excavating the royal residences of various ancient Egyptian rulers. Despite dating back to c.3200 BC (the Old Kingdom), the palette has survived intact, and was a votive (gift) offered up by King Narmer to the sun god Amun-Ra. What makes this artefact so significant historically is that it not only bears some of the earliest-known examples of Egyptian hieroglyphics but also commemorates a major event in ancient Egyp-

Both sides of the Narmer Palette

tian history - the unification of Lower Egypt and Upper Egypt into a single land, with King Narmer as the first ruler of both lands.

On one side of the palette, King Narmer is vividly portrayed as ruler of Upper Egypt smiting his Lower Egypt enemy, and facing his own incarnation as the falcon deity Horus, god of the sky. On the other side, there are various depictions celebrating Narmer's triumph after capturing the crown of Lower Egypt, thereby unifying Upper and Lower Egypt. The largest, most striking image on this side, however, does not feature Narmer at all. Instead, it portrays a pair of inordinately long-necked creatures whose flexible necks entwine around one another, forming a border round a central circular reservoir that some researchers believe may have been used to hold perfume, or to serve as a receptacle within which such cosmetics were manufactured in situ.

These extraordinary beasts are generally referred to as serpopards (though in at least one reference source they are termed mafedets), for good reason. For whereas their necks are decidedly serpentine in appearance, their heads are very leopard-like. As for their bodies: I have seen them likened variously to panthers, lions, and even baboons. After having finally witnessed the palette at first-hand (prior to then, I knew of its images only from various internet pictures of varying quality), I agree that their bodies, long limbs, and lengthy tails certainly possess a degree of simian similarity, more than I had previously realised when simply viewing pictures of them. But what were these serpopards meant to be - wholly symbolic, a purely legendary beast, perhaps a very distorted portrayal of some known animal, or something more than any of these options?

The reason why I was already familiar with the Narmer Palette is that in the past it has attracted a degree of cryptozoological speculation that the serpopards may conceivably represent a stylised or alternatively a distorted depiction of some mokele-mbembe-type species of surviving long-necked dinosaur that was alive at least at the time of King Narmer. Of course, this would not be the first time that ancient Middle Eastern records have inspired theories of historical dinosaur survival - the mushush or sirrush carved on King Nebuchadnezzar II's resplendent Ishtar Gate of Babylon (which I was privileged to view at first-hand when visiting the Vorderasiatisches Museum in 1983 in what was then East Berlin), the huge monster Behemoth referred to in the Holy Bible, and a reptilian mystery beast in a short book of the Apocrypha entitled 'Bel and the Dragon' all readily come to mind. Could the serpopard be another putative neo-dinosaur?

Much as I would like to admit it to this select crypto-company, after viewing the palette's serpopards up close and personal I was left in no doubt whatsoever that these necking entities were unquestionably mammalian, not even remotely reptilian. The serpopard head, complete with ears, is indeed leopard-like, not leonine as some have suggested, but the toes of the feet, posture of the body and limbs, as well as the limbs' relative lengths, and the shape and carriage of the tail all struck me as rather more monkey-like than feline. As for the palette pair's disproportionately long, impossibly flexible necks, it seems likely that they were intertwined not only to symbolise the union of Upper and Lower Egypt (as well as the eastern and western heavens?), but also for practical purposes - to fit neatly around the palette's central reservoir. Interestingly, each of the two depicted serpopards is held on a leash by a handler, who may be a slave, or a tribute, indicating perhaps that the serpopards were a gift to King Narmer, or possibly even domesticated?

Significantly, depicted serpopards are not restricted to the Narmer Palette. Another early Hierakonpolis palette, known as the Oxford or Two Dogs Palette and retained at Oxford University's Ashmolean Museum, also bears a pair of these striking creatures on one side, plus a single one on the other side. The paired serpopards on this artefact have even longer necks than those on the Narmer Palette, but this time they are not entwined - instead, they are held in a painful-looking zig-zag pose above their bodies, one on each side of a central reservoir. On the Two Dogs Palette (named, incidentally, after the two superficially canine - but quite possibly hyaenid - beasts comprising the upper section and outer sides of the palette, though the head of one is missing), the necks of the serpopards are striped, and there are stripes on their foreparts too.

Another serpopard-depicting palette is the Four Dogs Palette held at the Louvre, Paris, and there is also a preserved cylinder seal from Susiana, the high country of the ancient Persian civilisation of Elam, that depicts a series of very long-tailed neck-entwined serpopards. Clearly, therefore, the serpopards of the Narmer Palette were clearly not just an invention of its sculptor, devised merely as a decorative motif for bordering and highlighting the palette's central reservoir and/or as a symbol of King Narmer's unified Egypt.

A more conservative identity than a mokele-mbembe but no less intriguing is that perhaps the serpopards were poor representations of a giraffe (a species that once existed in Egypt), possibly based upon indirect descriptions of what this exceptional creature looked like rather than personal observations. If this were so, however, surely the giraffe's long legs would have been mentioned and described to the sculptor, not just its long neck. Yet although the serpopards' legs are fairly long, they are far shorter than one would expect for a giraffe, whereas their necks are much too long. In any case, it just so happens that there is absolute iconographical proof readily to hand to confirm that the serpopard and giraffe are totally discrete animals.

On the reverse side of the Two Dogs Palette, a wide range of creatures is depicted, including readily-identifiable lions, antelopes, goats, a hartebeest- or gnu-like ungulate - and not only a serpopard but also a clearly-recognisable giraffe, the latter beast complete with long inflexible erect neck, small horns as well as ears on its head, long giraffe-like legs, hoofed feet, and downward-pointing tail. Just above it, offering a perfect opportunity for direct comparison, is a flexible-necked, hornless, leopard-headed, shorter-legged, toe-footed, upward-tailed serpopard - indisputably a wholly different animal.

Equally worthy of note is that alongside portrayals of real animals on the Two Dogs Palette is not just a serpopard but a winged griffin too, depicted in traditional composite form with leonine body, eagle's head, and feathered wings. This provides immediate proof that ancient Egyptian sculptors carved real and fabled animals together, so the presence elsewhere of serpopards depicted alongside people and real animals cannot be taken as firm evidence that the serpopards themselves must also be real.

Accordingly, the most reasonable solution to the mystery of its identity is that the serpopard is nothing more than another composite (albeit exotic-looking) mythical beast, just like the griffin, as well as certain other Egyptian monsters such as the hippo-bodied crocodile-headed ammut, and the venomous winged snakes that reputedly swarmed across ancient Egypt each year like locusts (indeed, perhaps they were directly inspired by actual locust swarms). After all, not all beasts of legend are creatures of crypto-

The Colossi of Memnon; the right-hand Colossus is the former singing version *(Dr Karl Shuker)*

zoology in disguise - from centaurs and minotaurs to cactus cats and gooseberry wives, the inventive human imagination is more than sufficiently capable of summoning forth from its uncharted depths a veritable menagerie of wholly original monsters surpassing even the wildest excesses of Mother Nature.

MEMNON'S SINGING STATUE

I first documented the statue that sang in 1996, within my book *The Unexplained*, little realising then that exactly a decade later I would be standing quite literally at its feet, and at those of its fellow monolith. Once there, however, I realised just how apt was its official name, or at least part of it. Looming like petrified giants amid the ruins of Thebes on Egypt's West Bank, just across the Nile from Luxor, these two enormous edifices are known as the Colossi of Memnon, and as each stands 62 ft high, they are indeed colossal. Their connection – or lack of it – with Memnon, however, is another matter entirely.

Hewn from quartzite sandstone, the Colossi are gigantic representations of Pharaoh Amenhotep (=Amenophis) III, seated on his throne, and were erected in approximately 1500 BC. Nowadays battered by time, much-damaged, and faceless but still largely intact, originally they stood guard at the entrance to Amenhotep's memorial (mortuary) temple, of which only scattered ruins remain today. In 27 BC, the eastern Colossus was itself partially destroyed when a sizeable earthquake sent its upper half crashing to the ground, but this catastrophe was somewhat ameliorated by the statue thereafter exhibiting a highly unexpected talent for any statue – it could sing!

As first reported in 20 BC by eminent historian Strabo after visiting Thebes, each morning at daybreak this wrecked Colossus emitted a weird bell-like sound, soft yet perfectly audible. Others came to hear it, and confirmed Strabo's statement. Moreover, the Greeks subsequently claimed that the sound was the Achilles-slain Trojan War hero Memnon crying out each sunrise to his mother, Eos, the Greek goddess of the dawn (and, as a result, the twin statues thereafter became known, albeit wholly inaccurately, as the Colossi of Memnon). The belief that it brought good luck to those who heard it, and possessed oracular powers, ensured a steady stream of visitors to view – and listen to - this monolithic marvel. Even Oscar Wilde commemorated it in verse, the following lines appearing in his poem 'The Sphinx':

The once-vocal Colossus of Memnon *(Dr Karl Shuker)*

Still from his chair of porphyry gaunt Memnon strains his lidless eyes
Across the empty land, and cries each yellow morning unto thee.

Tragically, however, the song of this sirenesque statue was permanently, albeit inadvertently, stilled in 199 AD, when the Roman emperor Septimus Severus supervised the Colossus's long-awaited repair in a bid to win the oracle's favour. The fallen, shattered upper half was replaced with a newly-carved version, which did much to restore the statue's former majesty, but somehow rendered it mute – as it has never again been heard giving voice to its eerie, morningtide refrain.

As I noted in *The Unexplained*, several solutions to the riddle of the Colossus's erstwhile vocalisations have been offered:

These include: noises made by the escape of expanding sunrise-warmed air from inside exposed cracks in the statue; unequal expansion of various sections of the statue's lower half warmed by the sun, causing them to rub audibly against one another; conversion of former ultrasonic emissions into audible emissions by the earthquake's damaging effects; and even fake sounds created surreptitiously by Egyptian priests to gain prestige or money (or both).

Sadly, however, it is most unlikely that a conclusive answer will ever be found; and, just as it has now been for almost two millennia, the Colossus remains resolutely silent about the whole affair.

BRING ME THE HEAD OF OZYMANDIAS!

Percy Bysshe Shelley's celebrated poem 'Ozymandias' (quoted in full at the beginning of this chapter) is

The surviving head and shoulder of the
Ozymandias Colossus
(Steve F.E. Cameron/Wikipedia)

a succinct but extraordinarily powerful evocation of the folly of human vanity, and the stark image of an immense wrecked statue encompassed by an empty wasteland of desert sand as conjured forth by Shelley's strangely compelling lines has fascinated me ever since I first read them many years ago.

Only recently, however, during my visit to Egypt's West Bank, did I discover to my surprise and delight that Ozymandias, and his giant stone head, were much more than mere figments of a poet's imagination.

Ruling for 66 years and 2 months during the 13th Century BC, when ancient Egypt had attained the very zenith of its power and splendour, Pharaoh Ramesses II was possibly the most celebrated, and wealthiest, of all of this great country's ancient rulers, and he used his vast riches to create an unparalleled array of monuments glorifying his existence. None was more spectacular, however, than his memorial (mortuary) temple, located in Thebes on the West Bank, across the River Nile from present-day Luxor. It was originally known as "The House of Millions of Years of Usermaatra-Setepenra That Unites With Thebes-The-City in the Domain of Amon", but following his visit to its ruins in 1829, French archaeologist Jean-François Champollion coined a rather more concise yet no less fitting name for it – the Ramesseum.

Yet despite being dubbed the House of Millions of Years, in reality this architectural wonder lasted for a far shorter span of time, on account of its location. Tragically, it had been erected at the very edge of the Nile floodplain, and so its foundations were inexorably undermined by this mighty river's annual inundation, as well as by neglect and desecration in later ages. Eventually, together with the gargantuan statues that it contained, Ramesses II's magnificent temple collapsed, and in time became largely buried, with only scattered ruins to remind a modern world of its long-bygone glory.

Among its now-fallen statues was an immense stone giant dubbed the Ozymandias Colossus, which weighed over 900 tons and formerly stood as tall as Memnon's twin counterparts documented earlier in this chapter. Today, however, only its massive head, hands, and feet remain. These shattered relics, the pathetic remnants of Ramesses II's most grandiose attempt at self-glorification, are what inspired Shelley's famous sonnet, though the "two vast and trunkless legs of stone" were his own invention, as no such relics exist here today.

As for the memorable name, Ozymandias, this was actually a mis-transliteration from Egyptian into Greek - by 1st-Century-BC writer Diodorus of Sicily - of the first part of Ramesses II's throne name, Usermaatra Setepenra. This appears as a cartouche on the statue's shoulder, and can still be seen today as shown in the photograph here.

QUETZALCOATL'S SIMULACRUM AND CHIRPING PYRAMID

God of the wind, wisdom, and life, Quetzalcoatl was a major deity in Aztec mythology, and was represented as a magnificent feathered serpent. He had an exact counterpart in Mayan legends too, called Kukulkan, and at Chichén Itzá in Mexico there is an 1100-year-old Mayan pyramid dedicated to Kukulkan - and possibly even visited by him? Every year at the spring and autumn equinoxes, the pyramid's shadow moves along as the sun sets, and creates a striking illusion of a feathered serpent - an apt simulacrum invariably attracting many sightseers. Nor is that the only enigma featuring this pyramid.

If a person stands before it, at the base of its staircase, and claps their hands, the pyramid will emit a distinctive chirping echo - but not just any chirp. In 2002, Californian acoustical engineer David Lubman and a team of Mexican researchers revealed that it constituted a precise phonic replica of the call of the quetzal *Pharomachrus mocinno*, the sacred Mayan bird (*National Geographic Today*, 6 December 2002). Some researchers, such as Ghent University mechanical construction specialist Dr Nico F. De-

The stegosaurian glyph at Ta Prohm temple,
Angkor Wat, Cambodia
(John and Lesley Burke)

The Cambodian 'stegosaur' glyph seen in situ
with other glyphs at Ta Prohm

clercq, have since questioned whether this acoustical anomaly was intentional on the part of the pyramid's Mayan designers and architects (*Journal of the Acoustical Society of America*, December 2004), but if not it is surely a formidable coincidence.

A STEGOSAUR IN CAMBODIA?

Cryptozoological riddles can turn up in the most unlikely places, but few can be as unexpected as Cambodia's perplexing dinosaur carving. One of this country's most beautiful monuments is the jungle temple of Ta Prohm, created around 800 years ago, and part of the Angkor Wat temple complex. Like others from this time, it is intricately adorned with images from Buddhist and Hindu mythology, but it also has one truly exceptional glyph unique to itself. Near one of the temple's entrances is a circular glyph containing the carving of a burly, small-headed, quadrupedal beast bearing a row of diamond-shaped plates along its back - an image irresistibly reminiscent of a stegosaurian dinosaur!

This anachronistic animal carving is reputedly popular with local guides, who delight in baffling western tourists by asking them if they believe dinosaurs still existed as recently as 800 years ago and then showing this glyph to them. Could it therefore be a modern fake, skilfully carved amid the genuine glyphs by a trickster hoping to fool unsuspecting tourists? Or is it a bona fide 800-year-old artefact?

If so, perhaps it was inspired by the temple's architects having seen some fossilised dinosaur remains? After all, it surely couldn't have been based upon a sighting of a real-life stegosaur...could it?

THE BOOYA STONES

Prior to the coming of the Europeans, the priests of the Murray Islands, sited in the Torres Strait between New Guinea and

Queensland, were the keepers of a trio of mysterious rocks called the booya stones, which emitted an intense blue light. When any one of these stones was placed in a special holding device, this light was concentrated into a laser-like beam, and when aimed directly at anyone who had found disfavour, an x-ray effect was observed, followed invariably by the victim's death. Once the Europeans had reached the Murray Islands, however, their priests hid the booya stones in secret caves, and their location is now unknown. Australian historian Ion Idriess speculated that they were lumps of pure radium.

XANADU

It is widely but erroneously assumed that Xanadu is wholly fictitious, entirely confined to Samuel Taylor Coleridge's magnificent if incomplete poem, 'Kubla Khan', published in 1816, whose opening lines are: "In Xanadu did Kubla Khan a stately pleasure dome decree". In reality, however, Xanadu was Shang Tu, the Upper Capital and summer capital of Kublai Khan (1215-1294) - founder of the great Mongol (Yuan) dynasty - and located some 180 miles north of Beijing. Why Coleridge chose to delete the last letter of the real-life Emperor's name is unknown.

According to legend, and immortalised in Coleridge's poem, Kublai Khan built a vast garden of delights in Xanadu, with a hedonistic 'pleasure dome' at its heart. Here, drugged soldiers were lain, and were told upon awakening that they were in Paradise - so that they would happily fight to the death in the emperor's battles, in the tragically mistaken belief that they would then return to this dome of decadence forever. Whether such an edifice truly existed is a matter for conjecture, as much of Kublai Khan's life is shrouded in myth.

MYTHOLOGY AND MYSTERY OF THE GOLDEN FLEECE

The grim shadow of a raised sword fell across the pale, tensed body of Prince Phryxus as he lay upon the sacrificial slab, awaiting certain death at the hand of his own father, King Athamas of Orchomenus. This city in ancient Greece had once been rich and prosperous, but that year its crop of corn had failed, and its people were starving. Due to a false prophecy originating from Queen Ino, Phryxus's wicked stepmother who hated him, Athamas mistakenly believed that his city's crop failure was a curse inflicted by the gods and could only be lifted by the sacrifice to them of his beloved son, Phryxus. Happily, however, Ino's evil plan was known to the gods who thwarted it in spectacular fashion.

Suddenly, as Athamas's sword was poised to strike the fatal blow that would slay Phryxus, a golden stream of light poured forth from the clouds directly overhead. The king and the gathered congregation of spectators stared up in surprise, and as they did so the clouds parted and a magnificent winged ram with a shimmering fleece of pure gold appeared. The ram was Chrysomallus, sent by the messenger god Hermes to rescue the boy prince. As soon as Chrysomallus alighted on the ground, Phryxus ran to him and sat astride his broad woolly back. So too did Phryxus's young sister, Helle, whom Ino also hated.

Moments later, like a brilliant golden meteor, Chrysomallus was soaring speedily through the sky, journeying eastwards. Just as they were flying over the narrow stretch of sea dividing Europe and Asia, however, Helle looked down, and became so giddy that she fell off Chrysomallus's back, plummeting into the waves where she drowned. From then on, this sea was called the Hellespont. Chrysomallus, meanwhile, flew onwards with Phryxus until they reached the land of Colchis, in what is today the republic of Georgia. There, Phryxus was welcomed by Aeëtes, king of Colchis, after which Chrysomallus was sacrificed by Phryxus (how ungrateful!) to the sea god Poseidon, and his glorious golden fleece was hung in a sacred grove, guarded by a dragon. Many years later, the fleece would be the focus of an epic quest by a Greek prince called Jason and his bold crew, the Argonauts, sailing from Iolcus to Colchis via the Black

Sea aboard their famous ship the Argo.

Most people assume that this entire story is nothing more than a fanciful Greek myth, dating back 3000 years but with no basis in fact. Quite apart from the ostensible fairytale aspect of a winged golden-fleeced ram, there was no firm evidence to suggest that Greek ships could even have reached the Black Sea prior to the 7th Century BC, when they are known to have colonised this region. However, not everyone has dismissed the legend quite so readily.

In May 1984, a classics scholar called Tim Severin and a team of modern-day Argonauts set out aboard their own Argo, sailing from Volos (site of Iolcus) in Thessaly, Greece, to Vani (site of Colchis) in western Georgia. Their goal was to recreate Jason's alleged voyage - and thus prove that such a journey could truly have taken place in those far-off days. And in order to achieve as intimate a degree of verisimilitude as possible, Severin's specially-designed Argo was patterned on ancient Aegean vessels by naval architect Colin Mudie. It was then built by Greek shipwright Vasilis Delimitros, who crafted it from the same Aleppo pine used by Bronze Age Greek seafarers. When complete, it measured 54 ft in length.

Spanning 1500 nautical miles from the Aegean Sea through the Dardanelles (Hellespont), the Sea of Marmara, the Bosporus, and thence along the Black Sea's Turkish perimeter to his Georgian destination at its eastern limit, Severin's voyage in this new Argo took three months. It was often arduous, but it was also successful - thereby indicating that Jason's epic journey was indeed possible even back in those bygone times of ancient Greece. Moreover, as revealed in his fascinating book *The Jason Voyage* (1985), Severin also learned some interesting items of information relevant to the Golden Fleece.

This legend is very popular in Georgia even today - where it is taught in schools, and commemorated in product names, such as a 'Golden Fleece' brand of cigarettes. Also, western Georgia once harboured a thriving cult of ram worshippers that lasted from the middle Bronze Age into the modern era - as confirmed by such archaeological finds as a bronze ram's head totem dating from the 18th Century BC, and ram's head bracelets moulded from gold that date from the 4th Century BC. There is even an old Georgian folktale of a golden ram tethered by a golden chain in a mountain cave filled with golden treasure.

Severin also learned that the traditional method of prospecting for gold in Georgia, a method dating back countless centuries, is to anchor in a gold-rich stream bed the fleece of a sheep - because the fleece's wool is very effective at trapping particles of gold. After it has been left in the stream for a time, the result will be a fleece impregnated with gold - in other words, a golden fleece! True, it will hardly compare to its magnificent legendary counterpart, but it may have been sufficient to inspire such a legend long ago.

The possibility that this activity is the origin of the Golden Fleece account is certainly intriguing, but it is not a recent revelation. As far back in time as the 1st Century BC, the Greek geographer/historian Strabo had made the same suggestion. Similarly, Dr George Hartwig noted in *The Subterranean World* (1875) that modern-day gold prospectors still use sheep fleeces for this purpose in several different gold-bearing countries. Nevertheless, although this theory is undoubtedly compelling in its simplicity, it is not the only explanation on offer for the origin of the Golden Fleece legend.

In a paper published by the English scientific journal *Nature* on 13 April 1973, Dr M. L. Ryder and Dr J. W. Hedges from the ARC Animal Breeding Research Organisation at Edinburgh, Scotland, noted that the legend of the Golden Fleece may actually refer to fine wool - i.e. very valuable, lustrous wool composed of extremely fine fibres (generally obtained from Merino sheep nowadays). Moreover, in their paper they documented a sample of cloth composed of fine wool that had been obtained from a Scythian tomb in the Crimea, and which dated back to the 5th Century BC. If their reconciliation of the Golden Fleece legend with fine wool is correct, this Scythian sample is thus of particular significance. Its age and Crimean locality collectively confirm that fine wool was indeed associated with the Black Sea region, and at a time near to that of the Golden Fleece's appearance and Jason's quest.

An even more thought-provoking explanation for the Golden Fleece legend was proposed in an account by Dr G.J. Smith (*Nature*, 23 June 1987. A researcher in physical chemistry at Melbourne University, Smith recalled the Golden Fleece saga's early portion, featuring the famine suffered by the people of Orchomenus. He speculated that if this had indeed occurred, the city's farmers may have resorted to feeding their hungry sheep with leaves from the olive tree *Olea europaea*, which was widely cultivated in ancient Greece. Such an action would have great significance in relation to the legend of the Golden Fleece, because olive leaves contain considerable amounts of oleanolic acid - the precursor of a group of substances called pentacyclic triterpenoids. As Smith noted, when these substances (contained in the leaves of certain shrubs and other plants) are ingested by sheep, they damage the animals' liver in a manner inducing secretion of the bile pigment bilirubin and certain related pigments through their skin and into their fleece's wool - staining it a conspicuous yellow-gold colour!

Consequently, as outlined by Smith, if ingested in unusually large quantities (as might occur if sheep are fed principally upon olive tree leaves during a period of famine), oleanolic acid could conceivably create a veritable herd of golden-fleeced sheep. Such a bizarre sight might well be sufficient to initiate a folk legend of golden sheep - one that would ultimately be elaborated and exaggerated during numerous successive retellings over the years and the centuries, until the stirring saga of the Golden Fleece was ultimately born.

In a short letter published by *Nature* on 5 November 1987, Drs Patrick Moyna and Horacio Heinzen from the Faculty of Chemistry at the Universidad de la Republica in Montevideo, Uruguay, noted that oleanolic acid is also found in high concentrations within the epicuticular wax of grapes, but pointed out that this does not cause any liver damage to humans consuming bunches of grapes. They did not, however, provide any evidence to disprove Smith's theory regarding oleanolic acid's possible damage to the liver of sheep.

The concept of liver-damaged sheep with discoloured yellowish wool stained by bile pigments certainly fails to conjure forth the romantic image evoked by the stirring legend of the Golden Fleece, but the prosaic practicality of science is rarely able to match the imaginative wonder of illusory fable and fairytale.

A final scientific curiosity linking sheep and gold that is well worth mentioning here is the occurrence of reports from many parts of the world over the years concerning sheep that supposedly possess teeth plated with gold! These still appear spasmodically in the media even today, yet as far back as 25 August 1920 in a paper published by the *Proceedings of the Linnean Society of New South Wales*, Thomas Steel revealed that the golden colour was due merely to the reflection of light from the overlapping of thin films of encrusting tartar, deposited on the teeth by the animals' own saliva. As for the tartar - far from being gold, or even iron pyrites ('fool's gold'), this had been conclusively shown to be nothing more exciting (or valuable!) than impure calcium phosphate and organic matter.

It was Shakespeare who said: "All that glisters is not gold", and judging from the subjects examined here, he may well have been correct!

ZILPHION

Known to the Greeks and the Romans, who enjoyed eating its stem and young shoots, and whose sap and roots were condensed into a special syrup called laserpitium, zilphion (aka xilphion) was once a familiar, popular plant, and was even portrayed in various Didrachmai coins from classical Cyrene. Yet today, it is seemingly long extinct, and certainly long forgotten - why? According to Pliny the Elder, it grew wild in North Africa and earned great riches for those merchants shipping it across the Mediterranean to Europe. Yet even as Pliny was writing about it, zilphion was apparently becoming ever rarer. Three centuries later, it was totally unknown. Attempts to identify it with species familiar to modern-day botany, such as *Ferula asafoetida* and *Thapsia garganica*, have proven woefully unconvincing, leaving us to ponder anew over a few ancient descriptions and depictions - all that survives of the lost zilphion.

THE TOWER OF BABEL

One of the Bible's most intriguing mysteries is the Tower of Babel, described as follows in Genesis, chapter 11:

> And the whole earth was of one language, and of one speech. And it came to pass, as they [the descendants of Noah] journeyed from the east, that they found a plain in the land of Shinar [in present-day Iraq]; and they dwelt there. And they said one to another...let us build us a city and a tower, whose top may reach unto heaven; and let us make us a name, lest we be scattered abroad upon the face of the whole earth. And the Lord came down to see the city and the tower, which the children of men builded. And the Lord said, Behold, the people is one, and they have all one language; and this they begin to do: and now nothing will be restrained from them, which they have imagined to do. Go to, let us go down, and there confound their language, that they may not understand one another's speech. So the Lord scattered them abroad from thence upon the face of all the earth: and they left off to build the city. Therefore is the name of it called Babel; because the Lord did there confound the language of all the earth.

Thus was the evolution of separate languages explained in the Bible. But what of the Tower of Babel? It is often assumed that this spectacular edifice was wholly fictitious. However, as reiterated during a conference in September 1998 between Iraqi and Western scholars held in Baghdad, the general consensus is that its story is closely associated with a real monument. Namely, the magnificent ziggurat (pyramidal, many-tiered temple tower) dedicated to Marduk, a Mesopotamian sun god, that once stood in ancient Babylon - which has been identified as Babel by biblical scholars.

Towards the end of the 1800s, German archaeologist Professor Robert Koldewey began a series of extensive excavations at the site of what had once been Babylon, capital of the ancient Mesopotamian kingdom of Babylonia. He succeeded in unearthing Babylon's spectacular, long-buried Ishtar Gate, as well as the massive square base of Marduk's colossal ziggurat - also known as the ziggurat of Etemenanki ('House of the Foundations of Heaven and Earth').

Revealing the stupendous size of this erstwhile edifice, each of the four sides of its base measured 300 ft, thereby greatly exceeding the dimensions of all other ziggurats presently known.

However, the original Tower of Babel was probably not Marduk's ziggurat, because archaeologists believe that several monuments were successively erected and razed on this same site *before* the creation of this latter edifice. Indeed, the original tower may have been razed here as long ago as the 18th Century BC, during the reign of Hammurabi, the first king of a united Babylon.

As for Marduk's ziggurat, it appears to be this site's most recent monument - built and dedicated to Marduk by Babylon's King Nebuchadnezzar II (reigned 605-562 BC), after its immediate predecessor here had been destroyed over a century earlier by the Assyrians, as led by Sargon II, Sennacherib, and Assurbanipal.

Moreover, it was Nebuchadnezzar and his father, Nabopolassar (625-605 BC), who sought to restore it to

The Tower of Babel, as illustrated by Athanasius Kircher in 1679

its predecessors' former glory. Various ancient cuneiform inscriptions have been uncovered that describe these Babylonian rulers' reconstruction of what they hoped would compare favourably with the original Tower of Babel, Nabopolassar supervising the rebuilding of its foundations and lower tiers, and Nebuchadnezzar taking charge of the summit's creation.

Tragically, however, the sumptuous ziggurat that emerged from the rubble of its former incarnation did not last long. By the end of the 6th Century BC, Babylon had been conquered by Persia, and in 479 BC its Persian king, Xerxes, began razing the earlier Babylonian rulers' monument to Marduk.

Fortunately, its destruction did not happen overnight - indeed, more than two decades later, enough of the ziggurat remained for the Greek historian Herodotus, visiting it in c.458 BC, to be able to pen a detailed description.

According to Herodotus, it possessed eight towers, i.e. seven terraces and the summit, with a spiral path running upwards along the outside. At its summit stood:

> ...a great temple with a fine large couch in it, richly covered, and a golden table beside it. The shrine contains no image and no one spends the night there except (if we may believe the Chaldeans who are the priests of Baal) one Assyrian woman, all alone, whoever it may be that the god has chosen. The Chaldeans also say - though I do not believe them - that the god enters the temple in person and takes his rest upon the bed.

Each tier was a different colour, including gold, silver, blue, yellow, white, and black. The ziggurat's total height was just under 300 ft, and 58 million bricks were used in its construction.

The ziggurat of Marduk must have been a breathtaking sight, which is why, presumably, the Persian Xerxes yearned for its destruction - in order to wipe from sight an outstanding architectural example of the Babylonians' former superiority. For a time, however, it seemed that this grandiose monument may once more rise like a phoenix from the desert sands, because in 331 BC Alexander the Great was so impressed by the descriptions still existing of Marduk's ziggurat that he was eager to rebuild it. Few tasks had ever daunted this Macedonian leader, but when faced with the grim fact that 10,000 men working for 2 months were required merely to clear the land upon which the resurrected ziggurat would stand, even Alexander the Great ultimately conceded defeat. But what of its earliest predecessor - what did the biblical Tower of Babel itself look like? The sad truth is that we are unlikely ever to know. As noted by Turin University archaeologist Dr Giorgio Gullino, speaking at the earlier-mentioned conference in Baghdad: "We can imagine what the tower looked like. We can do it in our computers. But I am afraid we will be far from archaeological fact".

THE PORCELAIN TOWER OF NANKING

Most people can name at least some of the Seven Wonders of the Ancient World. Far less familiar, conversely, is the list that designates the Seven Wonders of the Middle Ages. Chronologically speaking, its medieval time-scale is interpreted very loosely, bearing in mind that among those monuments included in this list are Stonehenge, the Colosseum of Rome, and the Great Wall of China. Even so, the term 'wonder' is undeniably appropriate, not only to the three already named here, but also to their four fellow marvels. Namely, the Mosque of St Sophia in Constantinople (now Istanbul), the Catacombs of Alexandria, the Leaning Tower of Pisa, and - perhaps most incredible of all - the now-forgotten but incomparably exquisite Porcelain Tower of Nanking (=Nanjing).

Beginning in the year 1413 AD, this ethereal edifice's creation was commissioned by the Emperor Yung-Loh, as a peerless monument to the memory of his late mother. As one bygone chronicler eloquently wrote:

> He determined that its beauty should as far outshine that of any similar memorial, as the transcendent virtues of the parent, in her son's eyes, surpassed those of the rest of her sex.

Almost 20 years, and a fabulous sum

The Porcelain Tower of Nanking

of money (the equivalent of over a million pounds today), later, the emperor's Porcelain Tower was complete, and its magnificence surpassed even his own grandiose expectations. Standing more than 200 ft high and consisting of nine storeys topped by a lofty spire, it was faced from base to apex with the finest glazed, coloured porcelain. But that was not all.

At the summit of the tower's spire was a richly gilt sphere of brass, from which, like the tentacles of a gleaming metallic octopus, eight long iron chains extended to eight projecting points upon the roof - and from each chain a small bell was suspended, which hung over the tower's face. The same format was duplicated on each storey, yielding a uniquely delicate, fragile aspect contrasting pleasingly with the more formal upright outline of the tower itself. This fairytale effect was heightened further by the presence of lanterns in several specially-carved apertures round each storey's outer face. Quoting from the description penned by a long-demised Chinese chronicler, when these lanterns were lit:

> ...their light illuminated the entire heavens, shining into the hearts of men, and eternally removing human misery!

And as an act of veneration to the deities of Heaven, as well as a charm for warding away evil, two large brass vessels and a bowl were placed on top of the tower, and filled with priceless articles of countless kinds. Glittering precious gems, multicoloured pearls reputedly empowered with miraculous properties, quantities of gold and silver, and even a selection of silken wares, copies of ancient Chinese writings, and a box of the finest tea - all were present. The spectacular result was a radiant beacon of porcelain, emitting a dulcet paean of joy whenever the gentle breezes stirred its bright company of bells, and proffering a myriad of treasures beyond estimation in decorous obeisance to the deities in exchange for their divine benevolence. How richly indeed did this dream-like monument deserve its worldwide renown as an architectural wonder.

Tragically, however, like all dreams and fairytales, it could not last forever. Four centuries passed by without incident, but in March 1853 the city of Nanking was captured by the Taiping Rebellion - a major uprising against the Qing dynasty. Even so, such was its spectacular, awe-inspiring beauty that for three years the Porcelain Tower quelled even the rebels' destructive impulses, as they single-mindedly annihilated everything else appertaining to Nanking's imperial heritage. By 1856, however, their anti-Imperialist fervour had at last focused its impassioned attention upon the final symbol of this city's historic past - the Porcelain Tower. Having been created by an emperor, the tower was doomed. And so it was that this glorious triumph of inspirational human achievement was demolished - felled like a mighty oak tree, shattering its porcelain visage into an infinity of glinting fragments, stilling forever the tinkling laughter of its bells, and strewing its venerable offerings far and wide like cheap, discarded trinkets.

The victims of War have been many and multifarious, and one was the Porcelain Tower of Nanking. Truly a thing of beauty and a joy, if not forever, then certainly for several centuries - and perhaps even longer too, if we are able to recall and recapture its peaceful, timeless memory for a brief moment amid this hectic modern world.

Chapter 17

MEN WITH WINGS,
AND OTHER STRANGE THINGS

Inexplicabilia In Flight

For something is amiss or out of place
When mice with wings can wear a human face.

Theodore Roethke – *'The Bat'*

Many cryptozoologists are willing to countenance the existence of giant vulture-like birds frequently sighted in the skies over North America, and even in the reality of tantalisingly pterodactylian creatures reported from America, Africa, and even New Guinea. True, these may well seem controversial to orthodox zoologists, but they are creatures with which the cryptozoological community can feel comfortable, because they do at least correspond encouragingly with previously recorded (albeit supposedly extinct) animal species.

In contrast, there are certain other 'things with wings' from which even the most open-minded cryptozoologist prefers to avert his eyes - and for good reason.

Eyewitness accounts of bizarre flying entities resembling humans but with wings sprouting from their shoulders, or headless winged wonders with eyes set in their shoulders, or wolf-headed, bird-footed, sky-soaring sasquatches are ones that cannot even begin to be slotted into any mainstream zoological classification system. Indeed, these beings are assuredly from worlds, dimensions, or realities far-distant from our own – and may even be zooform entities.

WING OF BAT AND LEG OF FROG?

When it comes to winged encounters of the weird kind, the U.S.A. has probably experienced more than its fair share, but few have ever been weirder than the following example. On 12 September 1880, readers of the *New York Times* were startled by this amazing news item:

> One day last week, a marvelous apparition was seen near Coney Island. At the height of at least 1000 feet in the air a strange object was in the act of flying toward the New Jersey coast. It was apparently a man with bat's wings and improved frog's legs. The face of the man could be distinctly seen and it wore a cruel and determined expression. The movements made by the object closely resembled those of a frog in the act of swimming with his hind legs and flying with his front legs...When we add that this monster waved his wings in answer to the whistle of a locomotive and was of a deep black color, the alarming nature of the apparition can be imagined. The object was seen by many reputable persons and they all agree that it was a man engaged in flying toward New Jersey.

To my mind, this report's faintly jaunty wording does not encourage faith in its veracity, especially as none of its "many reputable persons" put their reputations on the line by coming forth to confirm it. Having said that, three years earlier, on 18 September 1877, a Mr W. Smith had allegedly seen "a winged human form" soaring over Brooklyn, and sent details in a letter to the *New York Sun*.

At 2.30 pm on 18 July 1953, yet another veritable 'bat man' paid what else but a flying visit to the United States, when it made an impromptu landing in the pecan tree of Mrs Hilda Walker, from Houston, Texas. She was standing with a teenage girl when this airborne apparition first came into view, and as it flew closer both women and two other observers too were clearly able to discern that its large bat-like wings sprouted directly from its back. Except for these aerial appendages, however, their uninvited visitor was fundamentally humanoid in form, and was wearing dark, tight-fitting clothes, but stood about 6.5 ft tall, and seemed to be emanating a glowing halo-like aura. After residing in the tree for about half a minute, he abruptly disappeared, and immediately afterwards an unidentified rocket-shaped object was seen, and heard, rising up over the rooftops and off into the sky.

SOME D.I.Y. FLYING MEN

Not all mystery flying men reported from the U.S.A. have been intrinsically winged - but that has not prevented them from becoming airborne. Take, for instance, the figure unexpectedly encountered on 6 January 1948 by Bernice Zaikowski, 61, in her own back yard at Chehalis, Washington. Walking outside to find out what was attracting so much attention from the children playing there, she heard a strange sizzling, whizzing sound, and to her amazement she saw what appeared to be a man, hovering upright over the barn roof!

Unlike those documented earlier in this chapter, however, this particu-

Flying men of the future from a 19th-Century sci-fi story

lar flying figure did indeed seem to be a man (as opposed to a man-like entity), but one who owed his unwonted aerial agility to the pair of long silver artificial wings fastened with straps across his shoulders, and whose movements were mediated by a panel of controls strapped to his chest. After treating his observers to a spectacular display of aerobatics, this anonymous (and anomalous) exhibitionist flew away, passing out of sight and into the history books. Mrs Zaikowski later approached army officials with her sighting, but they were thoroughly at a loss to explain it. Indeed, their only helpful(?) comment was: "It sounds like one of those saucer deals".

On 29 July 1880, a Kentucky newspaper called the *Louisville Courier-Journal* published an extraordinary account that could have been directly extracted from a Jules Verne novel - except for the fact that the people featured in it emphatically insisted that every word was true. The eyewitnesses in question, both local, were Ben Flexner and C.A. Youngman, who claimed that during the previous evening they had watched a man "... surrounded by machinery" fly through the sky. He appeared to be manipulating the machinery with his hands, and was borne aloft by fins or wing-like structures protruding from his back. Whenever he made these fins beat faster he ascended, and whenever they began to slow down he descended. Eventually, he flew out of sight, lost amid the thickening veil of twilight.

A DIY flying man from a French novel of 1879

The necessity of growing one's own wings for effective if inexplicable flight was also dispensed with by the trio of mystifying flyers wearing "dark drab flying suits", reported on 9 April 1948 by laundry worker Viola Johnson and janitor James Pittman in Longview, Washington. As they watched, the three figures circled the city at a height of around 250 ft, moving at about the same speed as a freight train, with unidentified apparatus attached to their sides, which may have been the source of the motor-like sound that accompanied their aerial circuit. As they drew closer, their two terrestrial observers could see that they were wearing helmets and were moving their heads around, as if monitoring their surroundings. As with the airborne visitor viewed by fellow Washington resident Mrs Zaikowski, the three flyers spied by Johnson and Pittman held themselves vertically when aloft.

Back on 14 April 1897, more than a hundred eyewitnesses had caught sight of a huge humanoid entity as it flew over Illinois's Mount Vernon. It was first spotted at approximately 8.30 pm, and remained clearly visible in the sky for the next half hour. Among its more eminent eyewitnesses was local dignitary Mayor B.C. Wells, who was fortunate enough to spy the bizarre being from his home's ensuite observatory. He stated that the humanoid moved as if swimming through the air, and bore what he described as "an electric light" on its back. Oddly, when this particular case was reported, no mention was made of wings, so perhaps its 'electric light' was some form of electrically-powered propulsion unit?

IS IT A BIRD? IS IT A PLANE? NO, IT'S BATSQUATCH!

Despite their wings, in other respects the humanoids recorded above did at least bear some resemblance to authentic humans. In contrast, no-one could ever accuse the entity spied by Brian Canfield of making even a token nod in the direction of such (relative) normality.

Batsquatch, illustrated by `Stigmata` (aka Gaz Stanley)

It was 9.30 am on 19 April 1993, and Canfield, an 18-year-old high school student, was driving a truck from Buckley to his home at Camp One in the foothills of Mount Rainier, above Lake Kapowsin, in Washington (what is it about this state that seems to attract such a diversity of winged wonders?). Suddenly, for no reason, his truck stopped, its lights died - and there, about 30 ft ahead in the road, was a creature of nightmare!

Standing at least 9 ft tall on its hind legs, covered in bluish fur, and remaining perfectly still, the horror in question had a wolf-like face, yellow eyes with half-moon pupils, a large mouth brimming with sharp white teeth (but no fangs), tufted ears, bird-like feet, a pair of powerful arms, and also a pair of huge wings, which began to flap as the monstrous being gazed at its unbelieving trucker eyewitness. After several minutes of this terrifying face-off, Canfield's grotesque visitor unfolded its wings, which were so enormous that they spanned the entire width of the road, and slowly took off into the air, stirring up considerable turbulence before soaring away towards Mount Rainier. Once it had gone, Canfield tried his truck's ignition again, and this time it fired without hesitation, enabling him to drive home.

Later that day, Canfield returned to the location of his incredible encounter, still greatly shaken but accompanied by his parents and a neighbour, seeking evidence of the entity's former presence that could substantiate his story, but nothing was found. Nevertheless, his abstemious lifestyle and reputation for honesty were such that no-one who knew him doubted that he had been telling the truth. Even columnist C.R. Roberts, who interviewed him for the *Tacoma News Tribune*, was convinced of his sincerity. And so it was that a decidedly novel newcomer was added to the chronicles of alien apparitions reported from Planet Earth, one whose surrealistic similarity to a bigfoot (sasquatch) incongruously sprouting the wings of a bat duly earned for it a highly apt, and memorable, media appellation - batsquatch.

MOTHMAN - POINT PLEASANT'S VERY UNPLEASANT VISITOR

Mothman,
illustrated by William Rebsamen

What I always say is that you know where you are with a monster, regardless of its shape or size, as long as it has a head - which is why mothman was such an unnerving critter.

On the evening of 15 November 1966, a car containing four people was travelling by an abandoned wartime explosives factory in Point Pleasant, West Virginia, when the car's occupants saw a strange man-shaped figure shuffling towards them out of the darkness. Standing 6-7 ft tall, it was grey in colour, with a large pair of folded wings. It also had a big pair of blazing red eyes, which seemed to be set in its shoulders - because it had no head!

The driver, Roger Scarberry, turned the car away from this spine-chilling entity and drove them away at high speed, but even as he did so, the same (or a second?) creature was seen standing by the road. As the car sped past, the creature opened its huge wings, rose up into the air, and flew after them, soaring overhead without flapping its wings yet effortlessly pacing the car and its panic-stricken occupants as they accelerated to 100 mph!

Throughout their terrifying journey, the people could clearly hear their winged pursuer emitting high-pitched squeaks. When they reached a sheriff's office, they reported the matter to him, but by then the creature had gone. During the next 13 months, many other reports of this eerie apparition were recorded from the Point Pleasant area, closely corresponding with the version given by Scarberry and his three companions, and the creature became referred to as mothman.

By the end of 1967, however, the sightings had ceased - mothman had apparently gone. Worth noting is that back in 1961, a very similar creature in the Point Pleasant area had allegedly frightened a car driver and her passenger - by standing in the middle of the road, opening its immense pair of wings (spanning 10 ft) as the car approached, then taking off vertically into the air, disappearing from sight almost instantly.

There is even the extraordinary possibility that during its reign of 1960s terror, mothman made at least one transatlantic flyover. During the evening of 16 November 1963, four teenagers were strolling down a country road at Sandling Park, near Hythe in Kent, when they saw a reddish-yellow light descend from the sky and sink behind a group of trees. As it vanished from sight, a bright golden light spanning 15-20 ft appeared in a nearby field, and floated 10 ft above the ground before it, too, disappeared behind some trees. Then, as the teenagers continued to watch, something dark and shambling came out from those trees, and began shuffling towards them. The teenagers lost no time in leaving their macabre visitor far behind - and little wonder, bearing in mind that it was as tall as a man, sported a pair of bat-like wings hanging down at its sides, plus a pair of webbed feet, but had no head.

CHICK-CHARNEY - THE BAHAMIAN BIRD-MAN

During 1984-1985, merchant seaman Curt Rowlett was working on Andros, the largest yet least-inhabited island in the Bahamas group, and while there he learnt of a peculiar leprechaun-like 'bird-man' said to inhabit the impenetrable mangrove swamps and vast pine forests in this island's uninhabited interior.

Known as the chick-charney (aka chickcharnie), it is described by the locals as half-human, half-bird, with a midget-sized humanoid body but the face of a featherless bird, with a pair of huge round eyes, and a long beak instead of a nose. Its feet each have three toes, it also has a tail, and is supposedly able to turn its head all the way round. According to native lore, the chick-charney lives in specially-created nests, which it constructs by bending over the tops of several pine trees and binding them together to produce a type of canopy. During his explorations of Andros, Rowlett actually discovered one such 'nest', which, as he could clearly see, had been formed by the tops of the pine trees intertwining and growing together. Rowlett wondered whether these peculiar configurations may have some natural origin, but he was unable to conceive how they could have become so intimately wound together. Moreover, as the 'nest' was in such a far-flung, rarely-visited location, and had been found purely by chance, there seemed no reason to suspect that it was a hoax.

Several centuries ago, Andros and other Bahamian islands were home to a large, 2-3-ft-tall species of flightless owl, *Tyto pollens*, the great barn owl, which is known from subfossil remains to have co-existed here with humans after Europeans and their slaves colonised the Bahamas during the 16th Century. It is possible, therefore, that modern-day native belief in the chick-charney is based at least in part upon distorted, exaggerated, distant memories of this very distinctive terrestrial bird. Like other barn owls, *T. pollens* could rotate its head very dramatically in either direction, and may have exhibited memorable territorial defence behaviour. It became extinct, however, after the islands' human colonisers began felling great swathes of its forest habitat's old-growth Caribbean pines.

BATTLING THE MAN-BATS OF RAYMONDVILLE AND LACROSSE

Sitting in his mother-in-law's backyard at Raymondville, Texas, on the evening of 14 January 1976, Armando Grimaldo suddenly heard a strange whistling, and a sound that reminded him of flapping bats' wings - a highly pertinent comparison, as it turned out. For just a few seconds later, he was attacked by a man-sized monstrosity with the face of a bat or monkey, a pair of large flaming eyes but no beak, dark, leathery, unfeathered skin, and a pair of huge wings yielding a massive 10-12 ft wingspan (i.e. twice that of any known species of bat). Swooping down at the terrified man, the creature snatched at him with its big claws, but, happily, Grimaldo was able to flee inside before his aerial attacker had inflicted any serious injuries. Nevertheless, his encounter was just one of several on file from this particular region of Texas during early 1976, all documenting sightings of a similar entity.

Another 'man-bat', 6-7-ft tall with a leathery 10-12-ft wingspan, clawed hands and feet, yellow eyes, well-delineated ears, and revealing a huge snarling mouth brimming with teeth, almost flew into the windscreen of a 53-year-old man, Wohali, as he was driving his truck (with his 25-year-old son as passenger) along a road near LaCrosse, Wisconsin, just after 9 pm on the evening of 26 September 2006. The creature then flew back up into the sky and vanished, but the shock of encountering it was such that both men became physically sick. They were convinced that it was a physical, tangible entity. This remarkable case has been documented by Linda Godfrey at http://www.cnb-scene.com/manbat.html

LaCrosse man-bat, illustrated by Richard Svenssen

Even more extraordinary is that a remarkably similar creature has been reported from the U.K., and is duly documented below for the first time in any book.

INTRODUCING BRITAIN'S BAT-WINGED MONKEY BIRD

The following winged wonder only became known to me in mid-October 2007, when Jan Patience, acting editor of the erstwhile British monthly magazine *Beyond* for which I contributed a major cryptozoology article each issue, brought to my attention a truly extraordinary email that she had just received from a reader. At that time, I was preparing a lead article on lesser-known British mystery beasts for the next issue of the magazine, so the email reached me in time for me to investigate it further and include a full account of the case in my article (*Beyond*, January 2008), and it is this account of mine that I shall now quote from here.

The email in question had been sent by Jacki Hartley of Tunbridge Wells, Kent, concerning a truly bizarre beast that she claims to have encountered on three totally separate occasions, and which she reported as follows:

> Back in 1969 I was 4 years old and travelling back from my auntie's house in London. Dad had been driving for about half an hour and we were going through the countryside. I was in the back of the car when I suddenly heard an awful, screeching scream. Mum and dad were in the front chatting and heard nothing. It was twilight and as I looked out of the back window into the trees, I saw what I could only describe as a monster.

> It had bat wings which it unfolded and stretched out before folding back up again, red eyes and a kind of monster monkey face with a parrot's beak and was about 3 feet in height.

> To my 4 year old mind it was terrifying and I had nightmares for weeks. I did not have a name for this thing in my vocabulary and so called it The Bat Winged Monkey Bird as it seemed to be such a weird mixture of animals. I saw it again late one night when I was eleven from the back window of the car on our way home from Hastings, I think we were travelling through Robertsbridge and I saw it for the third time last year.

> It was 4.30 in the morning and I was woken by the same horrible screeching sound. Thinking someone was being murdered in the street I jumped out of bed and ran to the window, catching the tail end of it as it flew past. I knew immediately what it was. . .the same horrible monster thing I had seen all those years ago, The Bat Winged Monkey Bird was back.

> I would be very grateful if you or any of your readers could shed any light on or put a name to this thing.

When Jan forwarded Jacki's email to me, I was naturally extremely intrigued, and replied directly to it, requesting any further information that Jacki could send. In response, I received this second email, together with two accompanying pictures, one of which is Jacki's own representation of what she saw, and the other a photo of her taken at the age when she experienced her first sighting of the monkey bird:

> My most recent sighting was at 4.30am on 19-10-06 outside my bedroom window, I saw the tail end of it fly past after the awful screeching noise woke me up, my address is in Tunbridge Wells in Kent.

> The time before that I was on my way back from Hastings in Sussex, having just passed through Robertsbridge. While the first time I saw it, we were coming back from London, heading toward Kent and had been travelling for about 20-30 minutes. The creature was sitting in a tree in a field off to my right.

> As for my parents, they just laughed and said I must have fallen asleep and had a nightmare; they weren't interested and simply dismissed it, even though I had nightmares about it coming to get me.

> Although the creature looked solid flesh and blood and made an awful sound, I think it could be paranormal in origin as I have never seen or read about anything that even vaguely resembles it, but if it is real, I would love to know if anyone else has seen it.

There is no doubt that this grotesque entity as described by Jacki does not correspond even vaguely with any known species of animal native to Britain. And even when venturing beyond the known into the realms of cryptozoology and zooform phenomena, there is little to compare with it, especially from the British Isles.

Perhaps the nearest is the Cornish owlman, a weird owl-human composite spied in the vicinity of Mawnan Church Cornwall, mostly during the 1970s, and discussed later here. Remarkably, however, as I discussed in my *Beyond* article, Jacki's creature is certainly reminiscent of the Raymondville man-bat(s), for which there has never been a satisfactorily explanation.

Jacki Hartley's sketch of the bat-winged monkey bird, and a photograph taken of her at the age when she first saw it

What is also very strange about Jacki's report is that she saw the creature in three very different geographical localities and over a considerable period of time, yet without any other eyewitnesses apparently having seen it – or have they? This is where you, gentle readers, come in.

Like Jacki, I would love to know if anyone out there has indeed seen this extraordinary creature, or something like it. So if you have, please get in touch with any details that you can provide. I now declare the case of the bat-winged monkey bird well and truly open for investigation!

ORANG BATI - THE 'FLYING MAN' OF SERAM

Another extraordinary 'man-bat' brought to public attention by me is one that was first documented in my book, *The Unexplained* (1996). Indigenous to the Indonesian island of Seram (Ceram) in the Moluccas group, and referred to by the native people there as the orang bati ('flying man'), it first became known to outsiders in 1986. This was when tropical agriculturalist Tyson Hughes spent several months working in Seram and amassed a great amount of data regarding the island's fauna - known and unknown.

According to Seram's coastal villagers, who live in terror of it, the orang bati has dark reddish skin, and is humanoid in form, standing 4-5 ft tall, but it possesses a pair of large black bat-like wings, plus a long thin tail. It emits a prolonged, mournful wail, and lives in large numbers within caves dotted along extinct volcanoes at the island's centre, which is encompassed by dense, scarcely-explored rainforests. Active at dusk, flocks of orang batis sometimes fly out across the island to raid coastal villages such as Uraur and reputedly abduct human babies and young children, carrying them away to their cave lairs, where they are presumably devoured as they are never seen again. Worth noting is that Seram does not have any monkey species, which may explain the orang bati's fondness for human infants?

It has been claimed that the orang bati are merely a normal human tribe with only professed flying powers. However, the testimony that Hughes gathered from villagers during his stay on Seram convinced him that their claims of a monstrous winged race were genuine.

VIETNAMESE BAT-WOMAN

Even more bizarre is the entity soberly spied one very early morning in August 1969 by Earl Morrison and two fellow American marines, while in Vietnam during the war. Stationed near Da Nang in South

Vietnam, they were on guard duty when, as Morrison later recalled:

> All of a sudden - I don't know why - we all three looked out there in the sky and we saw this figure coming toward us. It had a kind of glow and we couldn't make out what it was at first. It started coming toward us, real slowly. All of a sudden we saw what looked like wings, like a bat's, only it was gigantic compared to what a regular bat would be. After it got close enough so we could see what it was, it looked like a woman. A naked woman. She was black. Her skin was black, her body was black, the wings were black, everything was black. But it glowed. It glowed in the night - kind of a greenish cast to it. We saw her arms towards the wings, and they looked like regular molded arms, each with a hand, and fingers, and everything, but they had skin from the wings going over them. And when she flapped her wings, there was no noise at first. It looked like her arms didn't have any bones in them because they were limber just like a bat.

Before Morrison and his colleagues had a chance to make any further observations, however, their mystifying visitor took to the air and flew off overhead, just a few feet above them. Naturally, one could easily dismiss this strange tale as a hallucination caused by wartime trauma, but, if so, three separate soldiers all suffered precisely the same hallucination at exactly the same time

BIRDMAN...OR TENGU?

Stranger still was the entity encountered in or around 2002 by a Malaysian oil palm worker called Zainal. At that time, he was employed on one of the country's oil palm plantations recording the number of oil palm fruit bunches gathered each day by the plantation's harvesters, counting the number of bags of loose fruit collected by the pickers, and checking that no loose fruit had been left uncollected on the ground beneath the palm trees. One day, while checking for loose fruit in an old palm field containing very tall, mature trees, Zainal saw what he at first took to be a large white bird swooping from tree to tree. It reminded him of an eagle, but there is no species fitting that description in Malaysia, so when it perched on a tree Zainal moved towards it slowly to get a better look at it – and was stunned to discover that this 'eagle' had a human face!

Snow-white in colour, just like the creature's plumage, but totally bare, it had deepset human eyes, a hooked nose, thin mouth, and pointed chin. After the creature flew off, a shocked Zainal reported it to the plantation's staff. They in turn contacted a Western overseer, who investigated this incredible account further, and was very startled to discover that Zainal's birdman bore an amazingly close resemblance to a supernatural Japanese entity known as a tengu – half-man, half-bird, often seen in forests and usually the bringer of bad luck to those who encounter it. But that is just ancient folklore...isn't it? Presumably, therefore, it was merely a coincidence that shortly after seeing his birdman, Zainal met with a freak accident that almost cost him his sight in one eye.

LETAYUSCHIY CHELOVEK - THE 'FLYING MAN' OF FAR-EASTERN RUSSIA

The immense, forbidding taiga forests of the Primorskiy Kray Territory (Russian Far East) lays claim to the letayuschiy chelovek ('flying man'). One was briefly seen several years ago by hunter A.I. Kurentsov, when it flew over his fire on webbed, bat-like wings.

On 11 July 1908, Russian author V.K. Arsenyev's dog was tracking some unseen creature leaving humanoid footprints alongside a mist-enshrouded river in the Sikhote Mountains, near Vladivostok. Suddenly, Arsenyev heard the beating of giant wings, and saw something large and dark emerge from the mist and fly over the river. Whatever it was, it emitted a bloodcurdling cry that resembled a woman's scream but ended in a lugubrious howl. When he told the local Udeyan people, they affirmed that it had been one of these flying men.

Perhaps the most incredible report of all, however, came from Petropavlovsk, during the early 1990s, and was brought to public attention by Russian ufologist Alexander Rempel. On the tenth day after moving into a new house in this remote eastern Russian locality, the Ivanitzky family was horrified to discover a bizarre chirping beast under the bed. Attempts to dislodge this intruder by throwing slippers at it only incited it to grow larger, swelling up until it was almost three times its original size. Then, to their even greater alarm, it abruptly shot forth a long tentacle-like trunk from its nose, with which it tried to snare their legs!

Thoroughly panic-stricken by now, the Ivanitzkys sprayed the entity with household chemicals, and soon afterwards it rolled over to a far corner and lay there, apparently dead. When they cautiously examined it, they saw that in general body form it resembled a dog, but it had short bluish fur, two three-fingered paws, and a flattish humanoid face with very large eyes, a tiny lipless mouth, and a triangular hole instead of a true nose. Most striking of all, however, were its strong bat-like wings, which yielded a 4.5-ft span.

Frightened that they may have killed some exotic, State-protected species, they disposed of the body in a nearby ditch; but when they looked there a short while later, it had gone.

CHILE'S LOCUST-HEADED SNAKE-BIRD

In April 1868, a truly unique 'thing with wings' was spied flying over a mine in Copiapó, Chile, and three months later this incredible encounter was documented as follows in *The Zoologist*, a popular English journal, by one of its bemused eyewitnesses:

> ...at about five o'clock in the afternoon, when the daily labours in this mine were over, and all the workmen were together awaiting their supper, we saw coming through the air, from the side of the ternera, a gigantic bird, which at first we took for one of the clouds then partially darkening the atmosphere, supposing it to have been separated from the rest by the wind. Its course was from north-west to south-east; its flight rapid and in a straight line. As it was passing a short distance above our heads we could mark the strange formation of its body. Its immense wings were clothed with a grayish plumage, its monstrous head was like that of a locust, its eyes were wide open and shone like burning coals; it seemed to be covered with something resembling the thick and stout bristles of a boar, while on the body, elongated like that of a serpent, we could only see brilliant scales, which clashed together with a metallic sound as the strange animal turned its body in its flight.

Reading that description, I'm more tempted to identify this airborne anomaly as some bizarre form of flying machine, as indicated by the metallic nature of its scales and the fiery appearance of its 'eyes', rather than a living creature. However, any flying machine equipped with the head of a locust, not to mention feathery wings, would be as difficult to explain away in mechanical terms as a comparable corporeal beast would be in zoological terms!

CORNISH OWLMAN

Last, but by no means least, in this survey of phenomenal flyers is the Cornish owlman, whose bewildering history has been extensively researched by Fortean investigator Jonathan Downes in several editions of his book *The Owlman and Others*. On 17 April 1976, while holidaying with their parents near Truro in Cornwall, June Melling, aged 12, and her sister Vicky, 9, were greatly frightened to see a huge, feathered "bird-man", hovering over the tower of Mawnan Old Church. Their story was made public by Tony 'Doc' Shiels, self-styled surrealchemist and Wizard of the Western World, whose travelling theatre group was performing locally at the time. Shiels claimed that he had been blamed for the owlman's appearance by the girls' father, who had given him a sketch of this entity, drawn by June, which depicted a humanoid with pointed ears, a black beak, black pointed feet, and curved feathery wings instead of arms.

On 4 July 1976, two other girls, Sally Chapman and Barbara Perry (both aged 14), allegedly informed Shiels that they had seen the owlman on the previous day, standing among some trees near Mawnan Church before abruptly flying up and disappearing in their branches. Shiels asked the girls to sketch it, but whereas Barbara's picture recalled the quasi-human drawn by June Melling, Sally's was more reminiscent of a giant owl. Further sightings were reported from around Mawnan during the next two years. Curiously, however, all attempts by investigators to trace the Melling family and these other two girls failed.

Inevitably, therefore, as there did not seem to be any evidence for the owlman's reality that was undeniably independent of Shiels, rumours arose that it may be a pure invention. In 1995, however, a then-undergraduate science student publicly identified only as 'Gavin' (but whose true identity is known to me), who was completely unknown to Shiels, revealed to Jonathan Downes that during a visit to Mawnan woods as a 12-year-old in the late 1980s, he and his 13-year-old girlfriend had spied the owlman at close range.

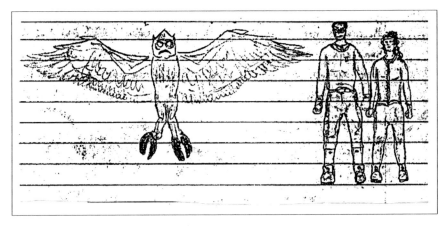

Sketch by 'Gavin' of himself, his girlfriend, and the owlman *(CFZ)*

Standing with wings raised, on a thick branch of a large conifer tree, it was grey and brown in colour, roughly 4 ft tall, with large glowing eyes, and two huge black toes or claws on each foot. When it saw them, the owlman's head jerked down and forwards, its wings lifted, and it jumped backwards, folding its legs. 'Gavin' was very unnerved by the sighting, and has always insisted upon public anonymity to protect his scientific credibility. Moreover, he avows that it was definitely not an owl, not even one as sizeable as the European eagle owl *Bubo bubo*, several escapee specimens of which have been recorded across Britain over the years (more recently, it has even bred in the wild).

SPECULATIONS AND SOLUTIONS - STILL UP IN THE AIR?

As we might expect, there has been no shortage of explanations and identities proffered over the years by these extraordinary airborne entities' many investigators.

Some have favoured cryptozoological solutions. Is it possible, for instance, that the orang bati and Vietnamese bat-woman comprise unknown species of giant bat - akin, perhaps, to the mysterious ahool long reported from western Java? American cryptozoologist Mark Hall has proposed that mothman might be an elusive species of enormous owl, pointing out that in the Pacific Northwest, the native American Indians and early Western settlers believed in the existence of a giant owl, which they called the big-hoot, great owl, or booger owl. Conversely, the more conservative 'official' explanation is that most mothman sightings probably featured a sandhill crane *Grus canadensis*, a very large bird that sports an extremely long neck – in stark contrast, of course, to the neck-less, and headless, mothman!

As John Keel revealed in *The Mothman Prophecies* (1975), however, at the time of this entity's appearances Point Pleasant was also experiencing a diverse spate of paranormal phenomena, including unexplained animal mutilations, UFO sightings, and visitations by sinister MIB-type figures, which indicates that something more than just a cryptozoological novelty was involved here. The same applies to the Cornish owlman. Another popular 'explanation' regularly aired by sceptics is mass hysteria or hallucination. Yet such a scenario is hardly tenable in relation to all of the cases discussed here. Equally, multiple-eyewitness sightings cannot be breezily discounted as hoaxes or misidentifications - 'solutions' favoured by the more cynically-minded.

Other notions that have been aired at one time or another include: visiting extraterrestrials; beings from other dimensions that have inadvertently slipped into our world through a 'window' or some such other interface; mind-generated entities akin to tulpas, but stimulated unconsciously by stress rather than deliberate intent; and elementals or nature spirits. Worthy of brief mention here is a claimed sighting in the Lake District of one such elemental, a deva, by Geoffrey Hodson, a Theosophist, who documented it in his book *Fairies at Work and at Play* (1925):

> My first impression was of a huge, brilliant crimson, bat-like thing, which fixed a pair of burning eyes upon me. The form was not concentrated into the true human shape, but was somehow spread out into the shape of a bat with a human face and eyes, and with wings outstretched over the mountainside. As soon as it felt itself to be observed, it flashed into its proper shape, as if to confront us, fixed its piercing eyes upon us, and then sank into the hillside and disappeared.

The simple truth seems to be that when dealing with such wholly alien entities as bat-men, bird-men, and their kind, we must accept that they are outside our world's normal reality (whatever that may be). Whether they hail from other dimensions, or planets in outer space, from alternate versions of this planet, or even from the inner space of our own minds, any attempt to categorise and label them as we would do with a new species of bird or butterfly is doomed to failure. The day that a mothman, owlman, or flying man peers out at us from an enclosure in London Zoo or any of our other wildlife parks will indeed be a long time coming - and perhaps that is no bad thing. There is much that we still do not understand in this life, and it can be dangerous to generalise, either in thought or deed.

Chapter 18

IT'S RAINING SPRATS AND FROGS!

Flotsam Flushed from the Super-Sargasso Sea

I accept that, when there are storms, the damndest of excluded, excommunicated things - things that are leprous to the faithful - are brought down--from the Super-Sargasso Sea - or from what for convenience we call the Super-Sargasso Sea - which by no means has been taken into full acceptance yet.

That things are brought down by storms, just as, from the depths of the sea things are brought up by storms. To be sure it is orthodoxy that storms have little, if any, effect below waves of the ocean - but - of course - only to have an opinion is to be ignorant of, or to disregard a contradiction, or something else that modifies an opinion out of distinguishability.

Charles Fort – *The Book of the Damned*

As far as I am aware, there has never been an authenticated case of cats and dogs raining down from the skies. However, there are countless reports of many other equally bizarre and unexpected items teeming forth upon an unsuspecting general public far below - everything from frogs and fishes, coins and crucifixes, star rot, angel hair, and exploding hailstones, to green rain, blue rain, electric rain, giant blocks of ice, cascades of stones, and even the (very) odd shower of knitting needles.

With his customary originality, anomalies archivist Charles Fort playfully postulated that perhaps, way above our own terrestrial abode, adrift in the upper atmospheric strata, there might be some form of aerial Super-Sargasso Sea (see this chapter's opening quote), where windborne items temporarily collect and separate into discrete type-specific aggregations before tumbling back down to earth, and where its stocks of items are being constantly renewed by the wind.

Sadly, no evidence has yet been obtained to support the existence of this aeolian sorting-house. Conversely, there is more than sufficient documentation on file regarding a wide variety of anomalous, mostly-unexplained sky-falls, as will be revealed here.

A METEOROLOGICAL MENAGERIE

If Fort's theory of an aerial Super-Sargasso Sea were to be believed, then it must contain a very substantial, constantly-replenished zoological assemblage, judging from the extraordinary diversity of animal life that at one time or another has dropped - and continues to drop - without warning from the heavens onto the startled heads of humanity beneath.

By far the most common examples on record are frog rains and fish rains, with cases dating back to the earliest times and new ones still being reported today. The *Deipnosophistae* is a historical anthology written in c.200 AD by the Greek historian Athenaeus, which contains the following pertinent passage:

> I also know that it has very often rained fishes. At all events Phenias, in the second book of his *Eresian Magistrates*, says that in the Chersonesus [peninsula] it once rained fish uninterruptedly for three days; and Phylarchus in his fourth book, says that people had often seen it raining fish, and often also raining wheat, and that the same thing had happened with respect to frogs. At all events, Heraclides Lembus, in the 21st book of his *History*, says: "In Paeonia and Dardania it has, they say, before now rained frogs; and so great has been the number of these frogs that the houses and the roads have been full of them; and at first, for some days, the inhabitants, endeavouring to kill them, and shutting up their houses, endured the pest; but when they did no good, but found that all their vessels were filled with them, and the frogs were found to be boiled up and roasted with everything they ate, and when besides all this, they could not make use of any water, nor put their feet on the ground for the heaps of frogs that were everywhere, and were annoyed also by the smell of those that died, they fled the country.

Frog rain, pictured on the front cover of an issue of *Fate (FPL)*

As the above report was not a personal eyewitness account, it may well be that the incidents described, which seem of near-biblical plague proportions, were grossly exaggerated. However, the same cannot be said of many more recent incidents that have been soberly recorded with statements from the eyewitnesses involved. In June 1979, for instance, Mrs Vida McWilliam of Bedford witnessed a heavy rainfall occurring throughout one Sunday, and the following day when she went outside she found numerous tiny frogs everywhere, with great numbers all over the lawn. Although she did not actually see the frogs falling from the skies, what she did find, which is of particular interest, was several of her garden's bushes liberally draped with batches of frogspawn. Needless to say, frogspawn is normally confined solely to ponds, so how could its unexpected presence hanging from bushes be explained if it had not dropped from above?

More recently, one Sunday in June 2005 the northwest Serbian town of Odzaci suffered a veritable frog deluge when thousands of tiny frogs apparently cascaded down upon it from the sky, even though there was no rain. Of especial note is that these frogs were different from those normally found in this area. Traffic was brought to a halt, and scared villagers fled inside their homes to escape from this bizarre downpour.

Nevertheless, a popular solution offered for the phenomenon of frog rain and other zoological precipitation is that the creatures did not actually drop earthward from the skies but were merely disturbed from their normal, concealed, terrestrial abodes by a heavy rain shower, thus appearing in numbers on the ground, and creating the impression that they must have fallen from above. However, I can state categorically that in at least certain cases, this proffered explanation is totally incorrect – thanks to a remarkable incident that happened to a member of my own family, as documented by me in my book *The Unexplained* (1996):

> In or around 1902, when she was then about eight years old, my grandmother Gertrude Timmins (née Griffin) was walking with her mother, Mary Griffin, across a field in what is now the town of West Bromwich, in the West Midlands. As they were walking it began to rain, so they opened their umbrellas, but a few moments later Gertrude felt a great number of quite heavy thumps on top of hers. When she peered out from beneath it, she saw to her amazement that the objects responsible for the thumps were small frogs, dropping down from above, hitting the top of her umbrella, bouncing off it, and falling to the ground around her feet. She became quite frightened, but her mother assured her that there was nothing to fear, informing her in a wholly matter-of-fact manner that it was merely a frog rain and that the frogs would stop falling soon – which they did☐This dramatic incident left such a vivid impression in my grandmother's mind that right up to her death in 1994, at the age of 99, she could still readily recall all of it.

Why this case is so significant is that the frogs had actually been cascading down on to the tops of the umbrellas, confirming that they had not merely emerged out of undergrowth on the ground when the rain began. Moreover, there were no buildings or trees nearby from which they could have dropped – always assuming that they could, or would, have found their way onto such structures anyway. In short, the only place where the frogs could have come from was the sky.

Such an explanation is even more valid in relation to falls of fishes and other totally aquatic species, as fishes clearly are not likely to have been lurking in holes or beneath bushes on land prior to a shower of rain. Over the years, many different species of fishes, freshwater and marine, have been identified in fish rains – from eels, perches, sardines, sprats, and gudgeons, to minnows, barbs, tench, smelts, and flounders, to name but a few.

Two gardens in Sheerwater, Surrey, experienced a downpour of sprats, a marine species, during the morning of 15 January 1993, which were found not only on the lawns but also on rose bushes and even on the roof of a shed. Yet the sea is almost 40 miles away and even the River Thames is more than 6 miles away. Dozens of living sardines, another marine species, fell into the streets of Ipswich, a town in Queensland, Australia,

The author's grandmother, Mrs Gertrude Timmins (and Patch), who witnessed a rain of frogs
(Dr Karl Shuker)

during a heavy rainstorm on 5 February 1989, even though Ipswich is 30 miles inland. And back on 29 May 1892, so many eels fell in a rain shower over Coalburg in Alabama that they were carried away in carts by enterprising farmers to be used as fertiliser. True, eels are known for being able to travel overland from one pond to another, but not in the kind of convoy required to explain away the quantities of eels featured in this remarkable incident.

Early engraving of a rain of fishes

Nor are animal falls restricted to frogs and fishes. Many other species have also been recorded, including snails and other molluscs, all manner of worms, starfishes, lizards, snakes, mice, crabs, insects, and even birds. The most popular theory for zoological falls is that the creatures in question have been lifted from their normal terrestrial or aquatic abodes by a whirlwind or waterspout, carried some distance through the air, and then dropped back down to earth. Initially, this seems quite a reasonable solution - until it is realised that these falls are almost always species-specific, i.e. a given animal fall normally contains only a single species.

Moreover, it does not usually contain any mud, vegetation, or any other substratum either - just the single animal species. Yet how could a waterspout or whirlwind be so extraordinarily selective, just lifting members of one species of fish or frog or snail, and leaving all other species in the same locality uncollected, and not even lifting up any substratum? It is this baffling specificity, this unexpected paucity of mixed showers, that poses such a problem when attempting to accept the vortex explanation for animal falls.

MANNA - FOOD FROM HEAVEN?

Manna is the name given to the miraculous 'bread' that God sent down from the sky in profuse quantities to sustain the Israelites during their 40-year journey through the wilderness to the Promised Land. But does manna really exist - and, if so, what can it be?

In fact, many contemporary accounts of manna exist, and it has been widely recorded in (though rarely beyond) Asia Minor, but appears somewhat variable in form, so that at least three biologically distinct types of manna are currently recognised.

The most familiar manna takes the form of small round bodies produced by lichens, most notably of the species *Lecanora esculenta* ('edible lichen'), which can be readily dislodged from their normal rocky substrata by gusts of wind and can then be transported aerially across great distances, thereby creating the illusion that they have fallen from the sky.

Early engraving depicting manna raining down from the sky *(FPL)*

A second well-known version may sound rather unpalatable but is prized as a delicacy by nomadic Bedouins in the Middle East. It comprises a sticky, sweet-tasting secretion excreted by cochineal-related insects when feeding upon the tamarisk tree *Tamarix mannifera* ('manna-producing tamarisk'). Once excreted, this insect-derived manna transforms into white, crystalline granules, recalling the Bible's description of manna as resembling coriander seed.

A third type of manna is the source of a crystalline carbohydrate termed manna sugar or mannite, and is exuded by the manna ash tree *Fraxinus ornus*, as well as certain seaweeds.

Yet none of these types of manna occurs in sufficient quantities, or contains sufficient nutrients, to sustain a large human population for 40 years. So unless the Israelites fed upon various other foodstuffs too, their survival upon manna alone remains a profound mystery.

ANGEL HAIR

Angel hair is the name given to one of the most enigmatic of all sky-falls - long, white, gossamer-like filaments that descend earthward often in vast quantities, cloaking meadows, streets, houses, or anything else that they land upon with their ethereal, silken strands. But what is angel hair - and where does it come from? Many eyewitnesses describe angel hair as resembling spider webs, and in most cases this is indeed what it probably is.

A fall of angel hair recorded from Ichinoseki City, Japan *(FPL)*

Very few reports of angel hair actually mention the presence of spiders amid the shroud-like sheets and threads drifting downwards or discovered festooning the ground, yet there is little doubt that this gauzy, filamentous material is merely an aggregation of threads produced by congregations of tiny money spiders in order to become airborne by a process known as ballooning.

Silken threads drawn out of their spinnerets when the spiders face a strong wind are lifted, together with the attached spiders, into the air by the wind and carried aloft, the spiders sometimes travelling great distances before finally gliding back to earth, where they simply abandon their threads, yielding spiderless, gossamer-like sheets called angel hair - as confirmed on several occasions by analysis of samples collected.

However, analysis has also shown that some angel hair is not spider gossamer. Certain examples appear to be dust particles held together by static electricity, which could explain incidents in which eyewitnesses who have gathered up angel hair have experienced a slight tingle or shock, as if it were electrically charged.

There is even the possibility that some angel hair may owe its origin to plasma phenomena, just as ball lightning apparently does too, because some strands of angel hair have been seen descending from spherical UFOs.

The most famous case of this nature was reported from the skies above Oloron-Sainte-Marie and other nearby villages in southwestern France on 17 October 1952, when about 30 domed saucer-like objects discharging angel hair were witnessed.

Despite its name, ball lightning can assume a range of different shapes, and this phenomenon's researchers are aware that when viewed at close range ball lightning is sometimes seen to produce filamentous extrusions. So perhaps angel hair is not a single phenomenon, but at least three totally different ones, yet all yielding superficially similar products.

PWDRE SER - THE ROT OF THE STARS

Equally strange, but far more grotesque, than angel hair is star rot or pwdre ser - names given to weird globules of gelatinous material sometimes discovered on the ground and elsewhere, usually after the sighting overhead of a falling star, and which have even been said by eyewitnesses to pulsate as if alive! In February 1980, several lumps of colourless star rot of the non-pulsating kind were found close together on his lawn by Philip Buller of Hemel Hempstead, but when a sample was analysed it appeared to be of amphibian origin and had perhaps been regurgitated by a predator. Other samples obtained in recent years have been unmasked as the semi-digested remains of a bird's meal, a genus of blue-green alga called *Nostoc*, and a type of fungus, *Tremella*, nicknamed devil's butter.

As for pulsating star rot, this appears to be an extraordinary kind of fungus known as a slime mould or myxomycete. It passes through a slug-like stage in its complex life-cycle, during which several individuals combine together to produce a mobile 'super-organism' termed a pseudoplasmodium, able to move across lawns and even climb telegraph poles. In the autumn of 2006, a profusion of star rot discovered by Mr Heywood in fields on his North Devon property, Duerdon Farm, was investigated by Richard Freeman and Mark North from the Centre for Fortean Zoology (CFZ). Photographs that they took of the cream, jelly-like aggregations observed there were tater shown to fungus expert Gordon Rutter who identified them as *Fuligo septica,* a slime mould

Yet if star rot truly has such earthbound origins as these, why is it popularly thought to originate from shooting stars? Worth noting is that star rot cases rarely if ever contain a complete scenario, i.e. the sighting of a shooting star, the discovery of the precise location of its landing, and the finding

Star rot at Duerdon Farm, Woolsery, autumn 2006 *(CFZ)*

there of star rot. In other words, the shooting star's occurrence and the finding in the same basic vicinity afterwards of star rot are most probably nothing more than coincidence. As with angel hair, it would seem that there is not a single origin or identity for star rot but several (see also Chapter 22).

GREEN RAIN AND PINK SNOW

In addition to traditional transparent rain and familiar white snow, all manner of decidedly unfamiliar, multicoloured precipitation has fallen from the skies upon countless localities worldwide down through the years. The snow witnessed by Milton S. Mayer drifting down over McCloud-McArthur Street in Dana, California, on 8 April 1953, for instance, was readily distinguished from normal snowfall by virtue of its bizarre bright green hue. Equally distinctive was the shower of green rain that fell upon Mobile, Alabama, in July 1948 - just 4 months after a viridescent rainfall had startled the town dwellers of Dayton, Ohio, on 26 March of that same year.

If you prefer blue rain, you should have been in Moline, Illinois, on 8 April 1954, when a heavy shower of unequivocally blue-hued rain unexpectedly fell - a shower made even more unexpected when tests conducted on samples of this rain discovered that it was radioactive. Additionally, showers of similar rain - blue and radioactive - also tumbled earthward elsewhere on that same day, including Davenport, Illinois, and Detroit, Michigan. Less than 2 months earlier, on 28 February, the town of Rensselaer in New York State had been doubly honoured, hosting not only a fall of blue rain but also a shower of pink snow. Back in July 1935, flurries of bright red snow had shocked the inhabitants of Montrose and Salida in Colorado, and long before that, in July 1881, a similar shower had been recorded from Colorado's Mount of the Holy Cross and documented on 10 July by the *New York Times*.

It seems most likely that, at least with respect to the green and red precipitations, their odd colouration is due to an organic inclusion of some kind, possibly algal. This is certainly true, for example, with regard to Greenland's famous Crimson Cliffs, whose red snow was described as far back as the early 1800s by Captain John Scott, and also to other polar examples. In many cases of yellow and brown rain, sand and pollen are also often shown to be the culprits. Perhaps the most bizarre (yet fully confirmed) explanation for an incident of green rain, when for 2 days in June 2002 the town of Sangrampur, east of Calcutta, India, was bombarded by an incessant lime-tinged deluge, is that these mysterious yellow-green droplets did not constitute a form of chemical warfare (as initially feared), but were rather the pollen-containing droppings defaecated by a huge swarm of migrating Asian honeybees.

As for blue rain, the presence of tiny cyanophytes (blue-green algae) may be responsible, and in certain cases where blue snow has been found on the ground but not seen falling it may even be an optical effect, caused by the snow particles scattering sunlight. Conversely, the reason why the blue rain that fell over various American towns on 8 April 1954 was radioactive remains unrevealed. It may have been some form of industrial pollutant, an explanation behind many eerie falls of black rain and snow, such as the ebony snowfall that descended upon Birmingham, Alabama, on 9 January 1949. However, the blue rain that fell over parts of northeastern Argentina on 2 September 1998 was later shown by scientists to have been caused by ash associated with fire activity in Bolivia and Brazil.

ELECTRIC RAIN AND EXPLODING HAIL

Less well-known but no less mystifying than coloured rain is electric rain - rain (and sometimes hail too) that sparks, emitting rays of light and crackling sounds, when it hits the ground. Anomalies chronicler William R. Corliss has documented several examples. During a heavy hailstorm that savaged Surrey on 16 July 1916, eyewitnesses claimed that when the hailstones hit windows, phosphorescent flashes of light could be clearly seen. In 1892, a particularly apposite witness, an electrical engineer, experienced a remarkable evening storm in Cordova, Spain, during which, following a flash of lightning, a shower of very large raindrops fell from some dark clouds overhead. And as each raindrop hit the ground, trees, or the walls of buildings, the engineer heard a faint crackle and saw a spark of light.

A comparable fall of electric rain occurred on the evening of 1 November 1844 in Paris, France, witnessed by a Dr Morel-Deville while crossing the court of the College Louis-le-Grand. Not only did he see the sparks and hear the crackles, he also detected a distinct smell of phosphorus. If such precipitation were indeed electrically-charged, this could explain the crackles produced, and the emissions of light may be due to triboluminescence (a phenomenon also responsible for light flashes seen when crushing sugar cubes).

Even more daunting than electric rain, however, must surely be an encounter with exploding hail. In such cases, when the hailstones hit the ground they fragment, and emit loud pistol-like reports as the fragments shoot off in all directions. Happily, only two such cases have currently been documented, and both were witnessed by the same person, W.G. Brown. Of these, he gave a particularly informative account of his second experience, which occurred during mid-afternoon on 11 November 1911, at Columbia, Missouri. Recalling the large hailstones that fell after a bout of thunder and lightning, Brown stated:

> ...on coming in contact with the windows or walls or pavement [the hailstones] in many instances exploded with a sharp report, so loud as to be mistaken for breaking window panes or a pistol shot. As the hail fell, the fragments sprang up from the ground and flew in all directions, looking like a mass of 'popping corn' on a large scale. The fall lasted two or three minutes, about half the hailstones being shattered, the ground in some places being nearly covered with stones and fragments.

Internal stresses accumulated within the hailstones may be responsible for these unusually explosive examples, but if so, why is this extraordinary phenomenon not more common? What was so special about the meteorological conditions on that occasion that only then (and on one single, earlier occasion) could explosive hail result? Also, it seems an astonishing coincidence that the only witness on both occasions was the same person.

ICEBERGS IN SPACE?

Quite aside from Fort's airborne Super-Sargasso Sea, one could be forgiven for believing in the existence of a veritable flotilla of stratospheric icebergs ominously soaring far overhead, bearing in mind the numerous reports of large (sometimes very large) blocks of ice abruptly dropping earthward from the skies up above that have been filed. Today, such objects, dubbed hydrometeors, are generally assumed to originate from aeroplanes flying overhead, but this cannot explain several well-documented cases that notably pre-date the invention of aircraft.

One of the most spectacular examples made its debut during an evening in August 1849, immediately after an exceptionally loud peal of thunder had been heard overhead, when a huge mass of ice fell near to the farmhouse of a Mr Moffat, on the Scottish estate of Ord. The ice block was estimated to measure almost 20 ft in circumference and of proportionate thickness, with a beautiful, mostly-transparent, crystalline appearance. A small portion of it could be seen to be an aggregation of unusually large hailstones, but the vast majority of it comprised a great quantity of diamond-shaped chunks of ice all intimately fused together. Nothing else had fallen with it, and no hail or snow was recorded anywhere in the vicinity. Although it is known that hailstones can sometimes fuse together, no tenable explanation for the creation by such fusion of ice blocks as big and heavy as the example cited above and many others on record has so far been offered.

STONES FROM THE SKIES

Until the reality of meteorites was scientifically established, any notion of stones falling out of the sky seemed totally preposterous - notwithstanding the numerous reports on file, dating back to antiquity. However, falls of stones that have nothing to do with meteorites are still occurring today, and still await a satisfactory explanation.

It was once believed that stones were in some way engendered directly during thunderstorms and hurled to the ground from the storm clouds. Accordingly, these meteorological missiles were dubbed thunderstones. This theory has long since been discounted, but the fact remains that falls of stones are most commonly reported during thunderstorms. On 12 June 1858, for example, a powerful thunderstorm over Birmingham, West Midlands, was accompanied by a shower of stones. And on 25 May 1869, in Wolverhampton, also within the West Midlands, a considerable number of pebble-like stones, each measuring approximately 0.75 in across, fell to earth during a severe thunderstorm. As a native Midlander, I can testify that both of these events are well-recorded in the local archives, and a meteorite identity was dismissed as a plausible explanation for them.

It is also unlikely, and for very obvious reasons, to be the solution for the stone fall that descended on the house of Steve Clark from Radpole, Dorset, during a weekend in December 1994, because these stones, small but numbering many dozens, were blue and red in colour. An eyewitness sitting in a car nearby watched them fall, and due to their large number the possibility that children had thrown them as a prank was ruled out. Moreover, they did not match any stones occurring locally, and their origin remains a mystery. So too does that of the countless dark-red pumice-like stones that fell in two showers like hailstones on Mrs Thomas Potter in San Diego, California, when she stepped outside during the gusty evening of 31 August 1969. She and her husband then watched them as they bounced off parked cars nearby, and later collected enough to fill an empty half-pint coffee jar.

The most reasonable theory for stone falls is that they are simply terrestrial stones that have been lifted into the air by swirling wind during a thunderstorm and then dropped back down to earth. However, the total weight of stones that have fallen in some instances is such that if this is indeed the correct explanation, the vertical wind velocity achieved during thunderstorms must be greater than currently anticipated. Of course, there is also the prospect in some cases that the stones have never fallen, i.e. they have merely been thrown around across the ground by low-altitude winds, but this theory cannot explain cases in which witnesses have personally seen - and felt - stones dropping down from on high. There is even the intriguing fact that in certain reports of poltergeist activity, stone-throwing has featured extensively. But that, as they say, is another story...

ARTEFACTS AHOY, AND OTHER AIRBORNE ANOMALIES

Some falls from the skies are so bizarre that they defy classification, let alone explanation. Take, for instance, the shower of small crucifixes recorded in 1503 by Lycosthenes. Could they possibly have been freak cruciform hailstones, but if so, how on earth (or should that be in Heaven?) had they been created? And what of the veritable cascade of nails that dropped out of the sky during the evening of 12 October 1888, striking the unfortunate wife of the lighthouse keeper at Point Isabel, Texas? After all, it is hardly likely that anyone was hiding on top of a lighthouse with a sackful of nails.

Even more astonishing was the hail of bullets that inexplicably dropped from the ceiling to the floor *inside* an office in Newton, New Jersey, on several separate occasions as reported in early 1929, and remains to this day one of the most famous if baffling events ever to have occurred there. Just as fa-

mous - but far more welcome, no doubt, to the witnesses - was a fall of pennies and half-pennies experienced by a group of children going home from school in Hanham, Bristol, one evening in September 1956, and the fall of cookies witnessed by Stanley Morris in May 1965 in Louisville, Kentucky.

Rather less beneficial was the fall of knitting needles experienced by F.W. Curry in Harrodsburg, Kentucky, in or around 1845, or the shower of soapflakes over Chickasaw, Alabama, in 1974 - though this last-mentioned precipitation was ultimately unmasked as residue from a water-treatment plant. But what about the deluge of mixed seeds - thousands upon thousands, which, when collected, weighed a hefty 10 lb - that fell on just three houses in a Southampton street during no less than 25 separate showers over several days in 1979, covering their gardens and roofs? The seeds were mustard, peas, beans, cress, and maize, and samples that were later planted all grew into healthy, normal crops - but where had they come from, and why were the falls so incredibly localised, and persistent?

Early engraving of a rain of crucifixes

Waterspouts, whirlwinds, optical illusions, electrical phenomena, and Fort's delightful if elusive aerial Super-Sargasso Sea - these and many other solutions have been offered down through the ages, but all are ultimately unsatisfying, and for the most part the mystery of anomalous falls from the skies remains unresolved. So next time you are out and a rainstorm begins, make sure that you have your umbrella open, because contrary to the popular song, you may well find that much more than just raindrops keep falling on your head!

Chapter 19

ONE OF OUR PLANETS IS MISSING!

Vulcan, Lilith, and Other Celestial Conundra

What grander idea can the mind of man form to itself than a prodigious, glorious and fiery globe hanging in the midst of an infinite and boundless space surrounded with bodies of whom our earth is scarcely any thing in comparison, moving their rounds about its body and held tight to their respective orbits by the attractive force inherent to it while they are suspended in the same space by the Creator's almighty arm! And then let us cast our eyes up to the spangled canopy of heaven, where innumerable luminaries at such an immense distance from us cover the face of the skies. All suns as great as that which illumines us, surrounded with earths perhaps no way inferior to the ball which we inhabit and no part of the amazing whole unfilled! System running into system, and worlds bordering on worlds! Sun, earth, moon, stars be ye made, and they were made!

Edmund Burke, praising the 'noble science' of astronomy

Today, astronomers acknowledge the existence in our Solar System of eight true planets, and some so-called dwarf planets. So whatever happened to the vanished fire planet Vulcan? And what of Phaeton – allegedly a planet formerly orbiting the sun where the major asteroid belt now exists?

And is there a hidden world, the cryptic Planet X, still awaiting discovery further away from the sun than even the icy Pluto? As revealed in this chapter, the curious case of the missing planets is as controversial today as it was more than a century ago.

VULCAN - THE LOST PLANET?

Named after the Roman god of fire by eminent French astronomer Prof. Urbain Leverrier shortly after its discovery, the most mystifying member of this tantalising trio must surely be the lost planet

---

Vulcan. It was aptly named, because with a distance of only 13,082,000 miles from the sun, Vulcan was even closer to it than Mercury - the nearest planet to the sun that is formally recognised today by science - which means that it would have been little more than an infernal wasteland of fire.

Leverrier had suspected the existence of an intra-mercurial planet for several years, after his studies

Prof. Urbain Leverrier

concerning the effects upon the solar planets' orbits of their gravitational fields had shown that Mercury's perihelion (its orbit's point of nearest approach to the sun) shifted position ('advanced') more rapidly than could be accounted for by traditional calculations of such shifts. He believed that this notable discrepancy between reality and mathematical prediction could be best explained by the presence of an unknown body (or bodies) inside Mercury's orbit, exerting a sizeable gravitational pull upon it.

Leverrier had initially favoured the existence of a series of asteroids, like those between Mars and Jupiter, rather than a single larger entity, as the elusive culprit - but then came the extraordinary claim of an amateur astronomer called Dr Edmond Modeste Lescarbault. A medical practitioner from Orgères-en-Beauce, near Orleans, who spent his spare time studying the heavens, albeit with only the most primitive of telescopes, Lescarbault seemed set to acquire astronomical immortality when for 75 minutes on 26 March 1859 he observed a round black object travel slowly across the upper section of the sun's incandescent face. Struck by this event's visual similarity to the transit of Mercury across the sun's face that he had watched 14 years earlier, Lescarbault was certain that the mysterious object must also have been a planet, but one that was closer to the sun than Mercury.

Lescarbault hesitated making public his discovery for some time, but eventually he wrote to Leverrier - who received his letter on 22 December 1859. Highly sceptical, but also well aware of the importance of his find if shown to be correct, Leverrier lost no time in visiting Lescarbault. After subjecting his equipment and methods of astronomical computation to an abrasive scrutiny, Leverrier returned home satisfied that Lescarbault had genuinely spied a hitherto undetected intra-mercurial planet, and he named it Vulcan.

Leverrier's calculations suggested that Vulcan's mass was only about 1/17th that of Mercury, which meant that its gravitational pull would be insufficient to account entirely for the mysterious discrepancy in Mercury's perihelion. However, he felt sure that there were other, smaller bodies also circling the sun inside Mercury's orbit whose collective gravitational pull would supply the missing amount required to balance the sums.

The most important step now was to obtain further sightings of Vulcan, and an ideal opportunity for this was not far away - the solar eclipse of spring 1860, when Leverrier's researches prophesied that Vulcan would pass in front of the sun's face, and when (due to the sun's greatly reduced brilliance during such an event) it should appear as a bright point of light closer to the sun than Mercury.

No sign of his newly-named fire planet was spied, however, either by Leverrier himself or by the numerous other scientific star-seekers eagerly scanning the skies for it. The same sad story was repeated in 1861, but at 8-9 am on 20 March 1862 a Manchester astronomer called Lummis observed a perfectly round object moving across the sun. Using Lummis's data, two French astronomers called Benjamin Valz and Rudolphe Radau were able to modify Leverrier's original estimates of Vulcan's orbit and distance from the sun, but their own respective estimates disagreed with each other. Clearly, more sightings were required, but once again they did not occur as rapidly as hoped for. Apart from a highly-disputed sighting on 4 April 1875, recorded by German astronomer Heinrich Weber in northeastern China and featuring a round dot on the sun, Vulcan stayed aloof this time until the solar eclipse of 29 July 1878.

This was when two American astronomers independently reported seeing a mysterious bright object that they assumed to be Lescarbault's prodigal planet. Michigan-based astronomer Prof. James Watson (observing from Rawlins, Wyoming) and New York nebula researcher Lewis Swift (observing from Pike's Peak, Colorado) had both spied a red disc near the star Theta Cancri, which reaffirmed Vulcan's scientific credibility for a time. However, all future attempts to trace Vulcan failed, so that by the 20th Century's commencement few astronomers believed in its existence - discounting Watson's observation as a sighting of Theta Cancri itself, and Swift's as that of a 7th-magnitude star close by.

Nevertheless, one major mystery remained unsolved - the anomalous orbit of Mercury, ostensibly influenced by Vulcan's gravitational pull. If Vulcan did not exist, how could Mercury's abnormal perihelion be explained? The answer came in 1915, courtesy of Albert Einstein and his famous Theory of Relativity, which revealed that the traditional methods of calculating the effects of gravitational pull upon bodies are inaccurate if strong pulls (such as the gravitational force exerted close to the sun) are involved. When his theory's method of calculation was utilised in predicting the shifting orbit of Mercury (the planet closest to the sun), to the delight of astronomers the result obtained coincided precisely with the actual rate of shift, accounting entirely for its hitherto perplexing perihelion - and thence eliminating any need for an elusive Vulcan as an explanation for the latter phenomenon.

Since then, Vulcan has been consigned to the history books (and to a certain TV and film series featuring a starship officer with decidedly pointed ears!). Yet it *is* possible that one or more Vulcans of sorts do exist, still eluding official scientific discovery. Mercury's surface is heavily pitted, suggesting that in bygone ages it experienced many wayward asteroid-type bodies, dubbed vulcanoids, colliding with it. Moreover, during a North American eclipse in 1970, Dowling College astronomer

Henry Courten photographed from three widely-spaced observation sites an intra-mercurial object near the sun that could conceivably comprise a vulcanoid. Although this would be much smaller than the estimated size of Leverrier's planet, any celestial body (however small) as close to the sun as that would be greatly deserving of the name Vulcan!

NEITH, A MISSING MOON OF VENUS; AND LILITH, EARTH'S DARK SECOND MOON

Equally mysterious, yet even less well-known than Vulcan, is Neith - the missing moon of Venus. According to modern astronomy, the planet Venus has no satellite. During the 1600s and 1700s, however, an enigmatic object resembling a bona fide Venusian moon was spied by several noted sky-watchers, including Giovanni Cassini - and in 1884, it was christened Neith by J.-C. Houzeau, director of the Royal Observatory of Brussels.

Even so, Neith had also acquired many sceptics, dismissing it as an optical illusion or misidentification of some other celestial body. In 1887, Neith received its death-blow, when astronomer Paul Stroobant presented a detailed paper to the Brussels Royal Academy of Sciences, in which he readily exposed Venus's 'moon' as a fixed star.

Remarkably, Earth was once thought to have a second, tiny moon, proposed in 1846 by French astronomer Frederic Petit of Marseilles to explain oddities in our planet's much larger, known moon's orbit. However, these were later shown to result from miscalculations rather than from any influences by an elusive unseen moonlet. Nevertheless, the prospect of a second moon orbiting Earth resurfaced when in 1918 astrologer Walter Gornold (aka Sepharial) claimed to have spied just such a body, and named it Lilith, after Adam's supposed first wife (prior to the creation of Eve). This corroborated an earlier statement in support of a second moon, by Hamburg scientist Dr Georg Waltemath, who had claimed in 1898 that this enigmatic satellite existed within an entire system of tiny moons orbiting Earth.

Gornold opined that Lilith escaped regular detection by being dark, and other supporters of the Lilith theory claim that it follows an orbit stationary to the opposing side of Earth's known moon, which again would render it permanently hidden from view except when crossing the sun (especially as its diameter is supposedly only a quarter of the known moon's). Hence it has been variously dubbed Dark Lilith, Black Lilith, and Ghost Lilith. However, the existence of this invisible moonlet is not recognised by mainstream science, who point out that it defies accepted gravitational laws and that it would certainly have been formally discovered by now, although it is still included by some astrologers in their horoscopes.

Incidentally, the asteroid 3753 Cruithne (discovered in 1986) is often referred to informally as 'Earth's second moon'. This is because although it orbits the sun, not Earth, its period of revolution around the sun is almost exactly equal to Earth's, which means that this asteroid and Earth give the appearance of following each other in their respective paths around the sun.

PHAETON - THE EXPLODED PLANET?

The famous asteroid belt between Mars and Jupiter remained wholly unknown to science until the 19th Century's onset, whereupon its members' progressive unveiling initiated another curious controversy. It stemmed from an intriguing (but now-discredited) mathematical dictum called the Titius-

Bode Law, which accurately predicted the respective distances from the sun of all of the Solar System planets - with one notable exception. The law predicted the presence of a planet between Mars and Jupiter - where no planet is known to exist. Consequently, astronomers set about scouring the heavens in search of what must surely be a hidden planet.

During the evening of 31 December 1800, Father Giuseppe Piazzi, observing the constellation Taurus from Palermo in Sicily, spied what he assumed to be a previously uncharted star, but over several successive nights he realised that it was continually changing its position. After he announced its puzzling existence, others sought it, and exactly one year after Piazzi's first sighting it was observed by German astronomer Heinrich Olbers. Once its possible identity as a comet had been examined and discounted, its status as a planetary body was formally accepted, and Piazzi named it Ceres - after the Greek goddess of agriculture. Even so, it could hardly be termed a planet - its diameter is only 620 miles!

Heinrich Olbers

On 28 March 1802, Olbers discovered a second, even smaller 'mini-planet' - Pallas - again orbiting in the gap between Mars and Jupiter where, according to the Titius-Bode Law, there should be a single, much larger world. As time went by, many other tiny planetary bodies were also discovered here, and became known as 'asteroids'; several thousand are now on record, with many more undoubtedly still awaiting detection.

The asteroid belt's presence led to speculation regarding whether it comprised the debris of a former planet that had somehow exploded or disintegrated – the Disruption Theory. During the 20th Century, this theory received particular attention in the former Soviet Union, where Russian meteorite expert Dr Yevgeny L. Krinov formally named this hypothetical lost world Phaeton (=Phaethon) - after the son of Helios, the Greek sun god, whose disastrous attempt to steer his father's fiery chariot through the sky so endangered the earth below that Zeus was forced to slay him with a thunderbolt.

Yet the Phaeton concept presents some profound dilemmas. For example, the collective mass of all of the currently known asteroids is very significantly less than the mass of even a very small planet. Also: if the asteroids had all originated from the same source, their orbits would intersect at the source's onetime location, but they exhibit no such intersection - proving that they do not share a common origin. Sensationally, some Soviet scientists have suggested that if Phaeton had existed, only an immense nuclear explosion could have destroyed it and yielded the asteroids in their near-circular orbits, thereby implying that it had been inhabited by advanced life forms. Yet studies of meteorites (which are themselves tiny portions of asteroids in many cases) have failed to uncover indications of the radioactivity that always accompanies explosions of this nature.

The current belief is that the asteroids are simply remnants of a planet that never completed its development – i.e. fragments of an embryonic planet, rather than debris from a destroyed planet.

Also of interest and relevance to the concept of a missing planet between Mars and Jupiter is the Planet V hypothesis, formulated by NASA scientists Drs John Chambers and Jack Lissauer, and formally presented by them during the 33rd Lunar and Planetary Science Conference in March 2002. According to their hypothesis, which is based upon computer simulations, a fifth planet, Planet V, existed between Mars and the asteroid belt approximately 4 billion years ago, exhibiting a highly eccentric, unstable orbit, but was destroyed when it plunged into the sun.

DWARF PLANETS, TNOS, AND PLANET X

The final controversy to be considered here remains as contentious today as it was over a hundred years ago - when the world's leading astronomers were engaged in heated dispute concerning the prospect that a major planet existed further away from the sun than the great gas planet Neptune, and whose gravitational pull was responsible for mysterious perturbations in the orbits of Neptune and Uranus. This unseen entity became known as Planet X, and was sought for decades, but without success - until February 1930, when a trans-neptunian planet was found by Lowell Observatory astronomer Clyde W. Tombaugh.

Nevertheless, this newly-exposed world, dubbed Pluto after the Roman god of the Underworld (in recognition of its gloomy existence further away from the sun than any other known planet in the Solar System), seemed too small to account entirely for the perturbations of Neptune and Uranus - even after its satellite, Charon, was discovered in 1978. Indeed, Pluto has now been officially demoted, reclassified in August 2006 as a dwarf planet.

So too has Eris (originally nicknamed Xena after TV's fictional warrior princess, but now officially named after the Greek goddess of discord). Another relatively diminutive trans-neptunian object (TNO,) although bigger than Pluto, Eris was discovered on 5 January 2005. The giant asteroid Ceres has also been reclassified in this way. Sedna, a TNO discovered on 14 November 2003 and named after the Inuit goddess of the sea, may be a dwarf planet too. Yet another enigmatic TNO is Quaoar, discovered on 4 June 2002, orbiting the sun in the Kuiper Belt (a Solar System region beyond Neptune that includes Pluto within its span), and named after a creation deity of the Native American Tongva people. All of these TNOs were, upon their respective discoveries, mistakenly deemed to be the elusive Planet X. However, the search for the real Planet X continues.

In 1992, what would appear to be a trans-plutonian asteroid was discovered. Currently named 1992 QB1, but with an estimated diameter of only 150 miles, it is a very far cry indeed from Planet X. Moreover, in January 1993 Dr Myles Standish of the Jet Propulsion Laboratory in Pasadena, California, announced that in his opinion the orbital deviations of Uranus and Neptune were due merely to observational errors, thereby eliminating any evidence for the presence of a really sizeable planet beyond Neptune. Others, conversely, such as Britain's most famous astronomer, Patrick Moore, remain optimistic that such a world does exist, conceivably awaiting detection in the southern sky.

To be continued...? Watch this space – and Outer Space!

Chapter 20

ON THE HORNS OF A DILEMMA

Do Unicorns Really Exist?

'What – is - this?' he said at last.

'This is a child!'...'We only found it today. It's as large as life, and twice as natural!'

'I always thought they were fabulous monsters!' said the Unicorn. 'Is it alive?'

'It can talk,'...

The Unicorn looked dreamily at Alice, and said 'Talk, child.'

Alice could not help her lips curling up into a smile as she began: 'Do you know, I always thought Unicorns were fabulous monsters, too! I never saw one alive before!'

'Well, now that we have seen each other,' said the Unicorn, 'if you'll believe in me, I'll believe in you. Is that a bargain?'

'Yes, if you like,' said Alice.

Lewis Carroll – *Through the Looking-Glass*

Its ethereal beauty, its imperious gaze, its evanescent presence, the magical properties of its spiralled horn, its fatal attraction to virginal maidens – all alluring, bewitching, memorable qualities that combine to render the unicorn one of the world's most famous, spellbinding creatures. Symbolically, it has been portrayed as representing Christ, Scotland, the moon, purity, royalty, strength, and chastity. Not bad for an animal that never actually existed – or did it? After all, in earlier days many travellers reported encountering this remarkable creature. Certainly, the unicorn is far more diverse in form, widespread in distribution, and tenacious in traditional lore for it to be wholly imaginary. So what exactly did they see?

HORNS OF PLENTY – A DIVERSITY OF UNICORNS FROM EUROPE AND THE EAST

It is widely but mistakenly assumed that even in mythology there was only ever one type of unicorn – the familiar milky-white equine beast bearing a slender spiralled horn at the centre of its brow. In reality, however, several very different versions upon this basic unicorn theme have been reported down through the ages across Europe and Asia.

Perhaps the earliest European documentation of the unicorn appeared in *Indica*, a work prepared by the 4th-Century-BC Greek historian Ctesias, who was appointed as court physician to Darius II, King of Persia. Ctesias stated:

> There are in India certain wild asses which are as large as horses, and larger. Their bodies are white, their heads dark red, and their eyes dark blue. They have a horn on the forehead which is about a foot and a half in length. The dust filed from this horn is administered in a potion as a protection against deadly drugs. The base of this horn, for some two hands'-breadth above the brow, is pure white; the upper part is sharp and of a vivid crimson; and the remainder, or middle portion, is black.

Edward Topsell's unicorn –
the most familiar unicorn type

Apart from their tricoloured horn and red head, these unicorns bear the greatest similarity to the modern-day image of this fabulous animal.

Very different was the monoceros, a hefty, bellicose beast again reported from India and first documented by one of Seleucus I's most eminent officials, Megasthenes, after being sent there by him on a diplomatic mission in c.300 BC. Megasthenes's description of the monoceros (which he referred to as the cartazon – a Sanskrit term translating as 'king of the wilderness') was recycled over three centuries later by the celebrated chronicler Pliny the Elder in his famous work *Naturalis Historia*:

> The Orsaeyan Indians hunt apes, which are white – but the wildest animal is the Monoceros, whose body is like a horse but which has the head of a stag, elephant's feet and a wild boar's tail. It utters a deep, growling sound, and a black horn, two cubits long, protrudes from the centre of its forehead. It is said that this animal cannot be captured alive.

This fearsome beast also appears in Arabian literature, where it is known as the karkadann, and has earned itself a formidable reputation, with one astonishing exception. This desert-dwelling unicorn would unhesitatingly attack anything, or anyone (including Alexander the Great himself in one memorable confrontation), that it encountered – unless that anything or anyone happened to be a ring dove. Amazingly, at the sound of this small pigeon's gentle cooing, the karkadann would slump to the ground, totally entranced, and would remain there for hours, soothed and lulled by the bird's singing. And if the dove alighted upon its horn, the karkadann would even refrain from moving, in case it disturbed its feathered troubadour and frightened it away.

17th-Century adaptation of a 1460 Persian manuscript depicting the karkadann

Similar unicorns were also documented by Marco Polo (1254-1324) during his sojourn at the court of Kublai Khan, describing examples witnessed in Burma as follows:

> There are wild elephants in the country, and numerous Unicorns, which are very nearly as big. They have hair like that of a buffalo, feet like those of an elephant, and a horn in the middle of the forehead, which is black and very thick...The head resembles that of the wild boar, and they carry it ever bent towards the ground. They delight much to abide in mire and mud. 'Tis a passing ugly beast to look upon, and is not the least like that which our stories tell of a being caught in the lap of a Virgin; in fact, 'tis altogether different from what we fancied.

Yet another variation upon the unicorn theme was the caprine equivalent reported from Germany's Harz Mountains (where it was termed the einhorn – 'one-horn') and elsewhere in Europe, and also from the mountainous lands of Tibet, Nepal, and Bhutan. Distinctly goat-like in form (complete with typical billy-goat beard) and cloven-footed, each foot comprising two hooves, it featured in numerous medieval European bestiaries, and is the unicorn type that was attracted by virginal maidens, as documented in a 13th-Century English bestiary:

> The unicorn...is a small beast not unlike a young goat, and extraordinarily swift. It has a horn in the middle of its brow and no hunter can catch it. But it can be caught in the following fashion: a girl who

is a virgin is led to the place where it dwells, and is left there alone in the forest. As soon as the unicorn sees her, it leaps into her lap, embraces her and falls asleep there. Then the hunters capture it and display it in the king's palace.

Alternatively, sometimes the hunters would cruelly slay the trusting unicorn, then crudely hack off its precious horn from its noble brow.

This story is the subject of the famous series of seven exquisite full-colour tapestries entitled 'The Unicorn Hunt', exhibited at the Metropolitan Museum of Art in New York, each of which is 12 ft high. They are believed to have been woven in c.1500 AD in Brussels to French designs. Equally famous, and beautiful, are the six glorious 'Lady with the Unicorn' tapestries woven in c.1490 AD that are now on show at the Cluny Museum, Paris. They depict the five senses, plus the virtue of dedication, and, intriguingly, portray the unicorn and the lion, traditionally sworn enemies, in harmony rather than in combat.

Larger, and much more deer- or antelope-like, than the European goat unicorns were the two cloven-footed unicorns reputedly observed in a park close to the Temple of Mecca by Roman traveller Lewes Vertomannus in 1503:

[They had] the head like a hart [stag], but no long neck, a thin mane hanging only on the one side. Their legs are thin and slender, like a fawn or hind. The hoofs of the fore-feet are divided in two, much like the feet of a goat. The outer part of the hinder feet is very full of hair. This beast doubtless seems wild and fierce, yet tempers that fierceness with a certain comeliness.

Larger still, and different again, was the mighty biblical re-em, a markedly bovine animal whose Hebrew-originating name was translated as 'unicorn' in several separate passages within an early Greek translation of the Bible's Old Testament known as the Septuagint, dating from the 3rd Century BC and commissioned by Ptolemy II of Egypt. It was clearly the same as the Assyrian rimi, a powerful ox-like beast.

Less familiar but equally formidable was the black unicorn recorded in the traditional lore of Siberia's Evenk people, which tells of a huge black bull-like creature with a single round, thick, tapering horn in the middle of its head. This belligerent beast would charge at any Evenk rider that it spied, tossing the unfortunate man into the air if it could reach him, and spearing him when he fell back down to earth until he died.

So what could all of these varied unicorns have been? It is clear that several totally discrete species of real animal inspired accounts of them, some of which, moreover, confused and combined one or more animal species to yield truly non-existent composites.

There is little doubt that the basis for the ass-like Indian unicorn documented by Ctesias is the Asiatic wild ass or onager *Equus hemionus*, a creamy-coated equine beast common in the East but which would have been unfamiliar to European travellers. Of course, it does not possess a horn, nor is its head red. However, it is interesting to note that the Indian unicorn's blue, red, black, and white colour combination matches the sacred colour combination popular in the East, so this may simply have been added to the onager description as a means of conveying a tradition rather than as a morphological reality.

Reports of the cartazon and monoceros undoubtedly refer to the rhinoceros, as does Marco Polo's description of Burmese unicorns, most probably the single-horned great Indian rhinoceros *Rhinoc-*

eros unicornis. True, Megasthenes distinguished the cartazon from the rhinoceros in his writings, yet his descriptions of both are extremely similar, indicating that they are indeed one and the same animal, with distorted accounts of rhinos filtering back to the Middle East no doubt explaining the legends of the Persian karkadann.

A Nepalese uni-ram

As for Siberia's black unicorn, veteran crypto-zoological writer Willy Ley boldly speculated that this creature may be drawn from folk-memories of late-surviving specimens of a huge long-legged prehistoric Siberian rhinoceros called *Elasmotherium*, whose massive single horn was borne not upon its nasal bones like those of modern-day rhinos but on its brow instead, as claimed for unicorns. Indeed, *Elasmotherium* is itself referred to informally as the giant unicorn.

The goat-like unicorns of montane Europe probably owe their origin to sightings of the chamois *Rupicapra rupicapra*, the famous goat-antelope peculiar to this continent. Similarly, the mountainous territories of Tibet, Bhutan, and Nepal can cite such creatures as the chiru or Tibetan antelope *Pantholops hodgsonii* and the domestic Nepalese goat as inspirations for their unicorns. Of course, as with the chamois such animals are normally two-horned, but individuals that through some accident had lost a horn might well have given rise to unicorn myths. Moreover, it is well known that the local Nepalese people can deliberately manipulate the horn development of infant rams to yield single-horned freaks.

The antelope- or deer-resembling unicorns of Mecca and elsewhere in the Middle East assuredly originate from sightings of the beautiful Arabian oryx *Oryx leucoryx*, a handsome creamy-white antelope whose two long spiralled horns are so close together that when the creature is viewed in profile it seems only to bear a single central horn. As for the re'em and rimi, these are believed by some scholars to have been based upon the now-extinct European wild ox or aurochs *Bos primigenius*, which became extinct in 1627.

Not all unicorns are so readily explained, however, as demonstrated by the following examples, recorded from some very unexpected global regions for unicorns.

UNICORNS, AND UNICORN CAVE PAINTINGS, IN TROPICAL AFRICA

As relatively recently as the mid-1800s, zoologists were still seriously speculating that the unicorn may one day be discovered alive, well, and very real in parts of tropical Africa, thanks to reports of certain mystery beasts whose identities remain opaque even today. One such creature was the abada, reported by Giovanni Cavassi from the Congo. In modern times, some zoologists have speculated that the abada may be a distorted account of the okapi *Okapia johnstoni*, the giraffe's smaller forest-dwelling cousin, which of course remained unknown to science until 1901, and which bears a pair of short horns.

Another African mystery unicorn was the a'nasa, referred to in 1848 by traveller Baron Von Müller from the Sudan, where it was said to be the size of a small donkey, with a thick body but thin bones, coarse hair, and a boar-like tail. On its brow was a single long horn, which it apparently let hang when alone, but erected immediately upon spying an enemy, and which it utilised as a formidable weapon.

Similar in description to the a'nasa was the ndzoo-dzoo, reported from a region lying northward of present-day Mozambique. Intriguingly, its horn, said to be 2-2.5-ft long, was claimed to be so flexible that the ndzoo-dzoo could coil it like an elephant's trunk, but it became erect when the creature was angered or threatened. Allegedly very fleet-footed and fierce, the ndzoo-dzoo would never fail to charge at any human that it encountered, and even if its target climbed a tree to escape its ire the ndzoo-dzoo would repeatedly charge the tree until it was felled - whereupon the enraged beast would then gore the helpless human. This behaviour is very reminiscent of a rhinoceros.

Most tantalising of all, however, is the presence of certain unicorn cave paintings in South Africa. Some are of striped zebra-like animals bearing a single horn on the brow. One, from the Bamboesberg area, depicts a yellow unicorn with black stripes, whose horn measures 10 in long and is covered in skin like a giraffe's horns. Others are dark brown horses but again with a brow horn, depicted in cave paintings discovered in South Africa's Drakensberg Mountain range. It has been suggested that these paintings may represent freak one-horned eland *Taurotragus oryx*, a huge striped antelope with spiralled horns.

UNICORNS IN THE NEW WORLD

It may come as something of a surprise to learn that creatures resembling unicorns have been reported in the Americas, but there is some tantalising evidence to suggest that such beasts have indeed existed here at one time or another.

In 1673, Sir Olfert Dapper made the following remarkable claim in his book *Die Unbekante Neue Welt* (*The Unknown New World*):

> On the Canadian border there are sometimes seen animals resembling horses but with cloven hooves, rough manes, a long straight horn upon the forehead, a curled tail like that of the wild boar, black eyes and a neck like that of the stag. They live in the loneliest wildernesses and are so shy that the males do not even pasture with the females except in the season of rut, when they are not so wild.

This description does not match that of any animal species known to exist today in North America.

In 1719, French explorer Bénard de la Harpe was leading an expedition along the Red River of Louisiana when, after joining a party of Nawdishe Indians near the confluence of the Washita, he saw to his amazement that they were roasting unicorns. He claimed that these extraordinary animals were the size of a common horse, with reddish hair as long as a goat's, rather thin legs, and a single horn in the middle of the brow that did not branch out into prongs or tines. Tasting their meat, he found it to be very palatable. A previous French explorer, Jean-Baptiste Le Moyne de Bienville, had been informed by the local Indians that the area bordering the Upper Washita River was frequented by unicorns, but unlike de la Harpe he had not personally seen – or eaten – any of them!

A third fascinating account of North American unicorns, this time in Florida, had been penned by Sir John Hawkins in the journal of his 1564 American voyage:

> The Floridians have pieces of Unicorns' horns, which they call Souanamma, which they wear about their necks, whereof the Frenchmen obtained many pieces. Of those Unicorns they have many; for they affirm a beast with one horn which, coming to the river to drink, puts the same in the water before he drinks. Of this Unicorn's horn there are some of our company that, having got them off the Frenchmen, brought them home to show.

This account readily echoes the European belief that before a unicorn drinks from a pool, it dips its horn into it in order to purify it, nullifying any toxins that it may contain.

Even more intriguing are the claims of Father Palacios, a Patagonian polymath interviewed by explorer Bruce Chatwin, as recorded in his famous book *In Patagonia* (1977). According to Palacios, unicorns once existed in this region of Argentina, but were hunted into extinction in the fifth or sixth millennium BC. However, he alleged that two ancient rock paintings (claimed to be around 10,000 years old) portraying unicorns exist at Lago Posadas. In one of these, the unicorn is depicted holding its horn erect. In the other, it is about to impale a hunter, and is aggressively stamping the pampas.

Chatwin subsequently visited Lago Posadas, and saw one of the paintings for himself, present on a huge dome of reddish basalt known locally as the Cerro de los Indios. The painting that he saw was of the unicorn holding its horn erect, and he noted that it had a thick neck and tapering body. Mystified by the painting, Chatwin wondered whether it really was as old as Palacios had claimed, and he concluded that if it were not, then the unicorn must surely be merely a bull in profile. But if it were truly ancient: "...then it had to be a unicorn".

THE CHINESE UNICORN OR KI-LIN

Just as the Chinese dragons are very different from their Western counterparts, so too is the Chinese unicorn - also known as the ki-lin, ching-ling, or qi-lin - fundamentally dissimilar from unicorns reported elsewhere in the world. Its name is actually a composite, a dual word derived from the male Chinese unicorn, known as the ki, and the female Chinese unicorn, called the lin. According to traditional Chinese lore, the ki-lin resembles a large scaly-bodied deer or calf, but with a single central branched horn whose tip is fleshy (demonstrating that this unicorn does not use it for combat purposes), the tail of an ox, the feet of a horse, and sometimes the head and beard of a dragon (this type of ki-lin is termed a king). Very ornate, it is decorated with the five sacred Chinese colours – red, yellow, blue, black, and white.

The king version of Chinese unicorn, as depicted in a Malaysian temple visited by the author during 2005 *(Dr Karl Shuker)*

The ki-lin

The saola or Vu Quang ox, painted by William Rebsamen

Chinese unicorns were noted for appearing at the births and deaths of great leaders or other significant persons, including, most famously, the eminent Chinese philosopher Confucius in the 6th Century BC. At his birth, a ki-lin reputedly appeared and dropped at the feet of his mother a jade tablet bearing the inscription: "The son of the mountain crystal, the essence of water, will redeem the fallen kingdom of Chu and become a king without a crown".

Seemingly the last sighting of a ki-lin was also the most unusual, when a unique pure-white specimen was spied during the 2nd Century BC by the Emperor Wu of the Han dynasty, appearing in the park of the royal palace. Nevertheless, the ki-lin has not been forgotten and is still a familiar, significant beast of Chinese legend today – unlike its direct Japanese counterpart, the kirin, which nowadays has largely passed into obscurity even in Japan itself.

Intriguingly, whereas Western unicorns have clearly been inspired by a wide range of real animals, there are few such animals in the Far East that comprise convincing models for the ki-lin – with one very noteworthy exception. In 1993, a dramatic new species of hoofed mammal was discovered by science in Vietnam – the saola or Vu Quang ox *Pseudoryx nghetinhensis*. This extraordinary creature resembles a lithe buffalo but with long antelope-like legs and a pair of extremely long, slender, deceptively oryx-like horns (hence its generic name, *Pseudoryx*). Indeed, when viewed sideways, its horns merge to give the creature a unicorn-like profile (as do the horns of the true oryx), which in turn has led some observers to speculate that perhaps this shy, rarely-seen species inspired the legend of the elusive ki-lin.

A ferocious Mongolian relative of the ki-lin is the poh, reputedly inhabiting this vast Asian country's plains, and said to have the appearance of a horse with a white body and black tail, but with a single central horn, and the teeth and claws of a tiger. According to ancient sources, the poh actually devours tigers and leopards, and emits a thunderous howl that sounds like a drum roll.

NEO-UNICORNS

Although they are undoubtedly the most famous ones, the unicorn and its burlier, rhinocerine cousin the monoceros are not the only members of the unicorn family, as demonstrated by this brief survey of some lesser-known relatives, or neo-unicorns.

Not all unicorns were paragons of purity and goodness. Some were decidedly duplicitous, sinister, and even downright evil. Perhaps the most infamous of these malevolent unicorns was the Persian shadhahvar. Usually portrayed as deer-like or antelope-like, it was distinguished by its extraordinary horn, which was elaborately curved and bore many hollow branches. When the desert wind blew through these flute-like structures, a paean of entrancing music would emerge, summoning forth any living thing within earshot to draw closer, mesmerised by this plaintive, evocative melody. As soon as a creature did approach within reach, however, the placid-looking shadhahvar would abruptly leap upon it and tear it apart, before feasting ravenously upon the hapless animal's flesh.

Early Persian depiction of the shadhahvar

Equally lethal was an even more incongruous neo-unicorn – the al-mi'raj. Native to an unnamed island in the Indian Ocean and often featured in Islamic poetry, this remarkable, rapacious beast resembled a large yellow rabbit or hare with long soft fur, but in best unicorn tradition it bore a spiralled horn, 2 ft in length and black in colour, upon its brow. Yet despite its innocuous appearance, all other animals sharing its island knew well to keep far away from the al-mi'raj, because just like the shadhahvar it would pounce upon anything that it could seize, and devour it straight away, even if its victim was several times larger than itself.

Traditional picture of the al-mi'raj

Far less dangerous, but no less peculiar, was the yale or yali, originally native to southern India, because this intriguing creature was somewhat of a contradiction in terms – a unicorn with two horns. Moreover, unlike the fixed horn of the true unicorn, the paired horns of the yale could be rotated in different directions, enabling it to aim them at any attacker approaching from any direction. Quite apart from its mobile horns, the yale was nothing if not memorable morphologically. The size of a horse, it sported the tusks of a boar, the tail of an elephant, and the body spotting of a leopard.

Engraving of monsters and monstrosities with an eale on extreme right

Not surprisingly, this startling beast was reputedly kept inside Indian temples to ward off evil spirits, and, perhaps rather more surprisingly, it ultimately entered British heraldry as one of the four heraldic beasts of the monarch. An earlier version of the yale was the eale, which shared the yale's Indian provenance, boar-like tusks, elephant tail, and moveable horns (though the eale's were far longer than the yale's). However, it was black or tawny all over, was much bigger than the yale, attaining the size of a hippopotamus, and was amphibious, able to live in the water as well as on land. Even more amphibious than the eale, however, was the camphor or champhur, a single-horned Ethiopian unicorn whose hind feet were webbed like a duck's, not hoofed.

The camphor, a web-footed unicorn

Less familiar than the yale and eale is yet another double-horned unicorn, the pirassoipi. Illustrated in a number of bestiaries down through the ages, and native to the Arabian lands bordering the Red Sea, it was normally depicted with two seemingly-fixed, forward-pointing horns, and, most distinctive of all, an extremely woolly, curly coat – giving this creature the appearance of a large ibex-like goat, upon which it may well have been based.

Possibly the least-known neo-unicorn of all, however, on account of its very limited distribution, was the baiste-na-scoghaigh, confined entirely to the Isle of Skye in Scotland's Inner Hebrides. This one-horned creature was more like the monoceros than the unicorn, as it resembled a huge lumbering rhinoceros.

THE ALICORN

The correct name for the unicorn's horn is the alicorn, and it is nothing if not interesting to note that whereas the unicorn is officially a wholly imaginary, non-existent creature, its horn is unequivocally real, and totally tangible. Not only that, it was very valuable in medieval times, due to its alleged ability to detect and neutralise poisons – dip the tip of an alicorn into a vessel suspected of containing a poison, and the latter would be instantly rendered harmless. Moreover, although such belief is no longer entertained today, alicorns remain just as valuable - in 1994, an exquisitely-carved example just over 3 ft long was sold at Christie's for £500,000, and is now on permanent display at Liverpool's World Museum.

Yet if the unicorn were truly unreal, what exactly are alicorns, and where did they originate? The answer is as fascinating as the objects themselves. In fact, an alicorn is the single, extremely long, spiralled ivory tusk (actually a left upper incisor tooth) of an

The pirassoipi, a woolly unicorn

The narwhal

A rare double-tusked narwhal

unusual species of whale called the narwhal or, aptly, the unicorn whale *Monodon monoceros*. Normally, only the male narwhal develops this remarkable structure, but occasionally a female will too, and even more rarely a male will develop a pair of them.

Up to 10 ft long, the narwhal tusk's purpose is still a subject for controversy, but due to the mass of nerve endings associated at its base, it is now believed to act as some form of sensory structure, possibly enabling the narwhal to detect changes in salinity, seawater composition, or even temperature or pressure.

D-I-Y UNICORNS

The bovine unicorn of Dr Dove

Lancelot the uni-goat
(Oberon Zell-Ravenheart)

The esteemed 19th-Century French zoologist Baron Georges Cuvier denounced the unicorn as an anatomical impossibility, stating that a single median horn could never develop from the paired frontal (brow) bones of a cloven-footed mammal's skull. In 1933, however, Maine University biologist Dr William Franklin Dove performed an extraordinary experiment that not only disproved Cuvier's claim but also wrote an entirely new chapter in the history of the unicorn. Dove removed the paired, lateral embryonic horn buds of a day-old male Ayrshire calf, trimmed their edges flat, then transplanted them onto the centre of the calf's brow, placing them side by side together, and waited to see what, if anything, would develop. What did develop was a long, massive, single white horn with a black tip, produced from the two buds fusing together, which projected forward from the centre of the calf's forehead – Dove had created his very own bovine unicorn!

However, unicorn verisimilitude was not limited to the calf's morphology. Astonishingly, as it matured into adulthood, this very striking, sturdy animal became such an accomplished fighter using its single horn that it was unhesitatingly recognised by all other cattle living with it as the herd leader – just as, according to traditional legend, the unicorn was always recognised as the leader of animals sharing its forest domain. This twin similarity between the mythical unicorn and its engineered real-life counterpart led to scientific speculation that perhaps primitive people had learnt this simple transplantation technique and had generated their own herd-leader unicorns, which in turn had been spied by travellers from other lands, thereby giving rise to the legend of the unicorn.

During the 1980s, a series of Angora goat unicorns, created

by Oberon (aka Timothy G) Zell-Ravenheart and his wife Morning Glory, using a patented surgical procedure based upon Dove's work, were exhibited at fairs, circuses, and sideshows across the United States. The most famous of these specially-created creatures was Lancelot, converted into a unicorn in 1980, who even appeared on television.

In Nepal, unicorn sheep (and also goats) have traditionally been created as highly sought-after curiosities, by allegedly branding with a red-hot iron the newly-sprouting horns of 2-3-month-old lambs, then treating with soot and oil the resulting wounds - which, when healed, yield a single horn sprouting forth from the centre of the brow.

A 21ST-CENTURY UNICORN

Occasionally, due variously to a genetic aberration or to injury, a deer will be encountered that has only a single antler instead of a pair. These are generally dubbed unicorn deer in press accounts, but this is incorrect because the lone antler is invariably sited either on the left side or on the right side of the deer's skull, not in the centre as in the fabled unicorn. However, a fully-confirmed, bona fide unicorn deer has now been recorded.

As reported in the *Daily Telegraph* and elsewhere on 11 June 2008, the one-year-old roe deer *Capreolus capreolus* in question was born in a park belonging to the Center of Natural Sciences in Prato, near Florence, Italy. What makes this individual – inevitably christened Unicorn - so special is the short, pointed, unbranched, horn-like antler growing from the very centre of his skull (albeit not, in strictest unicorn fashion, his brow). Scientists at the Prato-based research centre have so far been unable to explain this anomaly, its director, Dr Gilberto Tozzi, announcing that he has never encountered any previous report of such a deer, and that even Unicorn's own twin brother has normal, paired antlers

HORSES WITH HORNS

Intriguingly, the classical unicorn is not the only horned equine creature on file. Two remarkable steeds each sporting a pair of horns were recorded by South American explorer Felix de Azara in his *Natural History of the Quadrupeds of Paraguay and the River La Plata* (1837):

> I have heard for a fact, that, a short time ago, a horse was born in Santa Fé de la Vera Cruz, which had two horns like a bull, four inches long, sharp and erect, growing close to the ears; and that another from Chili [sic] was brought to Don John Augustin Videla, a native of Buenos Ayres [sic], with strong horns, three inches high. This horse, they tell me, was remarkably gentle; but, when offended, he attacked like a bull. Videla sent the horse to some of his relatives in Mendoza, who gave it to an inhabitant of Cordova in Tucuman, who intended, as it was a stallion, to endeavour to form a race of horned horses. I am not aware of the results, which may probably have been favorable.

As there is no indication that such a race of horned horses ever did arise, I can only assume that de Azara had been overly optimistic. However, I do know of at least one additional example on record. In 1929, a short account was published in the scientific journal *Anatomische Anzeiger* by German zoologists P.P. Winogradow and A.L. Frolow regarding a horse exhibiting lateral horn development on its brow, accompanied by a photograph of the horse's skull.

So, did the unicorn ever truly exist as a discrete entity, or was it never anything more than a montage of muddled sightings drawn from a diverse assemblage of known, zoologically-accepted species? In his recent book *Unicorns* (2007), Nigel Suckling offers the following wise words in answer to this most perplexing of questions:

> The true unicorn is a mystical creature endowed with wisdom and insight beyond the common human measure and able to communicate with the divine. Like an angel it can pass between worlds and dimensions.

> One medieval legend says that although unicorns chose to accompany Adam and Eve in their expulsion from Eden, once a year they were allowed to return there to refresh their spirits – because it had been their free choice to enter the mundane world, it had not been forced upon them. The message of the legend is that the unicorn never was a completely flesh and blood creature; it always still partly belonged to Paradise.

Amen to that, and long live the unicorn – whatever, and wherever, it may be!

By the light of the silvery moon – a spectacular
stained-glass unicorn

Chapter 21
WE THREE KINGS OF ALBION ARE
Who were King Lear, Cymbeline, and Old King Cole?

Who is it that can tell me who I am?
William Shakespeare – *King Lear*, Act I Scene IV

And I am something curious, being strange,
William Shakespeare – *Cymbeline*, Act I Scene VI

Old King Cole was a merry old soul
'Old King Cole' - Traditional English nursery rhyme

W hereas many of Britain's post-1066 monarchs rank among the world's most famous rulers, some of their much earlier antecedents are so extensively enveloped by mythology and misconception that it is often difficult to uncover the real leader - if indeed, in certain cases, there ever was a real leader - behind the legend. King Arthur is undoubtedly the best-known example, but there are plenty of others - three of the most interesting being King Lear, Cymbeline, and Old King Cole.

KING LEAR - FOUNDER OF LEICESTER?

There are a number of different theories regarding the origin and identity of King Lear, who has apparently undergone several transformations from antiquity to Shakespearian times. The earliest known equivalent would seem to be none other than a Celtic ocean god called Lir, the Irish Old Man of the Sea, whose children were transformed into swans by his evil wife, Aoife - thus changing his life from one that had previously been filled with happiness into one encompassed by sorrow, mirroring the grief experienced by Shakespeare's King Lear.

Engraving of King Lear

Another fabled counterpart is Llyr, who appears in *The Mabinogion* (a celebrated collection of eleven Celtic prose stories from various medieval Welsh manuscripts). A Welsh counterpart of sorts to Lir but associated with the Underworld, Llyr sired several Celtic deities, including the love goddess Branwen, the sea god Manawydan, and the famous hero Bran the Blessed. Llyr's link to King Lear derives from the 12th-Century Welsh scholar Geoffrey of Monmouth, who referred to him as King Leir in his major work *Historia Regum Britanniae*, and narrated the basic story familiar today to Shakespearian readers as that of King Lear and his three daughters - Gonorilla (renamed Goneril in Shakespeare's play), Regan, and Cordeilla (Cordelia).

Geoffrey of Monmouth also claimed that 'King Leir' founded Leicester. However, this scholar's writings are notorious for their inaccuracy (he had the regrettable tendency to make his work more entertaining by fleshing out historical events with storylines and characters from legend and folktale), and there is no factual basis for linking King Leir and Leicester. Indeed, the same can also be said for King Leir (or Lear) and his daughters, for they themselves seem to have been little more than a fictional amalgamation of mythical figures concocted by Geoffrey of Monmouth's fertile imagination, with no basis in reality.

Although Shakespeare's play *King Lear* is by far the most famous telling of this story, it is not the only one in existence. A version appeared in Edmund Spenser's *The Faerie Queene* (1590-96), and another featured in 16th-Century scholar Ralph Holinshed's *Chronicles of England, Scotland, and Ireland*, which was Shakespeare's source for his own version.

CYMBELINE - KING OF THE CATUVELLAUNI

In contrast, there is much more support for believing that another early British king documented by Geoffrey of Monmouth does have a foundation in fact. According to the latter, Cymbeline was king of much of southeastern England during the 1st Century AD, and although he paid an annual tribute to Rome, was greatly admired by the Romans. In his writings Geoffrey also claimed that Great Kimble in Buckinghamshire was named after him.

Unlike King Lear, King Cymbeline did exist but was known as Cunobelinus, King of the Catuvel-

Engraving of Cymbelline

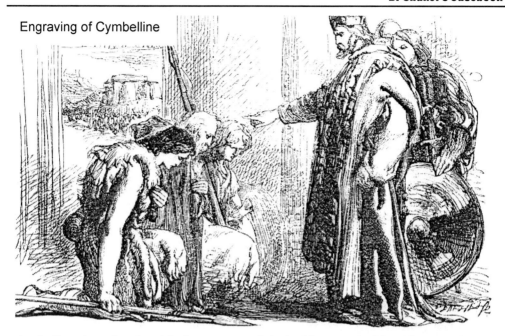

launi tribe, whose domain originally encompassed the land between the River Thames and Cambridgeshire. Under Cunobelinus's reign, however, which lasted more than 30 years, its area expanded until it covered much of southeastern England - and such was his reputation in Rome that the Romans dubbed him 'King of the Britons'. During his reign, the Catuvellauni conquered a neighbouring tribe called the Trinovantes, and Cunobelinus claimed their capital, Camulodunum, as his own. This later became known as Colchester.

Cunobelinus died in AD 40, and although Geoffrey of Monmouth's assertion that Great Kimble was named after him is not widely believed today, his name does live on in a local name for a series of small caverns in the chalk beds at Little Thurrock, Essex. These are referred to there as Cunobelin's goldmines.

Cunobelinus is also remembered as the father of another myth-enshrouded early British king - Caradoc, known as Caractacus to the Romans, who captured him and took him back to Rome in AD 51. Following his father's death, Caradoc led the Catuvellauni in a series of skirmishes against Claudius's Romans in Wales, but he was ultimately defeated after being betrayed by Cartimandua, queen of the Brigantes tribe.

WAS OLD KING COLE REALLY A MERRY OLD SOUL?

Every child has heard the famous nursery rhyme:

> Old King Cole was a merry old soul
> And a merry old soul was he;
> He called for his pipe and he called for his bowl
> And he called for his fiddlers three.

Every fiddler he had a fiddle,
And a very fine fiddle had he;
Oh there's none so rare, as can compare
With King Cole and his fiddlers three.

Yet even though this rhyme is believed to date back at least as far as 1708, how many children - or adults, for that matter - know who King Cole really was, or whether he even existed?

Once again, we have Geoffrey of Monmouth to thank for bringing this mysterious monarch to the attention of other scholars and chroniclers. Geoffrey stated that King Cole first came to fame as Coel, Duke of Colchester, but became King of Britain when he overthrew the 3rd Century incumbent to this title, a vehemently anti-Roman king called Asclepiodotus, which Coel killed in battle. After Coel's own death, and burial at Glastonbury, his daughter and sole heir, Helena, married a Roman senator called Constantius, stationed in Britain - a union that produced a son, Constantine, who ultimately became the first Christian emperor of Rome.

An interesting history, if true - but, as is so often the case with Geoffrey of Monmouth, it owed more to his talent for weaving memorable tales than to any factual storyline. First and foremost, as painstakingly revealed by Geoffrey Ashe in his excellent *Mythology of the British Isles* (1990), there is no evidence to suggest that Coel, or Cole, ever existed. Indeed, Ashe notes that Geoffrey of Monmouth may have based Cole upon a 5th-Century northern magnate. His supposed son-in-law, Constantius Chlorus (a Roman senator), was real, and he did marry a woman called Helena (later St Helena), but there is no record to link her to any Duke of Colchester. As for Coel's predecessor, the supposedly Roman-hating Asclepiodotus, whom he dethroned, he too existed - but not as a British king. On the contrary, Asclepiodotus was actually a Roman! Indeed, he was a praetorian prefect and later consul known in full as Julius Asclepiodotus, who assisted Constantius Chlorus in re-establishing Roman rule in Britain.

Old King Cole, from *The Baby's Opera*
by Walter Crane, c.1877

An alternative contender for the true identity of King Cole is Coel Hen, who lived around 350-420 AD, and may have been the last of the so-called Dukes of the Britons, who commanded the Roman army in Northern Britain. He may even have taken over the Northern capital of Eburacum, now York, at around the time when the Romans were withdrawing their forces from Britain.

Moreover, as part of a quaint tradition, for many generations the children of the North Devon village of Woodfardisworthy (=Woolsery) have been told that Old King Cole is buried in their village's church! The tomb in question, however, is actually that of local nobleman Richard Cole, whose worthy actions in defending the area against overseas invasion during the late 16th Century elevated him in the populace's eyes and esteem to that of a veritable king! As documented by J.T. Downes ISO (father of CFZ Director Jonathan Downes) in his book *Woolsery: The Village With Two Names* (1998):

> Despite the defeat of the Armada, the threat from overseas continued and Hartland Parish records show that in 1598 Richard Cole of Bucks Mills - whose tomb is in Woolfardisworthy church - paid for 100 lbs of gunpowder for transportation to the ships of the Royal Navy in Plymouth.

Cole's great patriotic gesture was rewarded by a handsome tomb and a measure of immortality that continues today - in 2007, local playwright Stuart Rickard produced a village pantomime telling a completely fictionalised story of Old King Cole's exploits

Tomb of Richard Cole, Woolsery's very own 'Old King Cole' *(CFZ)*

Indeed, regardless of who he was or whether he even existed at all, King Cole lives on in the great corpus of British traditional legends and lore - just like King Lear and (especially) King Arthur. And, again like these latter figures, his name is probably more familiar than a number of our country's real-life previous rulers.

All of which is a tribute of sorts to the story-telling skills of Geoffrey of Monmouth - as much a King-maker with the pen as ever Richard Neville, Earl of Warwick, was with the sword!

Chapter 22

SOMETHING IN THE AIR

Fishing for Sky Beasts in the Stratospheric Sea

Conceive a jellyfish such as sails in our summer seas, bell-shaped and of enormous size - far larger, I should judge, than the dome of St. Paul's. It was of a light pink colour veined with a delicate green, but the whole huge fabric so tenuous that it was but a fairy outline against the dark blue sky. It pulsated with a delicate and regular rhythm. From it there depended two long, drooping, green tentacles, which swayed slowly backwards and forwards. This gorgeous vision passed gently with noiseless dignity over my head, as light and fragile as a soap-bubble, and drifted upon its stately way...[I]n a moment, I found myself amidst a perfect fleet of them, of all sizes, but none so large as the first. Some were quite small, but the majority about as big as an average balloon, and with much the same curvature at the top. There was in them a delicacy of texture and colouring which reminded me of the finest Venetian glass. Pale shades of pink and green were the prevailing tints, but all had a lovely iridescence where the sun shimmered through their dainty forms. Some hundreds of them drifted past me, a wonderful fairy squadron of strange unknown argosies of the sky - creatures whose forms and substance were so attuned to these pure heights that one could not conceive anything so delicate within actual sight or sound of earth.

Sir Arthur Conan Doyle – *'The Horror of the Heights'*

Higher and higher the monoplane had soared through the sky's rarefied atmosphere, bearing its pilot, a pre-World War I airman called Mr Joyce-Armstrong, to altitudes never previously attained by any human. Suddenly, as his eyes gazed upwards even further, they beheld an astounding sight - surely one of Nature's most incredible, well-kept zoological secrets – which he described in the above quote.

In reality, of course, this entire account is wholly fictional - extracted from 'The Horror of the Heights', which is a science-fantasy short story written by Sir Arthur Conan Doyle and published in

1913. Yet what if its theme could be true?

Our planet has engendered life forms morphologically and physiologically specialised to occupy every conceivable ecological niche on land and in water - but what about the sky? True, certain species of bird and insect spend much of their time in flight. However, there is no type of animal that lives its entire life airborne in the sky, adapted to an exclusive existence amid the atmosphere encircling our planet - or is there? In reality, there are some tantalisingly thought-provoking reports and theories on file offering up for serious consideration the fascinating possibility that extraordinary creatures completely unknown to science - bona fide sky beasts - do indeed inhabit the cloud-dappled world above us, only rarely revealing themselves to amaze their perplexed eyewitnesses below.

SKY BEASTS AS LIVING UFOS - EARLY THOUGHTS

UFOs are many things to many people - from misidentified weather balloons or lenticular clouds to extraterrestrial spacecraft...or sky beasts. Over the years, several different authorities have soberly examined the prospect that at least some UFO reports are based upon sightings of sky beasts whose ethereal existence still awaits official recognition.

Certainly, Earth's all-encapsulating atmosphere is more than sufficient spatially to accommodate any number of aerial entities with ease, regardless of shape or size. After all, even the atmosphere's densest, lowest portion, called the troposphere, extends upwards for about 11 miles from our planet's surface. And exterior to the troposphere is the stratosphere (11-31 miles above the planet's surface), followed by the mesosphere (31-50 miles), ionosphere (50-400 miles), and exosphere (400 miles and beyond, stretching into Space).

Perhaps the earliest modern-day investigator to consider the sky beast notion was arch-iconoclast Charles Fort, who spent much of his life amassing, assessing, and documenting anomalous data and reports of all kinds. Musing in his book *Lo!* (1931), Fort framed this concept very succinctly: "Unknown, luminous things, or beings, have often been seen, sometimes close to this earth, and sometimes high in the sky. It may be that some of them were living things that occasionally come from somewhere else".

However, the first notable researcher to formulate a specific theory of sky beasts (or ideoplasms, as he termed them) was John Philip Bessor, a veteran Pennsylvanian investigator of paranormal mysteries. Bessor's sky beast speculations began in 1946, but acquired great relevance a year later, following the now-famous sighting by Idaho businessman Kenneth Arnold on 24 June 1947 of a veritable UFO flotilla that he had encountered while piloting a Callair aeroplane near Mount Rainier, Washington. Coincidentally(?), this is the very same location where the bizarre batsquatch entity would be seen 46 years later.

Arnold saw nine UFOs in total, and although it is this sighting that inspired the media to coin the term 'flying saucer' and beget the alien spacecraft identity as a highly popular explanation for UFOs, Arnold's own thoughts on the matter were very different. He believed that his UFOs were: "living organisms, sort of like sky jellyfish...I suspect they just come down to look us over. I believe they are harmless or we would have had trouble with them long ago".

Inspired by Kenneth Arnold's words, on 7 July 1947 Bessor submitted his sky beast theory for UFOs to the U.S. Air Force - and, far from discounting the idea, an officer from the Air Force's Press and

'Cosmic Leap' - Philippa Foster's gorgeous painting of sky medusae

Radio Section replied that it was: "one of the most intelligent theories we have received". In an article published by *Fate* in December 1955, entitled 'Are the Saucers Space Animals?', Bessor deemed UFOs to be: "...a form of space animal, or creature, of a highly attenuated substance, capable of materialization and dematerialization, whose propellant is a form of telekinetic energy...Some may be quite invisible, others translucent, others opaque, still others capable of changing, chameleon-like, from one colour to another, from one form to another, from visibility to complete invisibility, all in a moment".

Interestingly, Bessor also linked his ideoplasms with reports of gelatinous meteors. These mysterious objects,

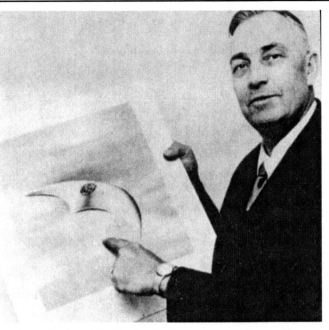

Kenneth Arnold with illustration of one of the UFOs seen by him *(FPL)*

bright and luminous in appearance, fall to earth like shooting stars, but instead of leaving behind any rocky residue (meteorites) at their landing sites as with true meteors, all that can be found is a puzzling jelly-like substance known as star rot or pwdre ser (see Chapter 18; also see my book *Mysteries of Planet Earth*, 1999). Yet if these gelatinous meteors had originated beyond our planet like true meteors, this jelly-like substance surely would have been totally incinerated while entering our atmosphere from Outer Space. Is it t.uly possible, therefore, as suggested by Bessor, that *some* star rot comprises the mortal remains of dead sky beasts that have fallen to earth from their airy abode in the sky?

FLYING FISHES IN A STRATOSPHERIC SEA?

While studying historical UFO reports, Bessor noticed that concentrations of sightings appeared to coincide with spasms of exceptional, freak weather. This correlation led him to speculate whether, just as fishes descend to less troubled, greater depths in the sea during surface storms (i.e. vertical migration), the sky beasts descend to lower atmospheric strata during severe weather to escape the harsh conditions prevalent at higher altitudes. Perhaps, as dubbed by veteran anomalies researcher Vincent H. Gaddis in his classic book *Mysterious Fires and Lights* (1967), we should think of sky beasts as fishes of the atmospheric sea.

There may be other reasons for the descent of sky beasts too. One of the most thought-provoking suggestions came from another prominent proponent of the sky beast scenario - Countess Zoe Wassilko-Serecki, an exceedingly wealthy, intelligent, liberal-minded thinker from Vienna. During the late 1950s, she published a number of articles collectively presenting a detailed appraisal of the sky beast concept, in which she claimed that the stratosphere is inhabited by enormous glowing organisms resembling gigantic bladders of colloidal silicones. Composed principally of pure energy but containing a central core of tenuous matter, they feed upon energy to sustain their luminous shapeshifting bodies, which are spherical when stationary but become fusiform (cigar- or spindle-shaped) when moving. In higher, rarefied strata of the atmosphere, however, such sky beasts might well be so diffuse that they are invisible - but when descending to lower, denser levels, their bodies may become sufficiently viscous to render them visible to the human eye.

Ufologists have often commented upon the notable frequency all over the world of UFOs sighted above localities containing energy-generating or energy-radiating sources, such as hydroelectric plants, and TV/radio transmitters. Those who believe that these airborne objects are extraterrestrial spacecraft may speculate that this is because their occupants are interested in learning more about our technology. Conversely, the Countess offered a much more biological explanation - if UFOs are sky beasts that feed upon energy, they would naturally be attracted to sources that regularly release considerable quantities of it. In the world of sky beasts, TV transmitters and other ground-based energy sources would no doubt be equivalent to a soup kitchen or even the proverbial free lunch!

The idea of organisms consuming pure energy is hardly unprecedented. A major component of photosynthesis in plants, by which they manufacture carbohydrate for nutritional purposes, is the harnessing of sunlight. Similarly, there are chemosynthetic bacteria ensnaring sulphur-released energy for much the same purpose at the base of a fairly recently-revealed biological foodchain within a sealed freshwater cave in Romania, as well as around hydrothermal rifts in the sea bottom near the Galapagos Islands and elsewhere.

SKY BEASTS AND THE STATUS QUO

The Countess's ideas have attracted interest from several scientists, including Ivan T. Sanderson. An American zoologist with a lifelong interest in cryptozoology and many other areas of the unexplained, Sanderson persuasively argued in support of the sky beast theory within one of his own books, *Uninvited Visitors: A Biologist Looks at UFOs* (1969).

During the 1960s, hydrophone inventor John M. Cage drew attention to the marked similarity between reports of UFOs pursuing aeroplanes and sightings of dolphins playfully following ships, and opined that these UFOs were sentient life forms which fed upon negative electricity: "They represent a life-form that diverged from the evolutionary process of this planet at such an early age that they do not fall into any predetermined category established for the classification of ordinary terrestrial lifeforms". Cage also substantiated another reason offered by the Countess for the occurrence of sky beasts in lower atmospheric zones. Perhaps they are simply curious about us!

For millions of years, the upper reaches of the sky have remained sacrosanct from contact by the realms of land and sea below. Here, according to the sky beast theory, these metrically vast but materially insubstantial entities have evolved in supreme isolation, unsullied and undisturbed by events on Earth. Since the early 20th Century, however, aeroplanes have roared on silver wings through the sky beasts' cathedrals of clouds, rockets have surged relentlessly upwards like volleys of scorching arrows, and nuclear explosions have spewed forth their deadly emissions in bilious contempt upon the once-immaculate roof of the world. Is it any wonder, therefore, as argued by Cage and the Countess, that UFO sightings are increasing? The sky beasts are cautiously venturing down, to discover what is happening in the frenetic zones far beneath their own languid kingdoms of air and space.

This reasonable hypothesis has been very eloquently expressed by a leading biologist from the Wright-Patterson Air Force Base in Dayton, Ohio, who prefers not to link his name in print with the possibility that sky beasts exist, but clearly has no doubt that they do:

> When you toss a pebble into a pool, you see the water animals - the nymphs, crawfish, minnows, water insects - streaking away in fright. But almost before the ripples have died in the pool they are back, investigating curiously. Perhaps events on this earth - atomic explosions, rocket flights into the stratosphere and so on - have been like the pebble tossed into the pool. They have disturbed the stratosphere, perhaps have sent ripples into the fabric of space itself. And the space animals, in curiosity, are coming down to investigate.

CONSTABLE AND THE CRITTERS - PHOTOGRAPHING SKY BEASTS

The most famous proponent of the theory that at least some UFOs are sky beasts is undoubtedly Trevor James Constable, who spent over 20 years pursuing evidence for the reality of such creatures, and wrote two books on the subject. The first of these, published in 1958, was *They Live in the Sky*; but the second one, first published in 1976 as *The Cosmic Pulse of Life* and abridged in 1978 as *Sky Creatures: Living UFOs*, is widely recognised as the definitive book on this sorely-neglected sub-discipline of ufology.

Christening them "critters", Constable proposed that sky beasts are unicellular in structure and reminiscent in form to amoebas, but possess a mica-like or metallic outer body surface. The remainder of their body is plasmatic - composed of plasma (the fourth state of matter), comprising an ionised gas containing approximately equal numbers of positive and negative charges. Critters range greatly in size, from half a mile long down to just a few inches.

Constable claimed that critters are seldom seen because they usually reflect infra-red light, which is invisible to the human eye. However, because they are able to change colour, sometimes they reflect visible light, and it is then that they are seen by human observers.

Notwithstanding this, evidence of their presence can apparently be obtained even when they are invisible because, according to Constable, critters can be photographed using infra-red cinefilm and an appropriate filter. In support of this remarkable claim, he included many photos allegedly portraying critters detected by him in the sky above California's Mojave Desert. Using Kodak IR 135 film that was loaded and unloaded in total darkness to prevent fogging, plus a Kodak Wratten filter over the lens (87, 87C, and 88A filters are all suitable), he and a colleague, James O. Woods, obtained numerous photos. The best were snapped just after dawn, and in the area of the Lucerne Valley between Victorville and Yucca Valley.

As the critters were invisible to the eye, Constable and Woods simply aimed their cameras at the sky and shot the entire horizon in overlapping pictures. Consequently, they never had any idea as to whether they had successfully photographed any critters until they actually developed their reels of film. Photos believed by the researchers to depict critters portray a variety of different forms. Some are fusiform and look like giant amoebas or enormous bladders (recalling the Countess's claims), but others are discoidal, and some of the biggest ones even have vaguely reptilian 'beaks'. Furthermore, as revealed by Constable's cinefilms, critters contract and expand as they speedily traverse the sky, and they are all luminous. Yet whereas some glow continually, others flash off and on. There are also a number of virtually transparent specimens, which appear almost two-dimensional (i.e. without any thickness).

Constable's curious images have never been exposed as hoaxes, and remain unexplained. Moreover, several other researchers have succeeded in obtaining similar 'critter photos' utilising Constable's methods. In May 1977, for instance, UFO enthusiast Richard Toronto visited the Mojave Desert and attempted to duplicate Constable's photographic work - with considerable success. He obtained a sizeable collection of photos depicting critter-like objects in the sky but which had not been seen by him while snapping the shots. When some of his films were examined by Kodak, however, they suggested that the critters were nothing more than drying spots and finger marks - a claim denied by Toronto. Conversely, when they were examined by psychical researcher D. Scott Rogo, he considered them to be 'thoughtographs' - images psychically imposed upon the film's emulsion by the thoughts of the photographer. Toronto, however, remains convinced that his pictures genuinely depict critters.

DRAGONS OF THE AIR

In Conan Doyle's story, 'The Horror of the Heights', aerial jellyfishes were not the only sky beasts encountered by the fictitious monoplane pilot:

> ...soon my attention was drawn to a new phenomenon - the serpents of the outer air. These were long, thin, fantastic coils of vapour-like material, which turned and twisted with great speed, flying round and round at such a pace that the eyes could hardly follow them. Some of these ghost-like creatures were twenty or thirty feet long, but it was difficult to tell their girth, for their outline was so hazy that it seemed to fade away into the air around them. These air-snakes were of a very light grey or smoke colour, with some darker lines within, which gave the impression of a definite organism...[but] there was no more solidity in their frames than in the floating spume from a broken wave.

In December 1917, however, an amazing instance (if genuine) of Nature mimicking Art occurred. For this was when the *Occult Review* published a remarkable report in which its author (who referred to himself merely as 'a philosophical aviator') claimed that one of his correspondents, a true-life airman, had actually encountered a very similar beast in reality!

According to the report, the true-life airman (unnamed) was a World War I pilot who had allegedly confronted at a considerable altitude a weird apparition that he likened to a colourful dragon, drifting towards him through the air at appreciable speed - an unnerving event that induced him to descend to earth straight away!

Assuming this bizarre report to be authentic, the pilot was probably suffering from oxygen deficiency - a hazard when flying at great heights, and capable of eliciting a wide range of optical hallucinations. However, the conspicuous absence of names for both the pilot and the 'philosophical aviator' unavoidably turns thoughts towards the possibility that the entire report was a hoax (perhaps even inspired by the Conan Doyle story). Yet even if this is true, how can we explain the many other reported sightings of dragonesque 'sky serpents' that have been recorded from around the world?

SKY SERPENTS, OR FIRE DRAKES?

Whatever a certain farmer in Bonham, Texas, expected to see when he looked up at the sky while working on his farm one day in June 1873, it certainly was not the gigantic yellow-striped serpent, writhing and thrusting, that floated overhead. A hissing sky snake was also reported during May 1888 in South Carolina's Darlington County, a Scandinavian equivalent was spied over southern Norway and Denmark in March 1935, and nine months later another one appeared over Cruz Alta in Brazil on two separate occasions.

Nor is Britain immune to overhead ophidians - the Devon town of Bideford was brightly illuminated by a twisting sky serpent for six minutes on 5 December 1762.

Perhaps the most dramatic example of all was the white, self-luminous 'sky eel' with flapping side fins that wriggled like a swimming fish across the sky above Crawfordsville, Indiana, for two successive nights on 4-5 September 1891. Viewed by hundreds of astonished local eyewitnesses, who were convinced that this 18-20-ft-long apparition was a living creature, the sky eel mostly hovered about 300 ft above the ground, emitting a loud wheezing sound, but occasionally it swooped down to within 100 ft, radiating hot air. Neither a head nor a tail could be discerned, but it did possess a fiery red eye-like structure.

What are we to make of these entities from the ether? Terrors in semi-tangible form, or merely a selection of unusual meteorological monsters, akin to aurorae and perihelia?

There is little doubt that certain aerial dragonesque or serpentine entities are indeed monsters of the meteorological kind. Indeed, in olden days they even had their own name – fire drake, though this is rarely used nowadays (even earlier names for them included the brenning and the dipsas). As far back as the reign of Elizabeth I, one chronicler, in a now-obscure tome entitled *Contemplation of Mysteries*, included the following concise, instructive explanation of this particular phenomenon:

> The flying dragon is when a fume kindled appeereth bended, and is in the middle wrythed like the belly of a dragon; but in the fore part, for the narrownesse, it representeth the figure of the neck, from whence the sparkes are breathed or forced forth with the same breathing.

A more expansive description concerning the skyborne appearance of fire drakes was published in *The World of Wonders* (1882), edited by A. Taffs:

> These strange, and indeed startling sights, have appeared under certain peculiar and favourable conditions of the atmosphere, and, philosophically considered, are easily and satisfactorily accounted for. When vapours of an inflammable kind collected in the air and ascended to a cold region, the vehement agitation thereby produced induced a flame. The highest part, being more subtle, assumed the singular form of what was presumed to be the dragon's neck, and then, having been made crooked by the repulse it received, formed the dragon's belly, while the hind part, turned upwards by the force of the same collision, represented the monster's tail. Then, with impetuous motion, it fled through the heavens – all ablaze, as it were – striking deadly terror into the hearts of the ignorant and superstitious.

More recently, John H. Parkinson explained that early sightings of aurorae certainly explain some bygone records of fiery sky dragons (*New Scientist*, 14 April 1983):

> These dramatic spectacles are caused by charged particles, which are accelerated in solar flares, being diverted towards the poles by the Earth's magnetic field where they enter the upper atmosphere causing ionisation and producing light. Aurorae appear as streamers or curtains, varying in colour from whitish-green to deep red, often changing minute by minute.

Nevertheless, there are a few sky serpent and aerial dragon reports on file that cannot readily be resolved by straightforward meteorological means. One of the most notable of these was first documented in my previous book, *Extraordinary Animals Revisited* (2007), and definitely warrants republication here:

> The [report] was given to me in early March 2001 by a exceedingly knowledgeable, well-respected, but thoroughly mystified British naturalist who, while willing for me to publish an account of his sighting,

was understandably reluctant to be publicly identified in relation to it, or to permit the precise location of his sighting to be publicly revealed until he has investigated it further himself. (I have full details on file.) However, he averred that he did indeed see what he described to me, and that he was stone-cold sober at the time.

His sighting had occurred a week before he contacted me, and had taken place while he and two colleagues were visiting a certain locality in the county of Powys, Wales. The reason why they were there is that the naturalist had been contacted by an inhabitant of that locality, asking him to come and see "something", but declining to say what that "something" was. Consequently, to avoid any chance of becoming the victims of a hoax, the naturalist and his two colleagues did visit this locality, spending three days there, but in secret - not having told the naturalist's informant that they were coming.

Their sighting occurred on the evening of Day 2. Standing at the edge of some woods by a quarry at around 6.00 pm, they suddenly spied an extraordinary entity. Measuring 2.5 ft or so in length, it resembled a serpentine dragon with four short limbs, but its head was shaped very like that of a sea horse, and it was airborne - undulating and wriggling as it flew about 10 ft above the surface of the quarry in a wide circle. They were unable to recall seeing any wings, but it had a long tail that terminated in a pair of horizontal, whale-like flukes. The entity was green in colour and shimmered somewhat, but appeared solid, not translucent or ethereal, and they watched it for 3-4 minutes, at a distance of roughly 50 ft, before it finally vanished into one of the numerous caves and large crevices pitting the quarry.

The naturalist had the distinct impression while watching it that this creature, or whatever it may be, was deliberately seeking to keep them at bay, warning them off from approaching further into its territory. Attempts by him to photograph it proved unsuccessful, as the early evening light was not bright enough for the camera's automatic system to function satisfactorily. He is convinced that it was neither an optical illusion nor a model, but was truly alive, although its appearance was so uncanny that he felt chilled by the encounter.

After their sighting, they then visited the naturalist's informant, who duly confirmed that this is what he had hoped they would see, as he had seen it several times himself and was totally perplexed as to what it could be.

Whatever it was, this was surely no mere artefact of atmospheric engendering, and its modern-day occurrence demonstrates well that sightings of sky serpents are far from over.

FOO FIGHTERS, GIZMOS, ANGELS, AND MIN-MINS

As previously noted in Chapter 14, while flying their aeroplanes during World War II various English, German, French, American, and Japanese fighter pilots alike encountered small but highly mystifying balls of white or red light that became known as foo fighters. These luminous spheres behaved in an uncannily intelligent manner - pursuing the planes, dancing over their wings, and changing course to remain alongside them if the planes altered their direction. Indeed, foo fighters gave every impression of playful curiosity, as if they were wondering what these metallic bird-like objects were that had begun to streak through their hitherto inviolate aerial domain.

Of particular interest is that foo fighters could not be detected by radar. Conversely, radar records contain many traces of what would appear to be similar entities, nicknamed gizmos or angels, and exhibiting comparable activity to foo fighters, but which often remain invisible to the naked eye. Sometimes flying singly, but on other occasions in small groups moving in precise formations, and at speeds of 30-60 mph in all weather and at all times of the day and night, gizmos produce radar traces that can be readily distinguished from those of all known airborne objects such as birds, ionised air masses, insect swarms, aircraft, etc. Attempts made by scientists monitoring gizmos to reconcile them with meteorological phenomena such as bubbles of water vapour or non-homogeneities in the refractive index gradients believed to exist in the atmosphere have failed to win widespread acceptance.

And then there are the Australian min-mins and eerie dancing orbs of light reported from elsewhere in the world (collectively termed spooklights, and not to be confused with ball lightning) that have tormented observers for centuries by persistently pursuing them or luring them onwards, just like mischievous airborne sprites playing tantalising games. More than once, a human victim has attempted to swat one of these luminous globes (also dubbed BOLs – balls of light) - only for the globe to dodge deftly out of harm's way, before returning to the 'game' again! (See also Chapter 14.)

Trevor Constable proposed that foo fighters and their kind are small atmospheric relatives of the giant critters, possessing a rudimentary consciousness that expresses itself in their profound inquisitiveness regarding humans and their desire to indulge in what they consider to be pleasurable games with them - a view not always shared by their human participants!

THE RIDDLE OF THE RODS

The most recent additions to the panoply of mysterious but conceivably living aerial entities on file are the rods.

Also termed skyfishes or solar entities and often small in size (just a few inches in length, though specimens up to 10 ft long have also been claimed) but hugely controversial in nature, these enigmatic objects initially came to attention during the mid-1990s in – of all places! – the New Mexico town of Roswell, a name forever linked with stories of UFO sightings and crash-landed aliens. First reported and filmed there by José Escamilla, rods have since been observed on film obtained by other investigators and in many locations worldwide – but most notably inside a deep cavern called the Cave of Swallows, near San Luis Potosi, Mexico, on a film shot there in 1996 by professional cameraman Mark Lichtle, which seems to show hundreds of them.

Moving through the air so swiftly that they are invisible to the naked eye, when captured on film they appear as cylindrical objects that seem to possess fins, wings, or a rapidly-vibrating, undulating diaphanous membrane of some kind running along each side of the body. Such is their speed that one 'slow' rod was captured on just 10 frames of videotape, thus comprising a total on-screen time of only 0.3 seconds. Indeed, flying speeds of up to 185 miles per hour have been alleged.

Sceptics have dismissed these filmed rods as camera artefacts or digital distortions, inanimate debris, insects, and even birds – though no known insect or bird can move at anything even approaching the speeds claimed for rods. Believers, conversely, have sometimes likened them to flying centipedes in overall morphology, claim that they demonstrate conscious thought as they seemingly change direction to avoid collisions, and state that they comprise a radically new type of life form. It has even been suggested by intrigued biologist and rod investigator Ken Swartz that they may represent an undiscovered, modern-day aerial descendant of *Anomalocaris*, an ancient membrane-fringed marine invertebrate from the Cambrian Period 400 million years ago, with no known present-day relatives.

Regardless of their taxonomy, however, if rods are indeed living entities their metabolic rate must be truly phenomenal in order to power their incredible flying speeds – needing, in Swartz's opinion, to consume their own weight in food every single day. Of course, surely the riddle of the rods could be conclusively solved if a specimen were to be captured - but how do you go about snaring something

'Guiding Lights', painted by Philippa Foster

that moves so quickly it cannot even be seen? Three years ago, some determined Chinese research staff provided the answer.

During May-June 2005, after their facility's surveillance cameras had filmed numerous examples of rods, the scientific research staff at the Tonghua Zhenguo Pharmaceutical Company in China's Jilin Province took matters into their own hands by installing huge nets. When their surveillance cameras confirmed that rods were indeed flying into these nets, the staff inspected their contents, only to find that they contained nothing more than ordinary moths and other familiar flying insects. Later investigations confirmed that the appearance of flying rods on the surveillance equipment's video was an optical illusion created by the camera's slower recording speed (done to save video space).

Moreover, an episode devoted to rods in the 'Monster Quest' television series, first screened in January 2008, showed a flying rod that had been filmed simultaneously by a traditional video camera and by a high-speed video camera. Tellingly, whereas the former showed a brightly-illuminated rod with multiple undulating wings, the latter revealed it to be nothing more extraordinary than a common moth in flight.

Faced with such damning evidence, the sceptics' view concerning rods, that they are merely optical artefacts, is by far the most convincing, and is effectively explained as follows in the online Wikipedia entry for these entities:

> Evidence points to the conclusion that they are mere tricks of light which result from how images (primarily video images) are recorded and played back. In particular, the fast passage before the camera of an insect flapping its wings has been shown directly to produce rod-like effects, due to motion blur, if the camera is shooting with relatively long exposure times. (In low-light conditions or even when pointed at blue sky, the automatic exposure programming of a video camera is likely to select the longest possible exposure time, which is 1/60th second per video field for NTSC format or 1/50th second for PAL format.) This criticism points to such video being physically unable to capture a clean image of something which moves so fast relative to the camera. In particular, the "membrane" in a video frame of a rod is effectively a time-lapse of the wings of the flying animal over several wingbeats that occurred during the field exposure time, while the central "rod" is a time-lapse image of the body, showing the full distance traveled during the field exposure time. The effect is especially pronounced with large, long-bodied insects which have broad wings and fairly slow wingbeats, such as mantises, grasshoppers, and katydids, or completely opaque wings such as moths. On video equipment which resolves the two interlaced fields of a single video frame (which are captured successively and then displayed as alternating horizontal lines), the "rod" effect can be seen to alternate from one field to the other, producing the distinctive gaps between successive images. Similar results can be produced using standard film, if there is a long exposure and/or a stroboscopic lighting effect which lasts more than a single wingbeat. This is the technical evidence, demonstrating that one can produce "rod" effects at will if one uses the right equipment, lighting, and subject.

In short, it would seem that rods are illusive rather than elusive. Whether the same can be said for other aerial anomalies documented here, conversely, remains to be seen. Perhaps we should look for clues beyond our own atmosphere...?

SKY BEASTS ON JUPITER?

In 1977, Cornell University astrophysicist Dr E.E. Salpeter and eminent astronomer Prof. Carl Sagan published a paper in the *Astrophysical Journal Supplement* outlining their thoughts on the possible presence of life on Jupiter, and the likely forms that it may take, existing amidst this gargantuan gas planet's clouds. They concluded that much of whatever life might indeed reside there may well parallel the ecology of Earth's ocean fauna - yielding Jovian equivalents to our plankton, fishes, and larger fish-eating marine predators (turtles, sharks, etc), and which the two scientists have termed sinkers, floaters, and hunters respectively.

Salpeter and Sagan postulated that they could be enormous sac-like organisms, filled with helium, and propelling themselves through the planet's atmosphere by controlled expulsion of this gas from their bodies. However, they did not consider these airborne balloon-like beasts to be wholly conjectural. In fact, they suggested that the presence of such creatures (perhaps even attaining a total diameter of many miles in the case of the Jovian 'hunters') may explain the frequent occurrence over the planet of clearly-perceived areas of red colouration.

The two scientists hoped that this prospect could be examined by the close-up cameras that would be focused upon Jupiter by the U.S.'s Mariner 11 and 12 space probes after their launch later in 1977, but no evidence to support the existence of these (or any other) Jovian life forms was found. Even so, the ideas of Salpeter and Sagan remain plausible relative to the likely nature of life on that huge gaseous world, should any truly exist there. Back in 1971, moreover, the scenario of colossal sky jellyfishes on Jupiter had been the basis for one of Arthur C. Clarke's most famous science-fiction stories, 'A Meeting With Medusa'.

FINAL THOUGHTS – CRITTERS AND CROPFIELD CIRCLES

Today, the notion of 'biological UFOs' - UFOs as unknown species of giant airborne animal native to our planet but physically aloof from the rest of our world's wildlife - is somewhat out of vogue. Yet according to one latter-day researcher, physical evidence for their reality is regularly reported by the media and has become well known to everyone but remains unrecognised for what it is. In his absorbing book *The Circlemakers* (1992), British psychical investigator Andrew Collins proposed that energy released by sky beasts of the critter variety when swooping groundwards from the heavens is responsible for the eyecatching but enigmatic cropfield circles. He also cited critters to explain various scorched circles found in fields in New Zealand's North Island during 1969 following a local spate of UFO sightings.

How ironic it would be if cropfield circles, surely among the most conspicuous of terrestrial curiosities, ultimately confirmed the reality of sky beasts - indisputably the least conspicuous of all life forms indigenous to planet Earth.

Nor should we forget the unresolved radar blips produced by gizmos, the unidentified star rot found at the landing sites of gelatinous meteors, and the unexplained objects captured by Constable on infra-red film. Perhaps the sky beast scenario is not so intangible and remote from our modern technological world after all.

Even so, whenever I think of sky beasts, the words that come to mind are from a time long ago. Quoted from *Divine Weekes and Workes* by the 16th-Century Gascon poet Guillaume de Salluste du Bartas, they were not inspired by our sky beasts - yet these ephemeral entities could surely never be more aptly described:

> None knowes their nest, none knowes the dam that breeds them:
> Foodless they live, for th'aire alonely feeds them;
> Wingless they fly; and yet their flight extends
> Till with their flight their unknown lives-date ends.

SELECTED BIBLIOGRAPHY
AND FURTHER READING

As the present book covers such an exceptionally diverse range of subjects, it would be impractical in the extreme to make any attempt at providing a comprehensive bibliography of all sources consulted by me during the preparation of the original articles upon which it is based. However, the following were of particular use, and are highly recommended for further reading.

– anon

-, 'The Unicorn' (*Scientific American*, vol. 4, 1848), p.150.

-, 'Unicorn Drawings' (*Nature*, vol. 6, 1872), p.292.

-, 'The Appearance of Mermaids...' (*Daily Kennebec Journal*, 24 June 1873).

-, 'A Cape Breton Mermaid' (*Cape Brooklyn Eagle*, 22 August 1886).

-, 'A Yarn From Seattle: Capture of a Singular Creature Resembling a Merman'
    (*Cape Brooklyn Eagle*, 3 November 1896).

-, 'Can a Cat Fly?' (*Strand Magazine*, vol. 18, November 1899), p.599.

-, 'Gold-Coloured Teeth of Sheep' (*Nature*, vol. 107, 21 April 1921), p.249.

-, 'Cat With Wings! Zoo Experts Puzzled By Freak That Can "Fly"'
    (*Sunday Dispatch*, 10 June 1933).

-, 'Aparece en Madrid un Gato con Alas' (*Informaciones*, Madrid, 23 May 1950).

-, 'Historia de "Angolina" – La Gata Con Alas'
    (*Informaciones*, Madrid, 25 and 27 May 1950).

-, 'Otra Gata Con Alas' (*Informaciones*, 12 June 1950).

-, 'Court Suit to Determine Possession of Winged Cat' (*Evening Star*, Washington DC, 2 July 1959).

-, 'The Unicorn Legend' (*Animals*, vol. 1, no. 5, 5 February 1963), pp.28-30.

-, *Stranger Than People* (Young World Productions: London, 1968).

-, 'A Classic Case of "Angel-Hair"' (*Pursuit*, vol. 3, no. 4, October 1970), pp.72-3.

-, 'The Cat With Wings' (*Manchester Evening News*, 11 August 1972).

-, 'Thomas Bessie Spends His 73rd Birthday in a Pub – Again'
  ([a Yorkshire newspaper, title unknown], 5 December 1973).

-, 'Florida's Mystery Sphere' (*APRO Bulletin*, vol. 22, no. 5, March-April 1974), pp.1,4.

-, 'A Cat That's Bats' (*The News* [now *Fortean Times*], no. 6, September 1974), p.3.

-, 'What a Cat Flap!' (*Weekend*, 12 November 1980).

-, 'Probe Into 'Eyes Closed' Puzzle' (*Sandwell Express and Star*, 29 March 1989).

-, 'Nothing Fishy About the Tale From Outback' [fish rain]
  (*Sandwell Express and Star*, 1 March 1994).

- [WELFARE, Adam], *The Mermaid's Grave, Nunton, Isle of Benbecula, Scotland. An Examination
  of a Putative Grave-Marker Upon the Dunes Fringing Culla Bay* (Heritage Site and Landscape
  Surveys Ltd, 1994).

-, 'Creation 'Fireballs' Made By Boffins' (*Sandwell Express and Star*, 6 June 1996).

-, 'Landing of Blobs After UFO Alert' (*Sandwell Express and Star*, 4 November 1996).

-, 'Miracle Claimed as Wafer Shines Out' (*Belfast Newsletter*, 18 April 1998), p.3.

-, 'The Tibetan Roswell' [re *Sungods in Exile*] (*Fortean Times*, no. 109, April 1998), p.39.

-, 'Babel Tower's Tall Order' (*Sandwell Express and Star*, 18 September 1998).

-, 'The Lost Planet Idea For Comets' (*Sandwell Express and Star*, 8 October 1999).

-, 'Great Balls of Fire May Be Just Fluff' (*Sandwell Express and Star*, 3 February 2000).

-, 'Winged Cat 'From Hell' Put To Death' (*MosNews*, Moscow, 29 July 2004).

-, 'Mysterious Amphibious Human-Like Creature Spotted in the Caspian Sea' [runan-shah]
  (*Pravda*, 25 March 2005).

-, 'Cat Grows Wings' (*Ananova*, 24 May 2007).

A

ADAMSKI, George, *Inside the Space Ships* (Abelard-Schuman: New York, 1955).

ADAMSKI, George and LESLIE, Desmond, *Flying Saucers Have Landed*
  (T. Werner Laurie: London, 1953).

ALEXANDER, Caroline, *The Way To Xanadu* (Weidenfeld and Nicolson: London, 1993).

ALWAY, Carol *et al.*, *Strange Stories, Amazing Facts* (2nd edit.) (Reader's Digest: London, 1984).

ARNOLD, Neil, *Monster! The A-Z of Zooform Phenomena* (CFZ Press: Bideford, 2007).

ASH, Russell *et al.*, *Folklore, Myths and Legends of Britain* (Reader's Digest: London, 1973).

ASHE, Geoffrey, *Mythology of the British Isles* (Methuen: London, 1990).

ATKINSON, B.W., 'Riddle of Ball Lightning'
  (*Geographical Magazine*, vol. 41, no. 3, December 1986), pp.204-5.

AZARA, Felix de, *The Natural History of the Quadrupeds of Paraguay and the River La Plata*
  (London, 1837).

B

BADER, Chris, *Strange Northwest* (Hancock House: Blaine, 1995).
BAHN, Paul G., 'Making Sense of Rongorongo' (*Nature*, vol. 379, 18 January 1996), pp.204-5.
BAHN, Paul G., *The Easter Island Enigma* (Weidenfeld and Nicolson: London, 1997).
BAINTON, Roy, 'The Spear Carriers' [Spear of Destiny]
        (*Fortean Times*, no. 175, October 2003), pp.48-52.
BALDWIN, Barry, 'Classical Corner: 5. Falling Frogs'
        (*Fortean Times*, no. 138, September 2000), p.14.
BARBER, Richard and RICHES, Anne, *A Dictionary of Fabulous Beasts*
        (Macmillan: London, 1971).
BARCLAY, David, *Aliens: The Final Answer* (Blandford: London, 1995).
BARING-GOULD, Sabine, *Iceland, Its Scenes and Sagas* (Smith, Elder: London, 1863).
BARING-GOULD, Sabine, *Curious Myths of the Middle Ages* (Longmans, Green: London, 1892).
BATES, Laraine, [Letter re Woolpit's green children] (*Daily Mail*, 2 July 1997).
BEARDEN, Thomas, *Species Metapsychology, UFO Waves, and Cattle Mutilations*
        (Privately published, 1977).
BEATTY, Bill, *A Treasury of Australian Folk Tales and Traditions*
        (Ure Smith: North Sydney, 1960).
BEER, Lionel, *The Moving Statue of Ballinspittle and Other Related Phenomena*
        (Spacelink: London, 1986).
BEER, Rüdiger R., *Unicorn: Myth and Reality* (Ash & Grant: London, 1977).
BENNETT, D.M., *A Truth Seeker Around the World* (Privately published: New York, 1882).
BENNETT, Jon, 'Where in the World: Mermaids' (*Beyond*, no. 8, September 2007), pp.71-3.
BENNETT, Jon, 'Where in the World: Weeping Statues' (*Beyond*, no. 12, January 2008), pp.65-7.
BENTLEY, James, *Restless Bones: The Story of Relics* (Constable: London, 1985).
BENWELL, Gwen and WAUGH, Arthur, *Sea-Enchantress: The Tale of the Mermaid and Her Kin*
        (Hutchinson: London, 1961).
BESSOR, John P., 'Mysterious Lights of Australia' (*Fate*, vol. 6, August 1953), pp.87-90.
BESSOR, John P., 'Are the Saucers Space Animals?' (*Fate*, vol. 8, December 1955), pp.6-12.
BIRCHBY, Sid, 'Winged Cats' (*Fortean Times*, no. 35, Summer 1981), p.47.
BLUNDELL, Nigel and HALL, Allan, *Marvels and Mysteries of the Unexplained*
        (Tiger Books International: London, 1993).
BORD, Janet and BORD, Colin, *Alien Animals* (rev. edit.) (Panther: London, 1985).
BORD, Janet and BORD, Colin, *Modern Mysteries of Britain: 100 Years of Strange Events*
        (Grafton: London, 1987).
BORD, Janet and BORD, Colin, *Modern Mysteries of the World: Strange Events of
        the 20th Century* (Grafton Books: London, 1989).
BORD, Janet and BORD, Colin, *Atlas of Magical Britain* (Sidgwick and Jackson: London, 1990).
BORD, Janet and BORD, Colin, *Life Beyond Planet Earth?* (Grafton: London, 1991).
BORN, Jaap Van Den, 'The Bati-Men' (*Fortean Times*, no. 169, April 2003), p.55.
BOULAY, R.A., *Flying Serpents and Dragons: The Story of Mankind's Reptilian Past* (rev. edit.)
        (The Book Tree: Escondido, 1997).
BOWEN, Charles, 'Crash-Landed UFO Near Mendoza'
        (*Flying Saucer Review*, vol. 11, no. 3, 1965), pp.7-8.
BRANDON, Jim, *Weird America* (E.P. Dutton: New York, 1978).
BRIGGS, Katharine, *A Dictionary of Fairies* (Allen Lane: London, 1976).

BROOKESMITH, Peter (Ed.), *The Unexplained: Mysteries of Mind, Space and Time* (13 vols) (Orbis: London, 1980-83).

BROOKESMITH, Peter (Consultant), *Bizarre Phenomena* (Reader's Digest: Pleasantville, 1992).

BROOKS, J[ohn]. A., *Ghosts of London* (2nd edit.) (Jarrold: Norwich, 1991).

BROOKS, John, *The Good Ghost Guide: A Gazetteer of Over a Thousand British Hauntings* (Jarrold: Norwich, 1994).

BROWN, Robert, *The Unicorn: A Mythological Investigation* (Longmans, Green: London, 1881).

BURNHAM, Owen, 'Guiafairo – The Fear That Flies By Night' (*Animals and Men*, no. 10, July 1996), p.33.

BURNHAM, Owen, 'Kikiyaon the Soul Cannibal' (*Uri Geller's Encounters*, no. 9, June 1997), pp.66-7.

BURNHAM, Owen, *African Wisdom* (Piatkus: London, 2000).

BUTLER, W.F., 'Fragility in the Skin of a Cat' (*Research in Veterinary Science*, vol. 19, 1975), pp.213-16.

C

'CARLOSOX', 'Man-Faced Eagle' (*UFO Digest*, 8 December 2006).

CARMICHAEL, Alexander, *Carmina Gadelica* (2 vols) (Norman MacLeod: Edinburgh, 1900).

CARPENTER, John, '"Reptilians and Other Unmentionables"' (*MUFON UFO Journal*, no. 300, April 1993), pp.10-11.

CHATWIN, Bruce, *In Patagonia* (Jonathan Cape: London, 1977).

CLARK, Jerome, 'Space Animals?' (*Fate*, vol. 44, April 1991), pp.40-4.

CLARK, Jerome, 'Rot From the Stars' (*Fate*, vol. 44, May 1991), pp.30-5.

CLARK, Jerome, 'Mermaid Mysteries' (*Fate*, vol. 47, November 1994), pp.37-40 and (*Fate*, vol. 47, December 1994), pp.34-6.

CLARK, Jerome, *Unexplained!* (2nd edit.) (Visible Ink: Detroit, 1999).

CLARK, Jerome and COLEMAN, Loren, 'Winged Weirdies' (*Fate*, vol. 25, March 1972), pp.80-9.

CLARK, Jerome and COLEMAN, Loren, 'Unidentified Flapping Objects' (*Oui*, October 1976), pp.95-106.

CLARK, Jerome and COLEMAN, Loren, *Creatures of the Outer Edge* (Warner: New York, 1978).

CLAYTON, Peter and PRICE, Martin (Eds), *The Seven Wonders of the Ancient World* (Routledge: London, 1988).

CLOUDSLEY-THOMPSON, John, 'Ball Lightning' (*Country-Side*, new series, vol. 27, no. 1, 1989), p.10.

COHEN, Daniel, *Encyclopedia of Ghosts* (Michael O'Mara: London, 1984).

COHEN, John, *Human Robots in Myth and Science* (George Allen and Unwin: London, 1966).

COLEMAN, Loren, 'The Strange Case of the Two Charlie Wetzels' (*Fortean Times*, no. 39, Spring 1983), pp.44-6.

COLEMAN, Loren, 'Creatures From the Black Lagoon' (*Fortean Times*, no. 40, Summer 1983), pp.43-7.

COLEMAN, Loren, *Mysterious America* (Faber and Faber: London, 1983).

COLEMAN, Loren, *Curious Encounters* (Faber and Faber: London, 1985).

COLEMAN, Loren, 'Other Lizard People Revisited' (*Strange Magazine*, no. 3, Winter 1988), pp.34,36.

COLEMAN, Loren, *Mothman and Other Curious Encounters*

(Paraview Press: New York, 2002).

COLEMAN, Loren and CLARK, Jerome, *Cryptozoology A-Z: The Encyclopedia of Loch Monsters, Sasquatch, Chupucabras, and Other Authentic Mysteries of Nature* (Fireside: New York, 1999).

COLEMAN, Tim, 'Serpents in the Sky' [rods] (*The X Factor*, no. 93, 2000), pp.2581-5.

COLLINS, Andrew, *The Circlemakers* (ABC Books: Leigh-on-Sea, 1992).

CONSTABLE, Trevor J., *Sky Creatures* (Pocket Books: New York, 1978).

CORRALES, Scott, 'Return of the Birdmen' (*Fate*, vol. 51, October 1998), pp.14-17.

CORLISS, William R., *The Unexplained: A Sourcebook of Strange Phenomena* (Bantam: New York, 1976).

CORLISS, William R., *Handbook of Unusual Natural Phenomena* (Sourcebook Project: Glen Arm, 1977).

CORLISS, William R., *Mysterious Universe: A Handbook of Astronomical Anomalies* (Sourcebook Project: Glen Arm, 1979).

CORLISS, William R., *Incredible Life: A Handbook of Biological Mysteries* (Sourcebook Project: Glen Arm, 1981).

CORLISS, William R., *Tornados, Dark Days, Anomalous Precipitation, and Related Weather Phenomena - A Catalog of Geophysical Anomalies* (Sourcebook Project: Glen Arm, 1983).

CORLISS, William R., *Anomalies in Geology: Physical, Chemical, Biological – A Catalog of Geological Anomalies* (Sourcebook Project: Glen Arm, 1989).

CORLISS, William R., *et al.* (Consultants), *Feats and Wisdom of the Ancients* (Time-Life: Alexandria, 1990).

CROSSWELL, Ken, 'Hopes Fade on Hunt For Planet X' (*New Scientist*, vol. 137, 30 January 1993).

CRUZ, Joan C., *The Incorruptibles* (Tan Books: Rockford, 1977).

CURRER-BRIGGS, Noel, *The Shroud and the Grail* (Weidenfeld and Nicolson: London, 1987).

D

DAILEY, Timothy, *Mysteries of the Bible: Exploring the Secrets of the Unexplained* (Publications International: Lincolnwood, 1998).

DALRYMPLE, James, 'As I Stared at the Shroud I Felt a Shudder. Despite What the Sceptics Say, I Will Not Forget Its Hypnotic Effect' (*Daily Mail*, 20 April 1998), p.22.

DAMON, E., *et al.*, 'Radiocarbon Dating of the Shroud of Turin' (*Nature*, vol. 337, no. 6208, 16 February 1988), pp.611-15.

DANCE, S. Peter, *Animal Fakes and Frauds* (Sampson Low: Maidenhead, 1976).

DAVID-NEEL, Alexandra, *With Mystics and Magicians in Tibet* (Bodley Head: London, 1931).

DECLERCQ, Nico F., *et al.*, 'A Theoretical Study of Special Acoustic Effects Caused By the Staircase of the El Castillo Pyramid at the Maya Ruins of Chichen-Itza in Mexico' (*Journal of the Acoustical Society of America*, vol. 116, no. 6, December 2004), pp.3328-35.

DENNIS, Jerry, *It's Raining Frogs and Fishes: Four Seasons of Natural Phenomena and Oddities of the Sky* (HarperCollins: New York, 1992).

DEVEREUX, Paul, *Earth Lights* (Turnstone Press: London, 1982).

DEVEREUX, Paul, *Earth Lights Revelation* (Blandford Press: London, 1989).

DOWNES, J.T., *Woolsery: The Village With Two Names* (Privately published, 1998).

DOWNES, Jonathan, 'Sky Beasts' (*Sightings*, vol. 2, July 1997), pp.10-15.
DOWNES, Jonathan, *The Owlman and Others: 30th Anniversary Edition*
    (CFZ Press: Bideford, 2006).
DOWNES, Jonathan, 'Star Rot' (*Beyond*, no. 7, August 2007), pp.62-5.
DUNN, Euan, 'Lead Kindly Light' [min-mins] (*BBC Wildlife*, vol. 13, June 1995), p.25.

E

EBERHART, George M., *A Geo-Bibliography of Anomalies: Primary Access to Observations of*
    UFOs, Ghosts, and Other Mysterious Phenomena (Greenwood Press: Westport, 1980).
EBERHART, George M., *Mysterious Creatures: A Guide to Cryptozoology*
    (2 vols) (ABC-Clio: Santa Barbara, 2002).
EDWARDS, Frank, *Stranger Than Science* (Ace Books: New York, 1959).
EDWARDS, Frank, *Strange World* (Lyle Stuart: New York, 1964).
ELSOM, Derek, 'Catch a Falling Frog' (*New Scientist*, vol. 108, 2 June 1988), pp.38-40.
EMSLEY, John, 'Graveyard Ghosts are a Gas' [will o' the wisps] (*New Scientist*, 19 June 1993).
ENGLISH, Rebecca, 'It's Raining Fish! Streets of Resort are Sprinkled With Sprats'
    (*Daily Mail*, 7 August 2000).
EVANS, Hilary and GRANT, Reg (Consultants), *UFO: The Continuing Enigma*
    (Reader's Digest: London, 1992).
EVANS, Hilary, SHUKER, Karl P.N., *et al.* (Consultants),
    Almanac of the Uncanny (Reader's Digest: Surry Hills, 1995).
EVANS, P.R., 'Unicorns' (*The Biologist*, vol. 32, no. 1, February 1985), p.6.

F

FAIRLEY, John and WELFARE, Simon, *Arthur C. Clarke's World of Strange Powers*
    (Collins: London, 1984).
FAIRLEY, John and WELFARE, Simon, *Arthur C. Clarke's Chronicles of the Strange*
    and Mysterious (Collins: London, 1987).
FITCH, Eric I., *In Search of Herne the Hunter* (Capall Bann: Chieveley, 1994).
FITCHIE, Peter, 'The Ehlers-Danlos Syndrome' [in a cat]
    (*The Veterinary Record*, vol. 90, 5 February 1972), p.165.
FLOYD, E. Randall, *Ghost Lights and Other Encounters With the Unknown*
    (August House: Little Rock, 1993).
FOLLAIN, John, ''Face of Christ' Relic Found in Monastery' [Veil of Veronica]
    (*Sunday Times*, 30 May 1999), p.9
FORT, Charles, *The Complete Books of Charles Fort* (Reprinted by Dover: New York, 1974).
FORTUNE, Dion, *Psychic Self-Defence* (Rider: London, 1931).
FREEDMAN, David N., *et al.* (Consultants), *Mysteries of the Bible:*
    The Enduring Questions of the Scriptures (Reader's Digest: Pleasantville, 1988).
FREEMAN, Margaret B., *The Unicorn Tapestries* (Metropolitan Museum of Art: New York, 1976).
FREEMAN, Richard, *Dragons: More Than a Myth?* (CFZ Press: Bideford, 2005).

G

GACHOT, Theodore, *Mermaids: Nymphs of the Sea* (Aurum Press: London, 1996).
GADDIS, Vincent H., *Mysterious Fires and Lights* (David McKay: New York, 1967).
GAMON, David, 'Sungods in Cuckoo Land' (*Fortean Times*, no. 75, June-July 1994), p.57.
GARLASCHELLI, Luigi, 'You Can Get Blood From a Stone'
        (*Chemistry in Britain*, vol. 30, no. 7, July 1995), p.534.
GARLASCHELLI, Luigi, 'Chemistry of "Supernatural" Substances'
        (*Journal of the Society For Psychical Research*, vol. 62, no. 852, 1998), p.417.
GARLASCHELLI, Luigi, *et al.*, 'Working Bloody Miracles' (*Nature*, vol. 353, 1991), p.507.
GAYNARD, T.J., 'Star Slime' (*Fortean Times*, no. 93, December 1996), p.58.
GILBERT, Adrian G. and COTTERELL, Maurice M., *The Mayan Prophecies*
        (Element: Longmead, 1995).
GILROY, Rex, *Mysterious Australia* (Nexus: Mapleton, 1995).
GOOCH, Stan, *The Neanderthal Question* (Wildwood: London, 1977).
GOOCH, Stan, *Creatures From Inner Space* (Rider: London, 1984).
GORDON, Stuart, *The Paranormal: An Illustrated Encyclopedia* (Headline: London, 1992).
GORDON, Stuart, *The Book of Miracles: From Lazarus To Lourdes* (Headline: London, 1996).
GOSSE, Philip H., *The Romance of Natural History* (Jams Nisbet: London, 1860).
GOTFREDSEN, Lisa, *The Unicorn* (Harvill Press: London, 1999).
GOULD, Charles, *Mythical Monsters* (W.H. Allen: London, 1886).
GOULD, Rupert T., *Oddities: A Book of Unexplained Facts* (rev. edit.)
        (Geoffrey Bles: London, 1944).
GOULD, Rupert T., *Enigmas: Another Book of Unexplained Facts* (2nd edit.)
        (Geoffrey Bles: London, 1946).
GOVE, Harry E., *Relic, Icon or Hoax? Carbon Dating the Turin Shroud*
        (Institute of Physics Publishing: Bristol, 1996).
GRAHAM-DIXON, Andrew, 'How Much Am I Bid For A Unicorn's Horn?'
        (*Daily Mail*, 17 June 1994), p.9.
GRANT, John, *Unexplained Mysteries of the World* (Quintet Books: London, 1991).
GRAY, Affleck, *The Big Grey Man of Ben MacDhui* (2nd edit.) (Lochar: Bankhead, 1989).
GREEN, Andrew, *Our Haunted Kingdom* (Wolfe: London, 1973).
GREENWELL, J. Richard (Ed.), 'New Guinea Expedition Observes Ri'
        (*ISC Newsletter*, vol. 2, no. 2, Summer 1983), pp.1-2.
GREENWELL, J. Richard (Ed.), 'New Expedition Identifies Ri as Dugong'
        (*ISC Newsletter*, vol. 4, no. 1, Spring 1985), pp.1-3.
GRIFFITHS, Jeanne (Ed.), *Unicorns* (W.H. Allen: London, 1981).
GUILEY, Rosemary E., *Harper's Encyclopedia of Mystical and Paranormal Experience*
        (Harper: New York, 1991).
GUILEY, Rosemary E., *The Guinness Encyclopedia of Ghosts and Spirits* (Guinness: Enfield, 1994).

H

HAFFERT, John M, *Russia Will Be Converted* (AMI International Press: Washington,
        New Jersey, 1950).
HAINING, Peter, *The Monster Trap and Other True Mysteries* (Armada: London, 1976).

HAINING, Peter, *Ancient Mysteries* (Sidgwick and Jackson: London, 1977).

HAINING, Peter, *A Dictionary of Ghosts* (Robert Hale: London, 1982).

HAMBLING, David, 'The Aquatic Ape Theory' *Fortean Times*, no. 175, October 2003), p.18.

HANLON, Michael, 'Why the Discovery of This Planet Will Change Our View of Space For
    Ever' [Xena/Eris] (*Daily Mail*, 1 August 2005), p.13.

HARDY, Alister C., 'Was Man More Aquatic in the Past?' (*New Scientist*, vol. 7, 17
    March 1960), pp.642-5.

HARPUR, Patrick, *Daimonic Reality: A Field Guide to the Otherworld*
    (Viking Arkana: London, 1994).

HARRIS, Paul, 'The Green Children of Woolpit: A 12th Century Mystery and Its Possible
    Solution' (*Fortean Studies*, vol. 4, 1998), pp.81-95.

HARRISON, Ted, *Stigmata: A Medieval Phenomenon in a Modern Age*
    (St Martin's Press: New York, 1994).

HARRISON, Ted, 'What a Way to Spend Easter' [stigmata]
    (*Fortean Times*, no. 96, March 1997), pp.34-8.

HARTWIG, George, *The Subterranean World* (Longmans, Green: London, 1875).

HATHAWAY, Nancy, *The Unicorn* (Avenel Books: New York, 1984).

HEALY, Tony and CROPPER, Paul, *Out of the Shadows: Mystery Animals of Australia*
    (Pan Macmillan Australia: Chippendale, 1994).

HECHT, Jeff and WILLIAMS III, Gurney, 'Smart Dinosaurs' [dinosauroid]
    (*Omni*, vol. 4, May 1982), pp.48-50,52,54.

HENBEST, Nigel, 'Pursuing Persephone' [Planet X]
    (*New Scientist*, vol. 100, 17 November 1983), p.527.

HENEGHAM, Kevin, [Letter re stigmata] (*Daily Mail*, 19 January 2001).

HEUVELMANS, Bernard, *On the Track of Unknown Animals* (Rupert Hart-Davis: London, 1958).

HEUVELMANS, Bernard, *In the Wake of the Sea-Serpents* (Rupert Hart-Davis: London, 1968).

HEYERDAHL, Thor, *Aku-Aku: The Secret of Easter Island* (Allen and Unwin: London, 1958).

HEYERDAHL, Thor, *Easter Island: The Mystery Solved* (Souvenir Press: London, 1989).

HIGBEE, Donna, 'Involuntary Spontaneous Human Invisibility'
    (*The Anomalist*, vol. 3, Winter 1995-6), pp.156-63.

HIND, Cynthia, 'A Mermaid or a Mere Maid?' (*Fate*, vol. 53, November 2000), pp.10-12.

HIPPISLEY COXE, Antony D., *Haunted Britain* (Hutchinson: London, 1973).

HITCHING, Francis, *The World Atlas of Mysteries* (Collins: London, 1978).

HODSON, Geoffrey, *Fairies at Work and at Play* (Theosophical Publishing House: London, 1925).

HOPKINS, Nick, 'The Healing Hand' [of St John Kemble] (*Daily Mail*, 22 July 1995), p.15.

HOPPER, Miriam, [Letter re ball lightning] (*Daily Mail*, 6 December 1995).

HORSWELL, Cindy, 'Leapin' Lizards! Is That Brute For Real?' [Lizard Man]
    (*Houston Chronicle*, 31 July 1988), pp.1,6.

HOUGH, Rebecca, 'Winged Cat' (*Fortean Times*, vol. 114, September 1998), p.50.

HUMPHREY, Doris M., 'Dancing Balls of Light' (*Fate*, vol. 45, January 1992), pp.53-6.

HUMPHREYS, Colin J., *The Miracles of Exodus: A Scientist's Discovery of the Extraordinary
    Natural Causes of the Biblical Stories* (Continuum: London, 2003).

HUYGHE, Patrick, *The Field Guide to Extraterrestrials* (New English Library: London, 1997).

I

IDRIESS, Ion L., *Drums of Mer* [booya stones] (Angus and Robertson: Sydney, 1941).

J

JERLSTROM, P., 'Live Plesiosaurs: Weighing the Evidence' [yarru]
(*Creation Ex Nihilo Technical Journal*, vol. 12, no. 3, 1999), pp.339-46.
JOSEPH, Rhawn, *Neuropsychiatry, Neuropsychology and Clinical Neuroscience*
(Williams and Wilkins: New York, 1996).

K

KEEL, John A., *Strange Creatures From Time and Space* (Fawcett: Greenwich, 1970).
KEEL, John A., *The Mothman Prophecies*
(E.P. Dutton: New York, 1975; rev. edit., IllumiNet Press: Lilburn, 1991).
KEITH, Jim (Ed.), *Secret and Suppressed* (Feral Horse: Portland, 1993).
KELLY, Jane, 'Miraculous Weeping Madonnas? Here's One I Made Earlier'
(*Daily Mail*, 10 July 1995).
KEY, Ivor, 'Experts Find DNA on the Shroud of Turin' (*Daily Mail*, 30 March 1998), p.11.
KOZICKA, Maureen, *The Mystery of the Min Min Light* (Privately published, 1994).
KUDLOWSKI, Mark, [Letter re Planet X] (*Daily Mail*, 21 October 1998).

L

LATERRADE, J.F., 'An Attempt to Prove the Existence of the Unicorn'
(*American Journal of Science*, Series 1, vol. 21, 1832), pp.123-6.
LAURENTIN, René, *The Apparitions of the Blessed Virgin Mary Today* (2nd edit.)
(Veritas: Dublin, 1991).
LEHN, Waldemar H. and SCHROEDER, I., 'The Norse Merman as an Optical Illusion'
(*Nature*, vol. 289, 29 January 1981), pp.62-6.
LEWELS, Joe, 'The Reptilians: Humanity's Historical Link to the Serpent Race'
(*Fate*, vol. 49, June 1996), pp.48-52.
LEWIS, Michael, 'Lizard Man: S.C.'s Answer to Bigfoot'
(*State*, Columbia, 15 August 1988), pp.17-18.
LEY, Willy, *The Lungfish and the Unicorn* (Viking Press: New York, 1941).
LEY, Willy, 'Earth's Extra Satellites' (*Galaxy*, February 1962), pp.55-60.
LITTLE, W.T., 'Hunting Ghosts in a Ghost Town Out West' [Silver Cliff spooklights]
(*New York Times*, 20 August 1967).
LOWENSTEIN, Jerold M. and ZIHLMAN, Adrienne L., 'The Leaky Logic of the Aquatic
Ape' (*BBC Wildlife*, April 1986), pp.182-3.
LYALL, Sutherland, *The Lady and the Unicorn* (Parkstone Press: London, 2000).

M

MacLEAN, Alasdair, [Letter re Planet X] (*Daily Mail*, 23 November 1998).

MacLEAN, Paul D. and KRAL, V.A., *A Triune Concept of the Brain and Behaviour* (University of Toronto: Toronto, 1973).

MacMANUS, Dermot, *The Middle Kingdom: The Faerie World of Ireland* (Max Parrish: London, 1959).

MAGEE, Mike, *Who Lies Sleeping? The Dinosaur Heritage and the Extinction of Man* (AskWhy!: Frome, 1993).

MAGIN, Ulrich, 'Scaly Horrors: Amphibious Aliens, Frogmen and UFOs' (*Fortean Times*, no. 63, June-July 1992), pp.40-3.

MAMONTOFF, Nicholas, 'Can Thoughts Have Forms?' (*Fate*, June 1960).

MANNING, Aubrey, 'Mythical Animals' (*The Biologist*, vol. 31, no. 2, April 1984), p.70.

MARSHALL, Richard, *et al.*, *Mysteries of the Unexplained* (amended edit.) (Reader's Digest: Pleasantville, 1988).

McCAFFERY, John, *The Friar of San Giovanni* [Padre Pio] (Darton, Longman and Todd: London, 1978).

McCLURE, Kevin, *The Evidence For Visions of the Virgin Mary* (Aquarian: Wellingborough, 1983).

McEWAN, Graham J., *Freak Weather* (Robert Hale: London, 1991).

McHATTIE, Grace, *Cat Tales: The Life and Times of Cats This Century* (Ebury Press: London, 1992).

MEGGED, Matti, *The Animal That Never Was: In Search of the Unicorn* (Lumen Books: New York, 1992)

MEREDITH, Peter, 'Min Min is a Mirage' (*BBC Wildlife*, vol. 13, December 1995), p.96.

MICHELL, John and RICKARD, Robert J.M., *Phenomena: A Book of Wonders* (Thames and Hudson: London, 1977).

MICHELL, John and RICKARD, Robert J.M., *Living Wonders: Mysteries and Curiosities of the Animal World* (Thames and Hudson: London, 1982).

MOORE, Malcolm, ''Unicorn' Born in Italy' (*Daily Telegraph*, 11 June 2008).

MOORE, Patrick, 'Planet 10' (*Omni*, vol. 5, February 1983), p.24.

MOORE, Patrick, 'Tiny Triumph For Smiley's People' [1992 QB1] (*Daily Mail*, 10 October 1992).

MOREL, M.-H., 'Note sur une "Vache à Sphéroïde" du Hoggar' (*Bull. Soc. Hist. Nat. Afrique Nord*, vol. 33, 1942), pp.223-6.

MORGAN, Elaine, *The Aquatic Ape* (Souvenir Press: London, 1982).

MORRELL, R.W., *The Angel Hair Problem in Ufology* (Nufois Press: Nottingham, 1981).

MOYNA, Patrick and HEINZEN, Horacio, 'Why Was the Fleece Golden?' (*Nature*, vol. 330, 5 November 1987), p.28.

MULLIN, Redmond, *Miracles and Magic: The Miracles and Spells of Saints and Witches* (A.R. Mowbray: Oxford, 1978).

MUNRO, William, [Letter re mermaid sighting] (*The Times*, 8 September 1809).

N

NAAKTGEBOREN, C., 'Unicorns: Immortal and Extinct'

(*World Magazine*, no. 41, September 1990), pp.70-6.

NAISH, Darren, 'Intelligent Dinosaurs' (*Fortean Times*, no. 239, August 2008), pp.52-3.

NEWTON, Michael, *Encyclopedia of Cryptozoology: A Global Guide* (McFarland: Jefferson, 2005).

NICKELL, Joe, *Looking For a Miracle* (Prometheus: Amherst, 1993).

NICKELL, Joe, *Relics of the Christ* (University of Kentucky Press: Lexington, 2007).

NYOONGAH, Mudrooroo, *Aboriginal Mythology* (Aquarian: London, 1994).

O

O'DONNELL, Elliott, *Haunted Britain* (Rider: London, 1949).

OPSASNICK, Mark and CHORVINSKY, Mark, 'Lizard Man'
   (*Strange Magazine*, no. 3, Winter 1988), pp.32-3.

P

PARKER, K. Langloh, *Australian Legendary Tales: Dreaming Stories From the Oral Tradition of the Aboriginal People* (Wordsworth: London, 2001).

PARKINSON, John H., 'Dragon Slaying' (*New Scientist*, vol. 97, 14 April 1983), p.95.

PATTERSON, Donald F. and MINOR, Ronald R., 'Hereditary Fragility and Hyperextensibility of the Skin of Cats' (*Laboratory Investigation*, vol. 37, no. 2, 1977), pp. 170-9.

PHILLPOTTS, Beatrice, *Mermaids* (Russell Ash/Windward: London/Leicester, 1980).

PICKNETT, Lynn, *The Encyclopaedia of the Paranormal: The Complete Guide to the Unexplained* (Macmillan: London, 1990).

PICKNETT, Lynn and PRINCE, Clive, *Turin Shroud: In Whose Image?* (Bloomsbury: London, 1994).

POVAH, Frank, *You Kids Count Your Shadows: Hairymen and Other Aboriginal Folklore in New South Wales* (Privately published: Wollar, 1990).

PRESTON, Frank, 'The Green Children' (*The Unknown*, no. 11, May 1986), p.49.

R

RADKA, Larry B., *Historical Evidence For Unicorns* (Einhorn Press: Newport, Delaware, 1995).

RAVENSCROFT, Trevor, *The Spear of Destiny* (Corgi: London, 1974).

REDFERN, Nick, *Man-Monkey: In Search of the British Bigfoot* (CFZ Press: Bideford, 2007).

REED, A.W., *Aboriginal Fables and Legendary Tales* (A.W. Reed: Sydney, 1965).

RENISON, Jessica, *The Secret History of Unicorns* (Kandour: London, 2007).

RHODES, John, '"Reptilian Aliens", Reptoids, Draco, Abductions, Dulce Base, UFOs and Saurian Underworlds' (http://www.reptoids.com first accessed 3 November 1998).

RICKARD, Robert J.M., 'Birdmen of the Apocalypse!'
   (*Fortean Times*, no. 17, August 1976), pp.14-20.

RICKARD, Robert J.M., 'A Reprise For *Living Wonders*' [winged cats section]
   (*Fortean Times*, no. 40, summer 1983), p.9.

RICKARD, Robert J.M., 'Seeing Red – The Blood Miracles of Naples'

(*Fortean Times*, no. 65, October-November 1992), pp.36-41

RICKARD, Robert J.M. and KELLY, Richard, *Photographs of the Unknown*
(New English Library: London, 1980).

RIDPATH, Ian, 'Mars May Be Dead But There's Still Jupiter'
(*New Scientist*, vol. 73, 10 March 1977), p.582.

ROBERTS, Ainslie and MOUNTFORD, Charles P., *The Dreamtime: Australian Aboriginal Myths*
(Rigby: Adelaide, 1965).

ROBERTS, Ainslie and MOUNTFORD, Charles P., *The Dawn of Time: Australian Aboriginal*
Myths (Rigby: Adelaide, 1969).

ROBERTS, C.R., 'Mount Rainier-Area Youth Has Close Encounter' [batsquatch]
(*News Tribune*, Tacoma, 1 May 1994).

ROBERTS, David, 'Mali's Dogon People' (*National Geographic*, vol. 178, no. 4,
October 1990), pp.100-26.

ROBIN-EVANS, Karyl, *Sungods in Exile* (Neville Spearman: London, 1979).

ROBINSON, Roy and PEDERSEN, N.C., 'Chapter 2. Normal Genetics, Genetic Disorders,
Developmental Anomalies and Breeding Programs', *in:* PEDERSEN, N.C. (Ed.),
Feline Husbandry (American Veterinary Pub.: Coleta, California, 1991).

ROGERS, Raymond N., 'Studies on the Radiocarbon Sample From the Shroud of
Turin' (*Thermochimica Acta*, vol. 425, 2005), pp.189-194.

ROGO, D. Scott, *Miracles: A Parascientific Inquiry Into Wondrous Phenomena*
(Dial Press: New York, 1982).

ROGO, D. Scott and CLARK, Jerome, *Earth's Secret Inhabitants* (Tempo: New York, 1979).

ROMER, John and ROMER, Elizabeth, *The Seven Wonders of the World*
Michael O'Mara: London, 1995).

ROWLETT, Curt, 'Chick-Charney: Bird-Man of the Bahamas'
(*Strange Magazine*, no. 7, April 1991), p.34.

RUSSELL, Dale A. and SÉGUIN, Ron, 'Reconstructions of the Small Cretaceous Theropod
Stenonychosaurus inequalis and a Hypothetical Dinosauroid'
(*Syllogeus*, no. 37, 1982), pp.1-43.

RUSSELL, Eric F., *Great World Mysteries* (Dobson: London, 1957).

RUTTER, Gordon, 'Whatever Happened To...17. The Spear of Destiny'
(*Fortean Times*, no. 140, November 2000), p.66.

RYDER, M.L. and HEDGES, J.W., 'Ancient Scythian Wool From the Crimea'
(*Nature*, vol. 242, 13 April 1973), p.480.

S

SAGAN, Carl, *The Dragons of Eden: Speculations on the Evolution of Human Intelligence*
(Hodder and Stoughton: London, 1978).

SANDERSON, Ivan T., *Uninvited Visitors: A Biologist Looks at UFOs*
(Neville Spearman: London, 1969).

SANDERSON, Ivan T., *Investigating the Unexplained*
(Prentice-Hall: Englewood Cliffs, 1972).

SANDERSON, Ivan T., *Invisible Residents* (Universal-Tandem: London, 1974).

SASSOON, George & DALE, Rodney, *The Manna Machine* (Sidgwick & Jackson: London, 1978).

SCOTT, D.V., 'Cutaneous Asthenia in a Cat, Resembling Ehlers-Danlos Syndrome in Man'
(*Veterinary Medicine*, vol. 69, October 1974), pp.1256-8.

SEIFE, Charles, 'Cracked It!' [ball lightning] (*New Scientist*, vol. 159, 26 September 1998), p.6.

SESSIONS, Steve, 'Claw Men From Outer Space'
    (*Fortean Times*, no. 119, February 1999), pp.38-44.

SEVERIN, Tim, 'Jason's Voyage – In Search of the Golden Fleece'
    (*National Geographic*, vol. 168, no. 3, September 1985), pp.406-20.

SEVERIN, Tim, *The Jason Voyage: The Quest For the Golden Fleece* (Hutchinson: London, 1985).

SHARP, Peter F., 'Angel Hair'
    (*Flying Saucer Review*, vol. 10, no. 1, January-February 1964), pp.14-15.

SHAY, V.B., 'Lights Without Flame' (*Fate*, vol. 4, August-September 1951), pp.96-7.

SHEARS, Richard, ''Miracle' of Christ's Face in the Outback'
    (*Daily Mail*, 16 December 1995), p.18.

SHEPARD, Odell, *The Lore of the Unicorn* (George Allen and Unwin: London, 1930).

SHUKER, Karl P.N., 'Meteorological Monsters and Mysteries'
    (*The Unknown*, no. 31, January 1988), pp.5-11.

SHUKER, Karl P.N., 'The Search For the *Real* Golden Fleece'
    (*Fate*, vol. 42, no. 9, September 1989), pp.46-52.

SHUKER, Karl P.N., *Mystery Cats of the World: From Blue Tigers to Exmoor Beasts*
    (Robert Hale: London, 1989).

SHUKER, Karl P.N., 'Gallinaceous Mystery Birds'
    (*World Pheasant Association News*, no. 32, May 1991), pp.3-6.

SHUKER, Karl P.N., 'Miiaooow!' [winged cats] (*Me*, 8 September 1993), pp.56-7.

SHUKER, Karl P.N., 'Sky Beasts. Science Fiction – Or Science Fact?'
    (*SCAN News*, no. 4, October 1993), pp.3-6.

SHUKER, Karl P.N., 'Cat Flaps' (*Fortean Times*, no. 78, December 1994-January 1995), pp.32-3.

SHUKER, Karl P.N., 'On a Wing and a Purr...' (*Cat World*, no. 210, August 1995), pp.14-15.

SHUKER, Karl P.N., 'High Flyers' [winged cats] (*Wild About Animals*, vol. 7, October 1995), p.13.

SHUKER, Karl P.N., *In Search of Prehistoric Survivors:*
    Do Giant 'Extinct' Creatures Still Exist? (Blandford Press: London, 1995).

SHUKER, Karl P.N., 'Wonderful Things Are Cats With Wings' (*Fate*, vol. 49, April 1996), p.80

SHUKER, Karl P.N., 'Increíble, Pero Cierto – Gatos Con Alas'
    (*Enigmas*, no. 5, May 1996), pp.32-9.

SHUKER, Karl P.N., *The Unexplained: An Illustrated Guide to the World's Natural and*
    Paranormal Mysteries (Carlton Books: London, 1996).

SHUKER, Karl P.N., 'Stars In Their Eyes' [Dogon]
    (*Uri Geller's Encounters*, no. 1, November 1996), pp.62-5.

SHUKER, Karl P.N., 'How Green Were My Children...'
    (*Uri Geller's Encounters*, no. 2, December 1996), pp.45-8

SHUKER, Karl P.N., 'Worldwide Webs and Rotten Stars'
    (*Uri Geller's Encounters*, no. 3, January 1997), pp.20-3.

SHUKER, Karl P.N., 'Flights of Fantasy?' [winged cats]
    (*All About Cats*, vol. 4, March-April 1997), pp.44-5.

SHUKER, Karl P.N., *From Flying Toads to Snakes With Wings: From the Pages of Fate Magazine*
    (Llewellyn: St Paul, 1997).

SHUKER, Karl P.N., 'In Search of the Gabriel Feather'
    (*Wild About Animals*, vol. 10, May-June 1998), p.38.

SHUKER, Karl P.N., *Mysteries of Planet Earth: An Encyclopedia of the Inexplicable*
    (Carlton Books: London, 1999).

SHUKER, Karl P.N., 'The Skull of a Wulver?' (*Fate*, vol. 53, November 2000), pp.16-18.

SHUKER, Karl P.N., 'The Green Children of Woolpit'
    (*Fate*, vol. 54, May 2001), pp.17-19.

SHUKER, Karl P.N., *The New Zoo: New and Rediscovered Animals of the Twentieth Century*
(House of Stratus: Thirsk, 2002).
SHUKER, Karl P.N., 'The Shroud of Turin: Still Shrouded in Mystery and Controversy'
(*Fate*, vol. 55, June 2002), pp.36-7.
SHUKER, Karl P.N., 'If Cats Could Fly...' (*Fortean Times*, no. 168, March 2003), pp.48-9.
SHUKER, Karl P.N., *Extraordinary Animals Revisited: From Singing Dogs To Serpent Kings*
(CFZ Press: Bideford, 2007).
SHUKER, Karl P.N., 'From Earth Hounds to Monkey Birds: In Pursuit of Britain's Lesser-Known
Mystery Beasts' (*Beyond*, no. 12, January 2008), pp.24-31,
SHUKER, Karl P.N. (Consultant), *Man and Beast* (Reader's Digest: Pleasantville, 1993).
SHUKER, Karl P.N. (Consultant), *Secrets of the Natural World*
(Reader's Digest: Pleasantville, 1993).
SIEVEKING, Paul, 'Loveland Frog Leaps Back' (*Fortean Times*, no. 46, Spring 1986), p.19.
SIEVEKING, Paul, 'Lizard Man' (*Fortean Times*, no. 51, Winter 1988), pp.34-7.
SIEVEKING, Paul, 'Virgin Blood Floods Italy' [blood-weeping Madonna statues]
(*Fortean Times*, no. 81, June-July 1995), p.11.
SIEVEKING, Paul, 'It's All in the Lap [milk-drinking statues]
(*Fortean Times*, vol. 84, December 1995-January 1996), pp.16-17.
SIEVEKING, Paul, 'How to Get Blood From a Stone' [weeping/bleeding statues]
(*Fortean Times*, no. 84, December 1995-January 1996), p.18.
SIEVEKING, Paul, 'Millions Line Up To See Jesus's [sic] Dirty Shirt' [Holy Tunic]
(*Fortean Times*, no. 91, October 1996), p.11.
SIEVEKING, Paul, 'Christ in the Outback' (*Fortean Times*, no. 91, October 1996), p.15.
SIEVEKING, Paul, 'Don't Bite – There's a Nun on That Bun' (*Sunday Telegraph*, 12 January 1997).
SIEVEKING, Paul, 'Queer Jelly Found After UFO Sightings'
(*Fortean Times*, no. 95, February 1997), p.19.
SIEVEKING, Paul, 'Virgin Vision in Clearwater' (*Fortean Times*, no. 97, April 1997), p.8.
SIEVEKING, Paul, [Liquefying blood of St Lawrence/Lorenzo]
(*Fortean Times*, no. 105, December 1997), p.17.
SIEVEKING, Paul, 'Tsar Icon Weeps Myrrh' (*Fortean Times*, no. 124, July 1999), p.21.
SIEVEKING, Paul, 'Bloody Strange' [bleeding/stigmatic statues]
(*Fortean Times*, no. 166, January 2003), p.8.
SIEVEKING, Paul, 'Wonder-webs!' (*Fortean Times*, no. 171, June 2003), pp.6-7.
SIEVEKING, Paul, 'Planet Number Ten?' [Sedna] (*Fortean Times*, no. 184, June 2004), pp.4-5.
SIEVEKING, Paul, 'Caspian Sea Anomaly' (*Fortean Times*, no. 197, June 2005), p.14.
SIEVEKING, Paul, 'New Planet – Perhaps' [Eris/Xena]
(*Fortean Times*, no. 202, October 2005), pp.26-7.
SIEVEKING, Paul, 'Holy Signs and Miracles' (*Fortean Times*, no. 208, April 2006), p.8.
SIEVEKING, Paul, 'Pope Visits Veil Relic' (*Fortean Times*, no. 217, December 2006), p.8.
SIEVEKING, Paul, 'The Statues That Like a Drink' (*Fortean Times*, no. 217, December 2006), p.10.
SIEVEKING, Paul, 'Wonder Plant of the Ancients' [zilphion]
(*Fortean Times*, no. 221, April 2007), p.77.
SIEVEKING, Paul, 'Winged Monsters Appear' [LaCrosse man-bat]
(*Fortean Times*, no. 223, June 2007), p.10.
SIMON, Robin, 'Can This Really Be the Hand of Colossus?' (*Daily Mail*, 7 July 1987).
SIMONS, Paul, *Weird Weather* (Little, Brown: London, 1996).
SINGER, Stanley, *The Nature of Ball Lightning* (Plenum Press: New York, 1971).
SKIDMORE, Mark, [Letter re foo fighters] (*Daily Mail*, 6 December 1995).
SMITH, G.J., 'Jason's Golden Fleece Explained?' (*Nature*, vol. 327, 23 June 1987), p.561.
SMITH, William R., *Myths and Legends of the Australian Aboriginals*

(George G. Harrap: London, 1930).
SMYTH, Frank, *Ghosts and Poltergeists* (Aldus: London, 1976).
SOX, H. David, *File on the Shroud* (Coronet: London, 1978).
SPENCER, John, *The UFO Encyclopedia* (Headline: London, 1991).
SPICKLER, Theodore, 'Another Mystery Sphere'
      (*APRO Bulletin*, vol. 23, no. 2, September-October 1974), pp.6-8.
STASHOWER, Daniel, *The Enchanted World: Magical Beasts* (Time-Life: Amsterdam, 1985).
STEEL, Thomas, 'On Dental Encrustations and the So-Called 'Gold-Plating' of Sheep's
      Teeth' (*Proceedings of the Linnean Society of New South Wales*, 25 August 1920).
STEIGER, Brad, *The Flying Saucer Menace* (Award: New York, 1967).
STEIGER, Brad, *Atlantis Rising* (Dell: New York, 1973).
STEIGER, Brad, *Monsters Among Us* (Para Research: Rockport, 1982).
STONEHILL, Paul, 'Return of the Flying Man' [letayuschiy chelovek]
      (*Fate*, vol. 45, November 1992), pp.48-53.
SUCKLING, Nigel, *The Book of the Unicorn* (Paper Tiger: Limpsfield, 1996).
SUCKLING, Nigel, *Unicorns* (FF&F: Wisley, 2007).

T

TAFFS, A. (Ed.), *The World of Wonders: A Record of Things Wonderful in Nature, Science, and Art*
(Cassell, Petter, & Galpin: London, 1882).
TEMPLE, Robert K.G., *The Sirius Mystery* (Sidgwick and Jackson: London, 1976).
TEMPLE, Robert K.G., 'In Defense of *The Sirius Mystery*' (*Fate*, vol. 33, October 1980), pp.83-8.
THOREAU, Henry David, *Walden; or Life in the Woods*
      (Boston, 1854; reprinted by Everyman's Library: London, 1972).
THORNTON, Jacqui, 'When Green Children Fell To Planet Earth'
      (*Sunday Telegraph*, 20 April 1997), p.18.
TOIBIN, Colin (Ed.), *Seeing is Believing: Moving Statues in Ireland*
      (Pilgrim Press: Mountrath, 1985).
TORONTO, Richard, 'Living UFOs' (*Fate*, August 1980), pp.50-5.
TRIVEDI, Bijal P., 'Was Maya Pyramid Designed to Chirp Like a Bird?'
      (*National Geographic Today*, 6 December 2002).
TROTTER, Spencer, 'Concerning the Real Unicorn' (*Science*, vol. 28, 1908), pp.608-9.
TRUZZI, Marcello (Consultant), *Into the Unknown* (amended edit.)
      (Reader's Digest: Pleasantville, 1988).
TRUZZI, Marcello, *et al.* (Consultants), *Mystic Places* (Time-Life: Amsterdam, 1987).

U

UMAN, Martin A., 'Ball Lightning Wild and Wonderful'
      (*Nature*, vol. 300, no. 5893, 16 December 1982), p.578.
UNDERWOOD, Peter, *A Gazetteer of British Ghosts* (rev. ed.) (Pan: London, 1973).
UNDERWOOD, Peter, *A Gazetteer of Scottish and Irish Ghosts* (Souvenir Press: London, 1973).
UNDERWOOD, Peter, *Dictionary of the Occult and Supernatural* (Fontana: London, 1979).
USHER, George A., *A Dictionary of Plants Used By Man* (Constable: London, 1974).
UTTON, Tim, 'Tenth Planet Added to the Solar System' [Quaoar]
      (*Daily Mail*, 8 October 2002), p.36.

V

VALLÉE, Jacques, *Passport To Magonia: From Folklore To Flying Saucers*
(H. Regnery: Chicago, 1969).

W

W., R.B., 'Flying Cat' (*The Naturalist's Note Book*, volume for 1868), p.318.
WAGNER, Roy, 'The Ri – Mermen and Mermaids?' (*Fate*, vol. 36, August 1983), pp.44-9.
WASSILKO-SERECKI, Zoe, 'Startling Theory on Flying Saucers'
(*American Astrology*, September 1955).
WATSON, Lyall, *Supernature* (Hodder and Stoughton: London, 1973).
WATSON, Lyall, *The Romeo Error: A Matter of Life and Death* (rev. edit.)
(Coronet: London, 1976).
WATSON, Lyall, *Heaven's Breath: A Natural History of the Wind*
(Hodder and Stoughton: London, 1984).
WATSON, Lyall, *The Nature of Things: The Secret Life of Inanimate Objects*
(Hodder and Stoughton: London, 1990).
WATSON, Nigel, 'Historical Aerospatial Anomalies'
(*Fortean Times*, no. 36, Winter 1982), pp.44-6.
WEISS, Suzanne E. (Ed.), *Great Mysteries of the Past* (Reader's Digest: Pleasantville, 1991).
WELFARE, Simon and FAIRLEY, John, *Arthur C. Clarke's Mysterious World*
(Collins: London, 1980).
WELFARE, Simon and FAIRLEY, John, *Arthur C. Clarke's A-Z of Mysteries:*
From Atlantis to Zombies (HarperCollins: London, 1993).
WESTWOOD, Jennifer (Ed.), *The Atlas of Mysterious Places*
(Weidenfeld and Nicolson: London, 1987).
WIGNELL, Edel (Ed.), *A Boggle of Bunyips* (Hodder and Stoughton: Sydney, 1981).
WILKINS, Harold T., *Mysteries of Ancient South America* (Rider: London, 1946).
WILKINS, Harold T., 'The Strange Mystery of the Foo Fighters'
(*Fate*, vol. 4, August-September 1951), pp.98-106.
WILKINS, Harold T., *Secret Cities of Old South America* (Library Publishers: New York, 1952).
WILKINS, Harold T., *Monsters and Mysteries* (James Pike: London, 1973).
WILLIAMS, Russ, 'Mutant Hunting' (*Fortean Times*, no. 126, September 1999), p.51.
WILSON, Colin and GRANT, John, *The Directory of Possibilities*
(Webb and Bower: London, 1981).
WILSON, Colin and WILSON, Damon, *Unsolved Mysteries: Past and Present*
(Headline: London, 1993).
WILSON, Ian, *The Turin Shroud* (Victor Gollancz: London, 1978).
WILSON, Ian, *The Bleeding Mind* (Weidenfeld and Nicolson: London, 1988).
WILSON, Ian, *Holy Faces, Secret Places* (Doubleday: London, 1991).

WILSON, Ian and SCHWORTZ, Barrie, *The Turin Shroud: The Illustrated Evidence*
    (Michael O'Mara: London, 2000).
WINOGRADOW, P.P. and FROLOW, A.L., 'Hörnerförmige Bildungen auf dem Stirnbein eines
    Pferdes' [horned horse] (*Anatomische Anzeiger*, vol. 68, 1929), pp.93-4.
WOOD, Gerald L., *Guinness Book of Pet Records* (Guinness Superlatives: London, 1984).
WORLEY, Don, 'The Winged Lady In Black'
    (*Flying Saucer Review Case Histories*, no. 10, June 1972), pp.14-16.

X

X, 'Vampire Cats' (*Pursuit*, vol. 9, October 1976), p.93.

Z

ZELL-RAVENHEART, Oberon and DeKIRK, Ash L., *A Wizard's Bestiary: A Menagerie of Myth,
    Magic, and Mystery* (New Page: Franklin Lakes, 2007).

Index

B

Briggs, Dr Katharine, 34, 86
Brighton Pavilion, ghosts, 41
Brooks, Lynda, 23
Brotherhood of the Rising Sun, The, 111
Brown, Adrian, 154, 159
Brown, W.G., 227
Bubble of light, luminous, 175
Bubo bubo, 216
Budden, Alfred, 65-66
Buller, Philip, 225
Bunyip, 65, 119
Burnham, Owen, 161-162
Burrunjor, 117-118
Burton, Robert, 87
Bury St Edmunds, 88-89
Bushbaby, giant mystery, 163
BVM visions and images, 141-143

C

Cage, John M., 261
Cain, 95
Caligula, Emperor, 37
Calne, Sir Richard de, 86-88
Calno, Ricardo da, 88
Cambodia, 196
Cameroon, 163
Camphor (=Champhur), 246
Camulodunum, 253
Canfield, Brian, 208
Cape Breton, 81
Cape York Peninsula, 120, 122
Capreolus capreolus, 249
Capsule, Mendoza mystery, 151
Caractacus (=Caradoc), King, 253
Carmichael, Alexander, 72
Carmina Gadelica, 72
Carpenter, John, 92, 97-98
Cartazon, 238, 240-241
Cartimandua, Queen, 253
Cassini, Giovanni, 234
Cat-a-mountain, 26
Catherine of Siena, St, 135, 140, 144
Catherine the Great, 155, 159
Cats, flying, 25, 27
Cats, vampire, 27
Cats, winged, 13-30
Catsounis, Pagona and Pagionitis, 127

Catuvellauni tribe, 252-253
Cat-wolves, 166
Cavassi, Giovanni, 242
Cave of Swallows, 266
Cecilia, St, 134
Celano, Thomas de, 140
Center of Natural Sciences, Prato, 249
Centre for Fortean Zoology (CFZ), 23, 112, 255
Ceres (dwarf planet), 236
Cerro de los Indios, 243
Chagrin, 37
Chakpuah, 166
Chambers, Dr John, 236
Chamois, 241
Champollion, Jean-François, 195
Chapman, Ethel, 137
Chapman, Sally, 216
Chares of Lindos, 186
Charlemagne, 131
Charon (moon), 236
Chatwin, Bruce, 243
Cheeroonear, 123
Chehalis, 64, 206
Chevrotain, water, 166
Chichén Itzá, 195
Chickasaw, 229
Chick-charney (=Chickcharnie), 210
Chile, 45-46, 51, 59, 215
China, 24, 145, 202, 233, 268, 276
Chiru, 241
Chlorosis, 86, 88
Chosi City, 149
'Christ in the Clouds' acheropite, 142
'Christ in the Outback' acheropite, 142
Christian of the Order of Saint Anne, Sister, 143
Chronicles of England, Scotland, and Ireland, 252
Chrysomallus, 197
Church, John (Mrs), 157
Circlemakers, The, 269
Circles, cropfield, 269
Cirripectes alboapicalis (=*variolosus patuki*), 60
Civitavecchia, 128
Clague, Amy, 16
Clark, Steve, 228
Clarke, Arthur C., 269
Claudius, Emperor, 253
Cloche, Dr Dei, 140
Cloudsley-Thompson, Prof. John L., 170

D

E

F

M

N

O

P

Patience, Jan, 211
Patterson, Dr Donald F., 28
Patuki, 60-61
Pauca billee, 25
Pavão, Dr Antonio, 172
Pentacyclic triterpenoids, 199
Perry, Barbara, 216
Petit, Frederic, 234
Petropavlovsk, 215
Pfeiffer, Prof. Heinrich, 133
Phaeton (planet), 231, 234-235
Phaeton (sun god's son), 235
Pharomachrus mocinno, 195
Phenias, 220
Phenomena: A Book of Wonders, 43
Philip Neri, St, 144
Phillips, Dr George, 78
Phryxus, Prince, 197
Phylarchus, 220
Piazzi, Father Giuseppe, 235
Picknett, Lynn, 131
Pile, Gordon, 100
Pirassoipi, 246-247
'Pithecanthropus' thought-form, 109-112
Pittman, James, 207
Planet V, 236
Planet X, 231, 236
Planets, dwarf, 231, 236
Plate, Lolladoff, 152
Plesiosaur, 118
Pliny the Elder, 167, 186 200, 238
Pluto (dwarf planet), 231, 236
Poh, 245
Point Isabel, 228
Point Pleasant, 209, 217
Polycarp of Smyrna, St, 144
Ponce, José, 70
Potato, sweet, 56
Potkoorok, 124
Potter, Thomas (Mrs), 228
Preston, Frank, 88
Priego, Juan, 18
Primorskiy Kray Territory, 214
Prince, Clive, 131
Prul (winged cat), 21, 24, 28
Pseudoplasmodium,225
Pseudoryx nghetinhensis, 244
Psychic Self-Defence, 109
Ptilopachus petrosus petrosus, 162
Ptolemy II, 186, 240

Pukao, 49, 52
'Puss-In-Boots', tulpoid, 111-112
Pwdre ser, see Star rot, 225, 260
Python, diamond, 116-117

Q

QB1, 1992, 236
Quaoar, 236
Quetzal, 133, 195
Quetzalcoatl, 97, 195
Quibell, James E., 190
Quinkin, 122, 124
Quinns light, 174

R

Rabbit, winged, 29
Radau, Rudolphe, 233
Radiant boys (enfants brilliant), 31, 34-35
Radpole, 228
Raggatt, John, 23
Rain, black, 226
　　blue, 219, 226
　　electric, 226-227
　　fish, 220-221
　　frog, 44, 220-221
　　green, 219, 226
Ralph of Coggeshall, 86
Ramesses II, Pharaoh, 195
Ramesseum, The, 195
Ramsey, Prof. Christopher, 132
Rano Kau, 57, 59
Rano Raraku, 45, 47, 49, 51-55, 59
Raymondville, 210, 212
Raynar, Robert, 76
Read, Herbert, 84
Rebsamen, William, 60, 209, 244
Rector of Greystoke, 35
Red River, 242
Redfern, Nick, 41
Re'em, 240

Wulver, 36-37

X

Y

Z

Acknowledgements

I would like to offer my sincere thanks to the following persons, societies, organizations, and publications for their very kind interest, encouragement, and most generous assistance during my researches for this book and its subsequent preparation.

Alien Encounters; Dr Simon K. Bearder; Ian Belcher; *Beyond*; Janet and Colin Bord; British Library; British Museum (Natural History); Lynda Brooks; John and Lesley Burke; Owen Burnham; the Centre for Fortean Zoology (CFZ); the late Mark Chorvinsky; Prof. John L. Cloudsley-Thompson; Loren Coleman; Paul Cropper; *Enigmas*; *Fate* (www.fatemag.com); Richard Forsyth; *Fortean Times* (www.forteantimes.com); the Fortean Picture Library; Richard Freeman; Phyllis Galde; Bill Gibbons; Craig Glenday; *Goblin Universe*; Paul Harris; Tim Harris; Jacki Hartley; Isabela Herranz; *History For All*; Mandy Holloway; Graham Inglis; Fox Israel; Karen, my Easter Island guide; Oll Lewis; my agent Mandy Little of Watson, Little Limited; Ivan Mackerle; Bob Michaels; Dr Ralph Molnar; Richard Monteiro; Sarah Moran; Rosario Navaza; Michael Newton; the late Scott Norman; Carina Norris; Heather Norris; Terry O'Neill; Jan Patience; Nina Pendred; *Prediction*; John Raggatt; the late Roy Robinson; Mary D. Shuker; Paul Sieveking; *Sightings*; Martine Smids; Society For Psychical Research; Gaz Stanley (aka Stigmata); *Strange Magazine*; Nick Sucik; Richard Svenssen; the late Gertrude Timmins; *Uri Geller's Encounters*; *The X Factor* (magazine); Oberon Zell; Zoological Society of London.

Finally, I wish to extend an especial vote of thanks to my friend and publisher Jonathan Downes, who made it possible via the CFZ Press for this first compilation volume of my non-cryptozoological writings to see the light of day in print; to fellow Midlands-born mysteries researcher Nick Redfern for his splendid foreword - bostin', ar kid!! (it's a Midlands thing!); and to Philippa Foster, Tim Morris, William Rebsamen, and Richard Svenssen for their wonderful illustrations that they kindly permitted me to include in this book. Particular thanks go to Philippa for allowing me to utilise her spectacular sky medusae painting, 'Cosmic Leap', for this book's front cover. Moreover, those illustrations of Tim and some of Richard's were prepared by them specifically for this book – so, many thanks indeed, once again!

I have attempted to obtain permission for the use of all illustrations and substantial quotations known or believed by me to be from material still in copyright. Any omission brought to my attention will be rectified in any future edition of this book.

MY WEBSITE

If you have enjoyed this or any of my other books and articles, and have any information or personal experiences/eyewitness accounts relating to any aspect of cryptozoology, animal anomalies, animal mythology, or any other mysterious phenomenon that you consider may be of interest to me and that you wish to share with me, please email me at: karlshuker@aol.com

For full details concerning my work and publications, please visit my website at: http://members.aol.com/karlshuker (it can also be accessed at: http://hometown.aol.com/karlshuker).

You can also check out my official entry on Wikipedia, the Net's most comprehensive online encyclopedia, which can be accessed at: http://en.wikipedia.org/wiki/Karl_Shuker

Many thanks indeed.

A LISTING OF MY ORIGINAL ARTICLES

All of the chapters in this book are variously based upon, expanded from, or inspired by the following previously-published original articles of mine, all of which I have been very kindly permitted to utilize in this manner by the publishers of the respective magazines in which they first appeared:

SHUKER, Karl P.N., 'Cats With Wings – And Other Strange Things'
(*Beyond*, no. 1, October 2006), pp.36-42.
SHUKER, Karl P.N., 'Tales of the Uninvited'
(*Uri Geller's Encounters*, no. 20, April 1998), pp.30-5.
SHUKER, Karl P.N., 'Herne the Hunter'
(*History For All*, vol. 1, no. 4, August-September 1998), pp.56-7.
SHUKER, Karl P.N., 'The Many Ghosts of Glamis'
(*History For All*, vol. 1, no. 2, April-May 1998), pp.24-5.
SHUKER, Karl P.N., 'Phantoms of the Pavilion'
(*History For All*, vol. 1, no. 1, February-March 1998), pp.22-3.
SHUKER, Karl P.N., 'Visiting the Land of the Stone Giants' (*Fate*, vol. 61, July-August 2008).

SHUKER, Karl P.N., 'Top 10 Strangest Aliens' (*Alien Encounters*, no. 14, August 1997), pp.56-60.
SHUKER, Karl P.N., 'Funeral For a Mermaid'
(*Uri Geller's Encounters*, no. 1, November 1996), pp.32-5.
SHUKER, Karl P.N., 'How Green Were My Children?'
(*Beyond*, no. 8, September 2007), pp.40-4.
SHUKER, Karl P.N., 'Reptaliens' (*The X Factor*, No. 59, 1999), pp.1652-6.
SHUKER, Karl P.N., 'Reptoids From the Black Lagoon' (*The X Factor*, No. 60, 1999), pp.1667-71.
SHUKER, Karl P.N., 'All in the Mind? Tulpas, Egrigors, and Other Thought-Forms'
(*Sightings*, vol. 2, July 1997), pp.62-7.
SHUKER, Karl P.N., 'Denizens of the Dreamtime' (*Fate*, vol. 52, May 1999), pp.52-6.
SHUKER, Karl p.N., 'Blood, Sweat and Tears – In Search of Religious Marvels and Miracles'
(*Uri Geller's Encounters*, no. 7, Easter 1997), pp.17-25.
SHUKER, Karl P.N., 'A Message From Mars' (*Alien Encounters*, no. 11, May 1997), pp.24-7.
SHUKER, Karl P.N., 'Seeing Double!' (*Uri Geller's Encounters*, no. 5, March 1997), pp.42-5.
SHUKER, Karl P.N., 'The Secret Animals of Senegambia'
(*Fate*, vol. 51, November 1998), pp.46-7, 49-50.
SHUKER, Karl P.N., 'Great Balls of Fire!' (*Sightings*, vol. 2, June 1997), pp.66-71.
SHUKER, Karl P.N., 'Can UFOs Swim?' (*Alien Encounters*, no. 9, March 1997), pp.18-21.
SHUKER, Karl P.N., 'The Colossus of Rhodes'
(*History For All*, vol. 1, no. 3, June-July 1998), pp.40-1.
SHUKER, Karl P.N., 'The Great Palette Palaver'
(*Fortean Times*, no. 216, November 2006), pp.58-59.
SHUKER, Karl P.N., 'Jason y el Vellocino de Oro' (*Enigmas*, vol. 3, November 1997).
SHUKER, Karl P.N., 'The Tower of Babel – A Monumental Biblical Mystery'
(*Fate*, vol. 54, September 2001), p.19.
SHUKER, Karl P.N., 'The Porcelain Tower of Nanking'
(*History For All*, vol. 1, no. 5, October-November 1998), pp.54-5.
SHUKER, Karl P.N., 'Men With Wings and Other Flights of Fancy...Or Fact?'
(*Sightings*, vol. 2, June 1997), pp.52-9.
SHUKER, Karl P.N., 'Things With Wings – Living Pterosaurs, Giant Bats and Other Airborne
Anomalies' (*Beyond*, no. 5, May 2007), pp.36-42.
SHUKER, Karl P.N., 'It's Raining Sprats and Frogs!' (*Beyond*, no. 2, December 2006), pp.24-31.
SHUKER, Karl P.N., 'Tantalising Trio' [re controversial planets]
(*Prediction*, vol. 60, February 1994), pp.13-15.
SHUKER, Karl P.N., 'Mystery of the Missing Planets' (*Goblin Universe*, no. 6, July 1997), pp.23-5.
SHUKER, Karl P.N., 'Vulcano y los Planetas Perdidos'
(*Enigmas*, vol. 13, no. 144, November 2007), pp.58-62.
SHUKER, Karl P.N., 'On the Horns of a Dilemma – Do Unicorns Really Exist?'
(*Beyond*, no. 10, November 2007), pp.24-31.
SHUKER, Karl P.N., 'We Three Kings of Albion Are'
(*History For All*, vol. 2, no. 9, June-July 1999), pp.56-7.
SHUKER, Karl P.N., 'Sky Beasts and Cloud Creatures – Fiction or Fact?' in: DOWNES, Jonathan
(Ed.), *CFZ Yearbook 1996* (CFZ: Exwick, 1995), pp.5-22.
SHUKER, Karl P.N., 'UFOs – Not Flying Saucers But Sky Beasts?' (*Uri Geller's Encounters*, no. 4,
February 1997), pp.16-20.

About the Author

AUTHOR BIOGRAPHY

Dr Karl P.N. Shuker BSc PhD FRES FZS is a zoologist who is internationally recognised as a world expert in cryptozoology, as well as in animal mythology and allied subjects relating to wildlife anomalies and inexplicabilia. He obtained a BSc (Honours) degree in pure zoology at the University of Leeds (U.K.), and a PhD in zoology and comparative physiology at the University of Birmingham (U.K.). He is now a freelance zoological consultant and writer, living in the West Midlands, England. The author of 13 books (translated into over a dozen foreign languages) and countless articles, Dr Shuker is also the official zoological consultant for *Guinness World Records*, and has acted as a consultant and/or contributor for many other publications and television programmes. Dr Shuker appears regularly on television and radio, has served as a question setter for the BBC's cerebral quiz show *Mastermind*, and has travelled widely throughout the world during the course of his researches. He is a Scientific Fellow of the prestigious Zoological Society of London, a Fellow of the Royal Entomological Society, a Member of the International Society of Cryptozoology and other wildlife-related organisations, he is Cryptozoology Consultant to the Centre for Fortean Zoology, and is also a Member of the Society of Authors.

AUTHOR BIBLIOGRAPHY

Mystery Cats of the World: From Blue Tigers To Exmoor Beasts (Robert Hale: London, 1989)
Extraordinary Animals Worldwide (Robert Hale: London, 1991)
The Lost Ark: New and Rediscovered Animals of the 20th Century (HarperCollins: London, 1993)
Dragons: A Natural History (Aurum: London/Simon & Schuster: New York, 1995; republished Taschen: Cologne, 2006)
In Search of Prehistoric Survivors: Do Giant 'Extinct' Creatures Still Exist? (Blandford: London, 1995)
The Unexplained: An Illustrated Guide to the World's Natural and Paranormal Mysteries (Carlton:

London/JG Press: North Dighton, 1996; republished Carlton: London, 2002)
From Flying Toads To Snakes With Wings: From the Pages of FATE Magazine
(Llewellyn: St Paul, 1997)
Mysteries of Planet Earth: An Encyclopedia of the Inexplicable (Carlton: London, 1999)
The Hidden Powers of Animals: Uncovering the Secrets of Nature (Reader's Digest: Pleasantville/
Marshall Editions: London, 2001)
The New Zoo: New and Rediscovered Animals of the Twentieth Century [fully-updated, greatly-
expanded, new edition of *The Lost Ark*] (House of Stratus Ltd: Thirsk, UK/House of Stratus Inc:
Poughkeepsie, USA, 2002)
The Beasts That Hide From Man: Seeking the World's Last Undiscovered Animals (Paraview: New
York, 2003)
Extraordinary Animals Revisited: From Singing Dogs To Serpent Kings (CFZ Press: Bideford, 2007)
Dr Shuker's Casebook: In Pursuit of Marvels and Mysteries (CFZ Press: Bideford, 2008)

Consultant and also Contributor

Man and Beast (Reader's Digest: Pleasantville, New York, 1993)
Secrets of the Natural World (Reader's Digest: Pleasantville, New York, 1993)
Almanac of the Uncanny (Reader's Digest: Surry Hills, Australia, 1995)
The Guinness Book of Records/Guinness World Records 1998- (Guinness: London, 1997-)

Consultant

Monsters (Lorenz: London, 2001)

Contributor

Fortean Times Weird Year 1996 (John Brown Publishing: London, 1996)
Mysteries of the Deep (Llewellyn: St Paul, 1998)
Guinness Amazing Future (Guinness: London, 1999)
The Earth (Channel 4 Books: London, 2000)
Mysteries and Monsters of the Sea (Gramercy: New York, 2001)
Chambers Dictionary of the Unexplained (Chambers: London, 2007)

THE CENTRE FOR FORTEAN ZOOLOGY

I have an idea for a project which isn't on your website. What do I do?

Write to us, e-mail us, or telephone us. The list of future projects on the website is not exhaustive. If you have a good idea for an investigation, please tell us. We may well be able to help.

How do I go on an expedition?

We are always looking for volunteers to join us. If you see a project that interests you, do not hesitate to get in touch with us. Under certain circumstances we can help provide funding for your trip. If you look on the future projects section of the website, you can see some of the projects that we have pencilled in for the next few years.

In 2003 and 2004 we sent three-man expeditions to Sumatra looking for Orang-Pendek - a semi-legendary bipedal ape. The same three went to Mongolia in 2005. All three members started off merely subscribers to the CFZ magazine.

Next time it could be you!

Project Kerinci, Sumatra - 2003
In search of the bipedal ape Orang Pendek

How is the Centre for Fortean Zoology funded?

We have no magic sources of income. All our funds come from donations, membership fees, works that we do for TV, radio or magazines, and sales of our publications and merchandise. We are always looking for corporate sponsorship, and other sources of revenue. If you have any ideas for fund-raising please let us know. However, unlike other cryptozoological organisations in the past, we do not live in an intellectual ivory tower. We are not afraid to get our hands dirty, and furthermore we are not one of those organisations where the membership have to raise money so that a privileged few can go on expensive foreign trips. Our research teams both in the UK and abroad, consist of a mixture of experienced and inexperienced personnel. We are truly a community, and work on the premise that the benefits of CFZ membership are open to all.

What do you do with the data you gather from your investigations and expeditions?

Reports of our investigations are published on our website as soon as they are available. Preliminary reports are posted within days of the project finishing.

Each year we publish a 200 page yearbook containing research papers and expedition reports too long to be printed in the journal. We freely circulate our information to anybody who asks for it.

Is the CFZ community purely an electronic one?

No. Each year since 2000 we have held our annual convention - the *Weird Weekend* - in Exeter. It is three days of lectures, workshops, and excursions. But most importantly it is a chance for members of the CFZ to meet each other, and to talk with the members of the permanent directorate in a relaxed and informal setting and preferably with a pint of beer in one hand. Since 2006 - the *Weird Weekend* has been bigger and better and held in the idyllic rural location of Woolsery in North Devon.

Since relocating to North Devon in 2005 we have become ever more closely involved with other community organisations, and we hope that this trend will continue. We also work closely with Police Forces across the UK as consultants for animal mutilation cases, and we intend to forge closer links with the coastguard and other community services. We want to work closely with those who regularly travel into the Bristol Channel, so that if the recent trend of exotic animal visitors to our coastal waters continues, we can be out there as soon as possible.

We are building a Visitor's Centre in rural North Devon. This will not be open to the general public, but will provide a museum, a library and an educational resource for our members (currently over 400) across the globe. We are also planning a youth organisation which will involve children and young people in our activities. We work closely with *Tropiquaria* - a small zoo in north Somerset, and have several exciting conservation projects planned.

Apart from having been the only Fortean Zoological organisation in the world to have consistently published material on all aspects of the subject for over a decade, we have achieved the following concrete results:

- Disproved the myth relating to the headless so-called sea-serpent carcass of Durgan beach in Cornwall 1975
- Disproved the story of the 1988 puma skull of Lustleigh Cleave
- Carried out the only in-depth research ever into the mythos of the Cornish Owlman
- Made the first records of a tropical species of lamprey
- Made the first records of a luminous cave gnat larva in Thailand.
- Discovered a possible new species of British mammal - the beech marten.
- In 1994-6 carried out the first archival fortean zoological survey of Hong Kong.
- In the year 2000, CFZ theories where confirmed when an entirely new species of lizard was found resident in Britain.
- Identified the monster of Martin Mere in Lancashire as a giant wels catfish
- Expanded the known range of Armitage's skink in the Gambia by 80%
- Obtained photographic evidence of the remains of Europe's largest known pike
- Carried out the first ever in-depth study of the *ninki-nanka*
- Carried out the first attempt to breed Puerto Rican cave snails in captivity
- Were the first European explorers to visit the `lost valley` in Sumatra
- Published the first ever evidence for a new tribe of pygmies in Guyana
- Published the first evidence for a new species of caiman in Guyana

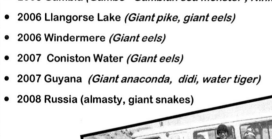

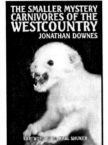

THE SMALLER MYSTERY CARNIVORES OF THE WESTCOUNTRY
Jonathan Downes - ISBN 978-1-905723-05-8

£7.99

Although much has been written in recent years about the mystery big cats which have been reported stalking Westcountry moorlands, little has been written on the subject of the smaller British mystery carnivores. This unique book redresses the balance and examines the current status in the Westcountry of three species thought to be extinct: the Wildcat, the Pine Marten and the Polecat, finding that the truth is far more exciting than the currently held scientific dogma. This book also uncovers evidence suggesting that even more exotic species of small mammal may lurk hitherto unsuspected in the countryside of Devon, Cornwall, Somerset and Dorset.

THE BLACKDOWN MYSTERY
Jonathan Downes - ISBN 978-1-905723-00-3

£7.99

Intrepid members of the CFZ are up to the challenge, and manage to entangle themselves thoroughly in the bizarre trappings of this case. This is the soft under-belly of ufology, rife with unsavoury characters, plenty of drugs and booze." That sums it up quite well, we think. A new edition of the classic 1999 book by legendary fortean author Jonathan Downes. In this remarkable book, Jon weaves a complex tale of conspiracy, anti-conspiracy, quasi-conspiracy and downright lies surrounding an air-crash and alleged UFO incident in Somerset during 1996. However the story is much stranger than that. This excellent and amusing book lifts the lid off much of contemporary forteana and explains far more than it initially promises.

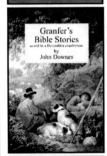

GRANFER'S BIBLE STORIES
John Downes - ISBN 0-9512872-8-1

£7.99

Bible stories in the Devonshire vernacular, each story being told by an old Devon Grandfather - 'Granfer'. These stories are now collected together in a remarkable book presenting selected parts of the Bible as one more-or-less continuous tale in short 'bite sized' stories intended for dipping into or even for bed-time reading. `Granfer` treats the biblical characters as if they were simple country folk living in the next village. Many of the stories are treated with a degree of bucolic humour and kindly irreverence, which not only gives the reader an opportunity to re-evaluate familiar tales in a new light, but do so in both an entertaining and a spiritually uplifting manner.

FRAGRANT HARBOURS DISTANT RIVERS
John Downes - ISBN 0-9512872-5-7

£12.50

Many excellent books have been written about Africa during the second half of the 19th Century, but this one is unique in that it presents the stories of a dozen different people, whose interlinked lives and achievements have as many nuances as any contemporary soap opera. It explains how the events in China and Hong Kong which surrounded the Opium Wars, intimately effected the events in Africa which take up the majority of this book. The author served in the Colonial Service in Nigeria and Hong Kong, during which he found himself following in the footsteps of one of the main characters in this book; Frederick Lugard – the architect of modern Nigeria.

CFZ PRESS, MYRTLE COTTAGE, WOOLFARDISWORTHY BIDEFORD, NORTH DEVON, EX39 5QR
w w w . c f z . o r g . u k

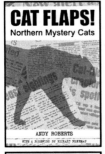

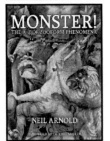

Other books available from
CFZ PRESS

CFZ PRESS

BIG CATS IN BRITAIN YEARBOOK 2008
Edited by Mark Fraser - ISBN 978-1-905723-23-2

£12.50

People from all walks of life encounter mysterious felids on a daily basis, in every nook and cranny of the UK. Most are jet-black, some are white, some are brown; big cats of every description and colour are seen by some unsuspecting person while on his or her daily business. 'Big Cats in Britain' are the largest and most active research group in the British Isles and Ireland. This book contains a run-down of every known big cat sighting in the UK during 2007, together with essays by various luminaries of the British big cat research community.

CFZ EXPEDITION REPORT 2007 - GUYANA
ISBN 978-1-905723-25-6

£12.50

Since 1992, the CFZ has carried out an unparalleled programme of research and investigation all over the world. In November 2007, a five-person team - Richard Freeman, Chris Clarke, Paul Rose, Lisa Dowley and Jon Hare went to Guyana, South America. They went in search of giant anacondas, the bigfoot-like didi, and the terrifying water tiger.

Here, for the first time, is their story...With an introduction by Jonathan Downes and forward by Dr. Karl Shuker.

CENTRE FOR FORTEAN ZOOLOGY 2003 YEARBOOK
Edited by Jonathan Downes and Richard Freeman
ISBN 978 -1-905723-19-5

£12.50

Edited by
Jonathan Downes and Richard Freeman

The Centre For Fortean Zoology Yearbook is a collection of papers and essays too long and detailed for publication in the CFZ Journal *Animals & Men*. With contributions from both well-known researchers, and relative newcomers to the field, the Yearbook provides a forum where new theories can be expounded, and work on little-known cryptids discussed.

CENTRE FOR FORTEAN ZOOLOGY 1997 YEARBOOK
Edited by Jonathan Downes and Graham Inglis
ISBN 978 -1-905723-27-0

£12.50

Edited By
Jonathan Downes
and Graham Inglis

The Centre For Fortean Zoology Yearbook is a collection of papers and essays too long and detailed for publication in the CFZ Journal *Animals & Men*. With contributions from both well-known researchers, and relative newcomers to the field, the Yearbook provides a forum where new theories can be expounded, and work on little-known cryptids discussed.

CFZ PRESS, MYRTLE COTTAGE,
WOOLFARDISWORTHY BIDEFORD,
NORTH DEVON, EX39 5QR
w w w . c f z . o r g . u k

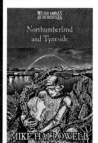

Printed in the United Kingdom
by Lightning Source UK Ltd.
132746UK00001B/85/P